GRAVITY AND
MAGNETICS IN
OIL PROSPECTING

**McGRAW-HILL
BOOK COMPANY**
New York
St. Louis
San Francisco
Auckland
Düsseldorf
Johannesburg
Kuala Lumpur
London
Mexico
Montreal
New Delhi
Panama
Paris
São Paulo
Singapore
Sydney
Tokyo
Toronto

L. L. NETTLETON
Geophysical Consultant
Adjunct Professor of Geology, Rice University

Formerly
Geophysicist, Gulf Research & Development Co.
Pittsburgh, Pennsylvania
Partner, Gravity Meter Exploration Co.
Houston, Texas

Gravity and Magnetics in Oil Prospecting

This book was set in Times New Roman.
The editors were Robert H. Summersgill and James W. Bradley;
the production supervisor was Charles Hess.
R. R. Donnelley & Sons Company was printer and binder.

Library of Congress Cataloging in Publication Data

Nettleton, Lewis Lomat, date
    Gravity and magnetics in oil prospecting.

    (McGraw-Hill international series in the earth and
planetary sciences)
    Includes bibliographies and index.
    1. Prospecting—Geophysical methods. 2. Petroleum.
I. Title.
TN271.P4N47   622'.18'282   75-17899
ISBN 0-07-046303-4

GRAVITY AND
MAGNETICS IN
OIL PROSPECTING

1 2 3 4 5 6 7 8 9 0 DODO 7 9 8 7 6

CONTENTS

PREFACE

It is 35 years since this writer's "Geophysical Prospecting for Oil" was published. That book covered oil exploration methods based on gravity, magnetic, and seismic principles. In the intervening years no new method based on a different fundamental principle has developed. In terms of relative effort, as measured by expenditures, the seismic method dominates the geophysical search for oil so much that less than 5 percent of the expenditures are used for gravity and magnetics, which are the methods that depend on the mathematical theory of the potential.

The writer has been continuously involved in the geophysical search for oil since 1928, largely in gravity, to a lesser extent in magnetic, and only incidentally in seismic applications. Therefore, in considering revision of the old book he did not feel competent to include the seismic method, with its very great later complications, particularly from the application of computer data processing techniques to increasingly complex field recording procedures.

Why bother with a book which is applicable to less than 5 percent of geophysical exploration for oil? For two reasons: first, the areas covered and the relative importance of the potential methods are much greater than their relative cost, and, second, because most of the practitioners in these methods feel that they would have greater and much more effective use if they were better understood and applied as aids to the interpretation of seismic data. Therefore the short "Part Three" is added to present some examples of "synergistic" interpretation in which the coordination of data of all three kinds reveals a total geologic result which is more detailed and deserves more confidence than could be derived from the analysis of the individual sets of data independently.

The presentation ranges rather widely in technical difficulty. This has been done deliberately in order to furnish the material needed for a textbook on the one hand and, on the other, to provide a guide and general reference for office and field personnel in day-to-day field work and interpretation. To that end, there are included simple graphical "tools" to permit quick, approximate methods to evaluate magnitudes and forms of expected gravity and magnetic anomalies. These can be used to compare effects expected from geologic interpretations with those observed.

Since his "retirement" from day-to-day geophysical activities, the writer has been dependent on friends and former associates for much help. Routine clerical assistance has been provided by the Geological Department of Rice University. Much drafting and other help came from Tidelands Geophysical Company, for whom the writer has been an independent consultant. Many former associates and friends in the Houston geophysical fraternity have answered questions and given other help.

Special acknowledgment is due Dr. Neal Jordan of Exxon, Houston, who read, criticized, and in some cases, added to the manuscript, particularly the gravity section.

Dr. Norman S. Neidel of Geoquest International, Houston, collaborated in the writing of Chapter 6, Anomaly Separation and Filtering, including preparation of the parts on Fourier transforms, wavelength filtering, etc.

Others who have read, criticized, and reread particular sections include Sigmund Hammer, Emeritus Professor of Geophysics at the University of Wisconsin, Robert J. Bean of Shell Research Laboratory in Houston, J. Lamar Worzel of the University of Texas Marine Biological Institute at Galveston, Robert E. Sheriff of Chevron Oil Co., Houston, and Robert E. West, Le Roy Brow, and Michael Alexander of Exxon, Houston, who divided Part One of the manuscript between them for criticism.

Part Three would not be possible without the seismic sections, the gravity and magnetic profiles, and the computer calculations used. Thanks are due Gulf Research and Development Co., Houston, and Joyce O'Brien for example No. 2 and Geophysical Services, Inc., Dallas, and Anthony Williams for example No. 3.

At Rice University, graduate students Alice Hickcox and Myrna Norvell were original and resourceful in bibliographic and editing help. Becky Lindig, Barbara Hawkins, and others shared the typing. Most of the new illustrations were drawn by Patricia Hill.

The material in this book, in addition to that from the writer's own experience, has depended a great deal, particularly for illustrations, on the published literature, and such sources are duly acknowledged. Both illustrations and limited text quotations have been taken from the writer's earlier publications. Particular acknowledgement is made to McGraw-Hill Book Company for material from "Geophysical Prospecting for Oil," to the Society of Exploration Geophysicists for free use of the 1954 paper "Regionals, Residuals and Structures" and the 1971 Monograph No. 1, and to the American Association of Petroleum Geologists for material from the 1962 paper on Gravity and Magnetics for Geologists and Seismologists.

To these and others too numerous to mention, including many probably forgotten during the 5 years of preparation of the manuscript who have contributed by personal discussion, published material, and in other ways, the writer is deeply grateful.

L. L. NETTLETON

INTRODUCTION

In its struggle to keep up with the ever increasing demand for its products, the oil industry is expending great effort and enormous sums of money to find new oil and gas reserves. To ensure its operation in the future, the industry must find new supplies to keep up with the rising rate of consumption. The growing dependence on imports makes American participation in domestic and foreign oil exploration more vital than in any time in the past.

Since their beginnings in about 1923, the various methods of geophysical prospecting have come to be an increasingly important part of the exploration branch of all the large oil companies and most of the smaller ones. From their initial extensive application and proved usefulness in the United States their use has spread to oil provinces throughout the world. The intensive search for new oil reserves has, to a large extent, tested by drilling those prospects which are evident from the surface. By the use of geophysical methods the search has been carried into areas where underground oil prospects have no visible surface expressions whatever. Their detection depends on artificial means for peering below the surface to see what the nature and attitude of the rocks may be. The various forms of geophysical prospecting are the tools for such seeing. Our vision is imperfect at best and often extremely hazy, so that different means may be called in to try to see the same picture. However, a very little light is extremely useful to one otherwise completely blind. Geophysical prospecting has proved valuable in the search for oil and is almost certain to be relied upon more and more to find the oil and gas for the ever increasing energy required for the transportation, power, heat, and chemical activities of the people of the world.

THE ACCUMULATION OF OIL

Geologists have known since comparatively early in the history of the oil industry that the accumulation of oil occurs under certain rather restricted underground conditions. An oil "pool" is not an open underground lake but a porous rock, more or less saturated with oil. Hence a "reservoir rock," usually a porous sandstone or limestone, is required. Oil will not accumulate unless there is a place for it to come from. Hence, there must be "source rocks," usually shale beds containing organic matter. Oil would not be held in a particular place unless there were something to prevent its further migration from that place. Hence there must be an impervious "cap rock" or other trapping condition above or adjacent to the reservoir rock.

The oil indigenous to the source rocks is not sufficiently concentrated to make an oil pool. The required accumulation necessitates migration of oil through porous rocks and some irregularity of the rocks which will arrest the migration in a comparatively local area.

The spaces between the grains of porous rocks are almost never void but are filled with fluid, which may be oil, gas, or water, either separate or mixed. As the fluids migrate through the porous rocks, the lighter ones tend to rise through and float on the heavier ones. If a trap of some kind exists within which the migrating fluids accumulate, they will tend to separate according to their relative densities, the lightest (gas) at the top, the oil next, and the heaviest (water) below.*

Traps that cause the local accumulation of oil are the result of a variety of geologic processes and rock deformations (Fig. I-1). Such deformations are usually called "structures." To serve as a trap it is generally considered necessary that the structure be "closed." This means that a contour on the cap rock or its equivalent is closed about a point from which the dip is downward in all directions. A simple structural dome may be considered as an upside-down bowl under which oil and gas accumulate over water, much as gas is collected over water with an inverted beaker in some simple chemical experiments.

When the requirements of source rock, reservoir rock, and cap rock are present in a general region, the area underlain by any suitable trap or closed structure may have a fair to high probability of being a potential oil field. Where the geologic movement that causes a trap is of such a nature or occurred at such a

---

* For a brief discussion of the fundamentals of the origin and accumulation of oil and gas, see Muskat (1937, pp. 43–54). References for Parts One, Two, and Three will be found at the end of the respective part.

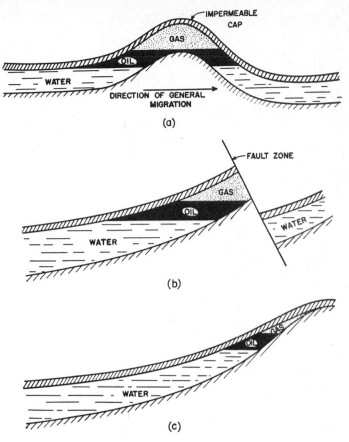

FIGURE I-1
Different ways in which structure may control oil accumulation.

geologic time that the rocks at the surface are affected, the presence of underground conditions favorable for the accumulation of oil may be inferred from a study of the nature and attitude of the surface rocks, i.e., by surface geologic mapping. However, in many cases, geologic deposits have been laid down subsequent to the geologic movement that caused the structure, so that there is absolutely no visible surface evidence of its presence. The problem and the *raison d'être* of geophysical prospecting are the finding of such buried structures.

Many kinds of geologic disturbances or conditions may form closed structures which are possible traps for the accumulation of oil. The means by which buried structures can be detected and the relative difficulty of their detection depend on the nature of the geologic disturbances forming the traps and the physical properties of the rocks involved. Also, the depth of the buried structure below the surface is always an important consideration.

GEOPHYSICAL METHODS OF PROSPECTING

Any conceivable property or process which can be measured or carried out at or above the surface and which is affected by the nature or attitude of rocks (or by oil itself) through a cover of hundreds to many thousands of feet of intervening rocks may be made the basis of a method of geophysical prospecting. Many principles have been suggested and tried [including some that are purely psychic (Blau, 1936)], but practically all the geophysical search for oil depends on a very few basic physical principles, a brief outline of which follows.

Gravitational Methods

Gravitational methods are based on the measurement at the surface of small variations in the gravitational field. Small differences or distortions in that field from point to point over the surface of the earth are caused by any lateral variation in the distribution of mass in the earth's crust. Therefore, if geologic movements involve rocks of differing density, the resulting irregularity in mass distribution will make a corresponding variation in the intensity of gravity. Sensitive instruments are used to measure relative values of gravity. The measured variations are interpreted in terms of probable subsurface mass distributions, which in turn are the basis for inferences about probable geologic conditions and the presence or absence of traps favorable for the accumulation of oil.

Magnetic Methods

Magnetic methods are based on the measurement of small variations in the magnetic field. This field is affected by any variation in the distribution of magnetized (or polarized) rocks. Most sedimentary rocks are nearly nonmagnetic, but the underlying igneous or basement rocks usually are slightly magnetic. A sensitive magnetometer is used to measure the relative or absolute values of magnetic intensity.

Variations in the magnetic field may be caused by inhomogeneities in composition of basement rocks or by structural or topographic relief of the basement surface. These variations may be measured at the surface or, more commonly, by suitable instruments carried in an aircraft.

The measured variations are interpreted in terms of the probable distribution of magnetic material below the surface, which in turn is the basis for inferences about the probable geologic conditions. A most important result of such interpretation is the determination of the depth to the basement rocks and therefore the thickness of sediments present. Furthermore, it is often possible to determine local relief of the basement surface capable of producing structural relief in the overlying sediments which could be favorable for the accumulation of oil. Very sensitive instruments now used can in some circumstances detect and map the much smaller effects from the very small magnetization of the sediments.

Seismic Methods

Seismic methods are based on the measurement of travel times of artificially induced elastic waves. Such waves, set up by explosives or by other sources of pulsed or continuous acoustic energy at or near the surface, travel in all directions from the source. Waves having paths in certain directions are refracted or reflected so that they come back to the surface. A series of sensitive seismic detectors at the surface and at various distances from the source is connected through suitable filters and amplifiers to a recorder. The resulting seismic record, which may be visual or on magnetic tape or both, provides the data for determination of the depth and form of certain reflecting or refracting "horizons" in the underlying rock series. In many cases, such seismic records show reflected waves which give very definite information about the depth to certain discontinuities in the lithologic character of the rock strata. Under favorable conditions a given geologic horizon many thousands of feet deep can be mapped very accurately by reflected seismic waves. In the refraction application of the seismic method the depths of different beds and their characteristic wave speeds can be determined by the travel times of waves that have been refracted through those beds. Under favorable conditions, seismic methods may give information that can be interpreted quite simply and directly in terms of geologic conditions. Under less favorable conditions the detection of reflected or refracted waves or their interpretation in terms of definite geologic conditions may be quite uncertain and leave many unanswered questions.

APPLICATIONS OF GEOPHYSICS TO PETROLEUM EXPLORATION

The different geophysical methods may serve quite different purposes in a general geophysical campaign for the exploration of a large area. For instance, a general preliminary or reconnaissance survey may be made by gravity or magnetic methods, or both. The aeromagnetic method is particularly suitable for a "first look" at any large and relatively unknown area. The interesting indications then can be selectively tested by the much more expensive, but usually more certain, seismic method or, in favorable circumstances, directly by drilling. In this way the more expensive exploration is concentrated on the areas that have higher probabilities of yielding favorable production possibilities. The results of one technique may provide information which makes it possible to apply another technique more intelligently. For example, seismic lines may be laid out with the proper orientation to test structural indications from gravity or magnetic surveys. Also, particularly in areas where the interpretation is difficult, the different methods provide different and often complementary information, which can give a much more certain answer than is obtainable by any one method alone. Illustrations from the combined interpretation of simultaneous magnetic, gravity, and seismic data at sea are given in Part Three.

The three basic physical principles of the gravity, magnetic, and seismic methods were first applied extensively to petroleum exploration on the Gulf Coast of the United States beginning about 1923. Since that time there have been great changes in instrumentation and advances in interpretation, but the same three physical principles are still employed. In spite of many attempts to use other principles, especially electrical, none has ever attained extensive field use, and current geophysical exploration is based on the same physical background that it started with 50 years ago.

COST OF GEOPHYSICAL SURVEYS

An airborne magnetic survey can be carried out for less than $10 per mile of line. Gravity stations can cost from less than $10 if made along roads in flat country to over $100 in jungle or rugged areas without roads. Seismic surveys cost from less than $100 per mile for some marine surveys to over $4000 for land surveys, depending on the accessibility of the area, the degree of detail, and the amount of data processing involved.

As a very crude order-of-magnitude rule of thumb, the costs per mile of line of magnetic, gravity, and seismic exploration stand in ratios of 1:10:100. This simple economic fact is the reason this book is devoted to those "potential" methods which, properly applied, can reduce overall exploration costs.

The relative simplicity of its basic concepts and its almost "picture book" results in many applications have made the seismic method much the most common petroleum exploration system. Because of this and its much greater cost, seismic operations use over 95 percent of the geophysical budget for oil exploration.

This book is concerned with the gravity and magnetic methods only. These are potential methods which depend on forces acting at a distance in a manner defined by the mathematics of potential fields. This background makes these methods much more obscure and difficult than the seismic methods to relate to geology.

Lack of understanding and appreciation of the usefulness of gravity and especially magnetic methods has resulted in limitations on their application and consequent great increase in overall exploration costs by the use of expensive methods where not needed or not applicable. It is hoped that this book may lead to much better understanding of the potential methods, i.e., "the other 5 percent," so that, even after over 50 years of application, they can assume their proper place in the economics of petroleum exploration.

PART ONE

Gravitational
Methods

1

FUNDAMENTAL PRINCIPLES AND UNITS

GRAVITATIONAL ACCELERATION

The law of universal gravitation was conceived by Newton from a study of Kepler's empirical laws of the motion of the planets. To explain these motions, Newton deduced the law that every particle of matter exerts a force of attraction on every other particle that is proportional to the product of the masses and inversely proportional to the square of the distance between them; i.e.,

$$F = G \frac{m_1 m_2}{r^2}$$

where F = the force between two particles of mass m_1 and m_2
r = distance between them
G = constant with dimensions $L^3 M^{-1} T^{-2}$ and a numerical value depending on units used

We all know that a body on the earth has weight. The weight of a body is the force acting on it due to the gravitational attraction of the earth. If we consider the body as one of the masses (say m_1), the force (weight) is the attraction on this

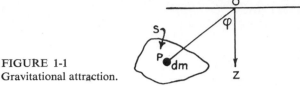

FIGURE 1-1
Gravitational attraction.

body of another body (m_2) which is the whole earth (which we may consider as concentrated at its center) at a distance r, the radius of the earth.

The force acting on the body m_1 also may be considered as defined by Newton's second law of motion; that is, $F = m_1 a$, where a is the acceleration that would be caused by the gravitational attraction of the earth if the body were allowed to fall. Thus the force on the body m_1 is exactly the same as if it were being accelerated at a rate

$$a = \frac{F}{m_1} = G\frac{m_2}{r^2}$$

Thus the attraction of the earth may be considered as a force per unit mass and therefore as exactly equivalent to an acceleration. Therefore, we often speak of the "acceleration" of gravity rather than the "force" of gravity, and the concept of acceleration rather than force will be used throughout this book.

Without reference to the earth we may consider that the gravitational field at a point O (Fig. 1-1) which is associated with any elemental particle of matter of mass dm at point P is equivalent to an acceleration of magnitude

$$a = G\frac{dm}{r^2}$$

directed toward the elemental particle dm.

The component of this acceleration in any other direction Z, making an angle φ with the line $\overline{OP}$, is

$$a_Z = G\frac{dm}{r^2}\cos\varphi$$

Finally, the component of acceleration at O in the direction Z due to a continuous body S is the sum of the effects of all particles dm within the body, or

$$(a_Z)_S = G\int_v \frac{dm}{r^2}\cos\varphi$$

where the integration is performed throughout the volume v.

THE GRAVITATIONAL CONSTANT

The constant G of the foregoing section is one of the fundamental constants of nature. It cannot be determined from astronomical observational data such as led Newton to the general law of gravitation. However, if all the quantities except G can be measured experimentally, the value of G can be calculated.

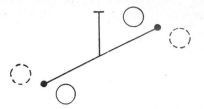

FIGURE 1-2
Principle of the Cavendish balance.

The earliest workers in this field were interested primarily in Δ, the average density of the earth, rather than the gravitational constant G. However, the two quantities are closely related, so that the determination of one evaluates the other. This may be shown as follows.

If we assume the earth to be a sphere of mass M and radius R, the gravitational acceleration g at its surface is

$$g = \frac{GM}{R^2} = \frac{G}{R^2}\left(\tfrac{4}{3}\pi R^3 \Delta\right) = \frac{4\pi R}{3}G\Delta$$

$$G\Delta = \frac{3g}{4\pi R}$$

Thus, if G and R are known, the determination of G* and Δ are equivalent problems, as the measurement of either determines the other.

Probably the earliest attempt to evaluate Δ was made by Bouguer, a French geodesist, in about 1740. He attempted to measure the deviations of a plumb line produced by the gravitational attraction of a high mountain in Peru. This deviation is a measure of the relative attractions of the earth and the mountain. From an estimate of the shape and density of the mountain mass, the density of the earth could be computed. Bouguer's value was much too large because of the effect, unknown at that time, of isostatic compensation. Other experiments based on the same principle were made later in Great Britain and gave approximately correct values for the earth's density.

The earliest direct laboratory measurement of Δ was made by Cavendish in England in 1797 and 1798, using an apparatus suggested by Rev. John Mitchell. In principle, the commonly called "Cavendish experiment" consists of measuring the deflection of a horizontal beam, carrying two small weights, which is produced by the gravitational attraction of two larger weights. One method, shown in Fig. 1-2, consists of interchanging the positions of the large weights, as indicated by the dotted lines, and measuring the resulting rotation of the beam or compensating the deflection by rotating the fiber support. From the torsion constant of the suspending fiber and the length of the beam (or the period of torsional vibration of the suspended system), the force of attraction between the masses is computed.

---

* Since only gravitational forces are considered, this expression applies to a nonrotating earth; therefore the value of G used must be the total value at the surface plus the normal outward component of centrifugal acceleration. The calculated theoretical value for $G\Delta$, based on a spherical nonrotating earth with mean radius 6.381×10^8 cm and gravity 982.0 cm/sec², is 36.797×10^{-8} sec⁻².

Then if the distances between the weights and their masses are known, all the quantities in the first equation on page 11 except G are measured, and since it can be computed, Δ can also be determined.

Boys in 1889 made a careful determination of G by a modified Cavendish balance. He used a balance with a very short beam and with the weights suspended at different heights to separate them enough to permit each of the large weights to act appreciably on only one of the smaller weights.

Many other measurements of G and Δ have been made by various modifications of the Cavendish torsion balance or by other means. The principal measurements are summarized in Table 1-1 and in Fig. 1-3.

Figure 1-3 shows the various experimental values in chronological order. The generally accepted value for the recent past has been about 6.667. The recent International System of Units (SI) (Mechtly, 1969) gives a value of 6.6732 with a probable error of 460 ppm. This is by far the largest uncertainty of any of the 32 basic physical constants of the system.*

* A more recent value is 6.6720, recommended in September 1973 in the report of the Task Group on Fundamental Constants (E. R. Caper, Chairman) to the Committee on Data for Science and Technology (CODATA) of the International Council of Scientific Unions (ICSU).

Table 1-1   MEASUREMENTS OF G OR Δ*

Name	Year	Method	G, 10^{-8} cgs units	Δ, g/cm³
Cavendish	1797–1789	Torsion balance	6.754	5.448
Reich	1837	Torsion balance	6.70†	5.49
Baily	1841–1842	Torsion balance	6.485†	5.674
Reich	1849	Torsion balance	6.594†	5.58
Cornu and Baille	1870	Torsion balance	6.618†	5.56
Jolly	1878–1881	Common balance	6.465	5.692
Wilsing	1886	Metronome balance	6.596	5.579
Boys	1895	Torsion balance	6.6576	5.5270
Braun	1896	Torsion balance	6.6579	5.5275
Eötvös	1896	Torsion balance	6.65	5.53
Richarz and Krigar-Menzel	1898	Common balance	6.684	5.505
Burgess	1899	Buoyant torsion balance	6.64	5.55
Poynting	.....	Ordinary balance	6.6984	5.4934
Heyl	1930	Torsion balance	6.670	5.517†
Zahradnicek	1932	Resonance	6.659	5.526†
Heyl and Chrzanowski	1942		6.673	5.514†
Rose et al.	1969	Rotating table	6.674	5.513†

* Tabulated largely from Poynting and Thompson (1927, pp. 36–44) and Boys (1923).
† Calculated by the $G\Delta$ product from one reported value.

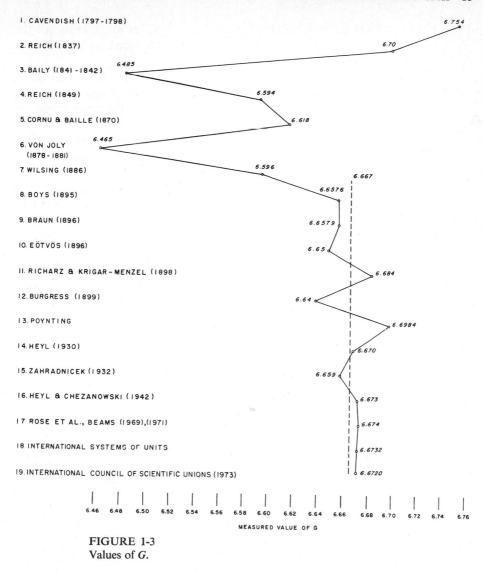

FIGURE 1-3
Values of *G*.

The value 6.6732 has been used throughout for the calculation of numerical coefficients which involve a value for *G*. The value for *G* in cgs units* is the attraction in dynes between two (spherical) masses of 1 g each when their centers are separated by a distance of 1 cm.

* The SI uses meters per second squared for acceleration. This unit, of course, is two orders of magnitude larger and numbers are 100 times smaller than when the older units (centimeters per second squared) are used. The SI system is used in much scientific literature, but at this time (1974) is not widely used in geophysics. See for instance, Markowitz (1973).

UNIT OF ACCELERATION

In most textbooks on physics, no special name has been generally used for the unit of acceleration. However, the name "gal"* is commonly used in the literature of geophysical prospecting. Thus

$$1 \text{ gal} = 1 \text{ cm/sec}^2$$

Since the gal is a rather large unit in terms of the magnitudes usually of interest in geophysical prospecting, the milligal (mgal) is more commonly used; i.e.,

$$1 \text{ milligal} = 1 \text{ mgal} = 0.001 \text{ gal} = 0.001 \text{ cm/sec}^2$$

Since the normal gravitational acceleration at the surface of the earth is about 980 gals, 1 mgal is approximately 1 part in 1 million of the normal gravity of the earth.

The "gravity unit" (gu) equals 10^{-4} gal or 0.1 mgal and is often used in calculation of field gravity observations and on commercial gravity maps. In this writer's opinion this use of two units differing by only one order of magnitude is regrettable since it often leads to confusion. In connection with tidal gravity measurements, the microgal (μgal $= 10^{-6}$ gal) is a convenient unit for the very small gravity changes involved.

---

* According to Jeffries (1962), the first use of the word was in Germany in a report by Wiechert of the Göttingen seismological station in 1909. Since then it has been gradually adopted and now is used widely in geophysical literature. The same gal is generally accepted as being in honor of Galileo because of his late sixteenth-century experiments on falling bodies from the leaning tower at Pisa.

2

GRAVITY OF THE EARTH

All gravity measurements necessarily are made in the gravitational field of the earth. Therefore, a knowledge of this field is required so that proper allowance for it can be made in reducing any kind of gravity measurement to a form useful for indicating geologic structure.

VARIATION OF GRAVITY WITH LATITUDE*

The earth is not a perfect sphere. To a rather close approximation, its shape is that of a perfect fluid for which a balance is maintained between the gravitational forces tending to make it spherical and the centrifugal forces of rotation tending to flatten it. As a result of this balance the equatorial radius is about 21 km greater than the polar radius.

* An extensive discussion of the corrections mentioned in this section is given in Heiskanen and Vening Meinesz (1958, chap. 6). A good mathematical treatment of the relation between the shape of the earth and its gravity field is found in Caputo (1967), and a very readable account is found in Garland (1965).

Flattening means that the acceleration of gravity is 5.17 gals greater at the poles than at the equator. This change can be broken down into three components as follows (Hammer, 1943):

1 The outward centrifugal acceleration at the equator, which is absent at the poles, makes an increase of 3.39 gals.

2 A point at the pole is nearer the center of mass of the earth, making an increase of 6.63 gals.

3 Because of the mass-shape factor, the attraction of the whole earth is greater at the equator than at the poles, making a decrease of 4.85 gals.

The total change, or sum of these effects, is 5.17 gals.

The exact shape of the earth has been a matter of great interest and scientific study since the time of Newton.* It has been derived from certain geodetic and astronomic measurements and from a study of the variation of gravity over the surface, together with a knowledge of the radius of the earth.†

The shape of the earth is expressed in terms of the dimensions of an ideal *spheroid* of reference. The dimensions usually given are the equatorial and polar radii *a* and *b*, respectively, or the equatorial radius together with the flattening; i.e.,

$$f = \frac{a - b}{a}$$

When the spheroid is determined from gravity measurements, a gravity formula giving the variation of gravity with latitude also is determined (by Clairaut's theorem†).

---

* For a very readable history of such studies, see Lambert (1936).

† This is shown by Clairaut's fundamental theorem, which may be written (to first-order approximation) as

$$f = \frac{5C}{2} - \beta$$

where *f* is the flattening of the earth expressed in terms of the equatorial radius *a* and the polar radius *b* as

$$f = \frac{a - b}{a}$$

C is the ratio of the centrifugal acceleration to the gravitational acceleration g_e at the equator; that is, $C = a\omega^2/g_e$, where ω is the angular velocity of rotation of the earth; β is the coefficient in the equation $g_0 = g_e(1 + \beta \sin^2 \varphi)$, in which g_0 is the gravity at latitude φ. β also can be expressed in terms of the gravity g_e at the equator and g_p at the poles as

$$\beta = \frac{g_p - g_e}{g_e}$$

The values of g_e and β are determined from an adjustment of gravity values determined at points widely distributed over the surface of the earth. From the values so determined and the equatorial radius, the flattening, which defines a spheroid giving the shape of the earth, can be calculated.

Gravity Formulas

Several slightly different spheroids and corresponding gravity formulas have been determined from time to time as the amount and precision of gravity and geodetic information have increased. Those which are of greatest interest with regard to the reduction of gravity surveys are discussed below.

Helmert's 1901 formula   This formula (Helmert, 1910, pp. 89–96) is

$$g_0 = 978.030(1 + 0.005302 \sin^2 \varphi - 0.000007 \sin^2 2\varphi)$$

where φ is the latitude. This corresponds to $a = 6{,}378{,}200$ m, $b = 6{,}356{,}818$ m, $1/f = 298.2$. This formula has been used extensively in connection with older gravity surveys.

The 1917 U.S. Coast and Geodetic Survey formula   For this formula see Bowie (1917, p. 134); it reads

$$g_0 = 978.039(1 + 0.005294 \sin^2 \varphi - 0.0000007 \sin^2 2\varphi)$$

The constants for this formula were based on the adjustment of 216 gravity values in the United States, 43 in Canada, 17 in Europe, and 73 in India. This formula has been used for the reduction of extensive gravity surveys made in the United States by the U.S. Coast and Geodetic Survey. The corresponding value for $1/f$ is 297.4.

The 1930 "international formula"   This formula (Cassinis, Doré, and Vallarin, 1937) is

$$g_0 - 978.049(1 + 0.0052884 \sin^2 \varphi - 0.0000059 \sin^2 2\varphi)$$

which corresponds to $a = 6{,}278{,}388$ m, $b = 6{,}356{,}909$ m, $1/f - 297.0$.*
   This formula was adopted by the 1930 General Assembly of the International Association of Geodesy and now is commonly used for the reduction of gravity measurements. A table of values of "normal" gravity, calculated from this formula, is given in the Appendix to Chap. 4.†
   Similar calculations can be made under the assumption that the earth is a triaxial ellipsoid. In this case, the gravity formula shows a dependence on the longitude λ, as in the formula (Niskanen, 1945)

$$g_0 = 978.0468[1 + 0.0052978 \sin^2 \varphi - 0.0000059 \sin^2 2\varphi$$
$$+ 0.000023 \cos^2 \varphi \cos 2(\lambda + 4°)]$$

---

* Values to 10^{-4} gal for each minute of latitude can be found in Theoretical Gravity at Sealevel for each Minute of Latitude by the International Formula, *U.S. Coast Geod. Sur., Publ.* G53, Washington, 1942.
   † Recent measurements indicate that the Potsdam value is too high and that the first term in the formula should be about 14 mgals lower (see page 58).

Several formulas of this type are listed by Caputo (1967, pp. 38–40), who shows that they imply a variation along the equator of around 200 m in elevation and only a few tens of milligals in gravity values. Such small variations in the gravity are hard to separate from those caused by lateral density variations, and the imagined accuracy to be gained by considering the triaxiality of the earth does not justify the increased complexity of the geodetic and gravimetric calculations.

APPLICATION TO GEOPHYSICAL PROSPECTING

For the application to geophysical prospecting, the interest in the gravity formulas lies in their use in making corrections for the normal northward or southward gravity increase. The maximum effect in middle latitudes (near 45° north or south) amounts to about 1.4 mgals/mile; therefore, a latitude correction must be made (see page 87). The minor differences between the various formulas are negligible, and the table calculated from the 1930 formula (Appendix to Chap. 4) is generally used for latitude corrections.

The first term (978.049) of the international gravity formula is the value of gravity at sea level at the equator. Its value depends both on a systematic adjustment of relative gravity measurements over the earth's surface and on an absolute value of gravity at some one place, to which the relative measurements are referred. The value given is based on a gravity value at Potsdam (located in East Germany with geodetic coordinates 52°22.86′N and 13°04.06′E and some 86¼ m above sea level) of 981.274, which was determined by absolute gravity measurements made in 1906 with a series of six Kater reversible pendulums of different mass. Recent new measurements by falling-body methods at other locations, referred to Potsdam by previously established relative gravity connections, have shown that the value there is about 13.8 mgals too high (see page 58). Any recognized correction to the Potsdam value would also apply to the first constant of the gravity formula. Thus, for the international gravity formula, the value would be changed from 978.049 to 978.035. For exploration uses of gravity, which are based on relative rather than absolute measurements, this small change is of no consequence.

The Normal Spheroid and the Geoid

If the earth were a perfect fluid with no lateral* variations in density, its surface would correspond to an ideal ellipsoid of revolution, the so-called normal spheroid represented by the gravity formula. This would be a level surface, and the direction of gravity everywhere would be perpendicular to this surface. Actually, the earth is not uniform, and there are departures of the level surface from the normal, or reference, spheroid. The actual level surface may be considered as that of the oceans (if free from disturbing tidal and wind forces) and the oceans as extended

---

* Darwin (1910) pointed out that the vertical variations in density from surface to center of the earth cause a very small depression, which reaches a maximum of 3 m at latitude 45°.

across land areas by imaginary deep canals. The level surface so defined is the "geoid." An ordinary spirit level indicates a surface parallel to the geoid (and a plumb line gives a direction normal to the geoid). The actual level surface is deformed or warped by the irregularities in density within the earth and the topographic irregularities of its surface. Evidence for this warping of the geoid is given by small deviations of the plumb, or vertical, from the direction it would have if it were perpendicular to the normal spheroid. These deviations (deflections of the vertical) can be determined by certain geodetic and astronomic measurements and also by surface integration of the gravity field around the point of calculation.

The very detailed measurements of artificial satellites have given an independent means of measuring the gross irregularities of the earth's surface. A geoid surface derived from satellite orbits given by Stacey (1969, p. 53) shows variations of continental dimensions and with relief from about $+70$ to -80 m.

The very local warping of the geoid is the source of the local variation of horizontal gravity measured by the torsion balance (see page 66). Further attention will be paid to the geoid surface in the section on isostasy (page 278).

VARIATION OF GRAVITY WITH ELEVATION (FREE-AIR EFFECT)

Gravity varies with elevation* because a point at a higher elevation is farther away from the center of the earth and therefore has a lower gravitational acceleration than one at a lower elevation. The rate of this normal vertical variation, or the vertical gradient of gravity, can be calculated quite accurately from the gravity formula and the radius of the earth. This can be shown approximately as follows.

The gravity at a point on the surface of a spherical earth is

$$g = \frac{GM}{R^2}$$

* Direct measurements of the vertical gradient of gravity were attempted in the past by very careful weighings with sensitive balances by means that permitted changing the vertical positions of the masses. Practically all such measurements contained systematic errors which gave values lower than the theoretical value. A review of these measurements and a new series of experiments made by direct measurements of gravity at different elevations in high buildings are given by Hammer (1938). Similar measurements (Kuo, Ottaviani, and Singh, 1969) were made in several tall structures at and near Columbia University. The results, after careful corrections for local effects, particularly from basement excavations for the buildings and from local topography, indicate actual variations in the vertical gradient of the order of ± 1 percent from the theoretical value. These variations are almost certainly caused by local irregularities of mass distribution in the earth's crust in the vicinity of the point of measurement. The same mass irregularities also cause local distortions of the gravity field. Theoretically, the anomalous vertical gradient can be calculated from the field. An attempt to compare calculated with theoretical vertical gradients by Hammer (1938, pp. 79 and 80) was only partially successful. Similar attempts by Kuo, Ottaviani, and Singh (1969) gave qualitatively similar indications, but actual calculations were not made.

where M is the total mass of the earth and R its radius. The vertical gradient is

$$\frac{dg}{dz} = \frac{dg}{dR} = -\frac{2GM}{R^3} = -\frac{2g}{R}$$

If we take for the mean radius of the earth $R = 6.367 \times 10^8$ cm and for the theoretical value of gravity at sea level and at $45°$ latitude $g = 980.629$ gals, then

$$\frac{dg}{dz} = -\frac{2 \times 980.629}{6.367 \times 10^8} = -0.3086 \times 10^{-5} \text{ gal/cm}$$
$$= -0.3086 \text{ mgal/m} = -0.09406 \text{ mgal/ft}$$

There is a small second-order term which is appreciable only at high elevations. According to Heiskanen and Vening Meinesz (1958, p. 149), this term amounts to only 0.07 mgal for an elevation of 1000 m, 0.3 mgal for 2000 m, and 1.7 mgals for 5000 m, or about $0.07h^2$, where h is in kilometers. This is nearly always neglected, and the free-air correction is calculated as 0.3086 mgal/m or 0.9406 mgal/ft.

If a proper correction for this elevation effect were not made, a gravity map would be strongly affected by differences in elevation between different points of measurement. Therefore, a correction for the elevation (including the Bouguer correction, page 89) is always made before mapping gravity measurements made for gravitational prospecting.

The simple correction for elevation, using the constant given above, is called the free-air correction because it is calculated as if the elevated point of measurement were suspended free in the air without any regard for the effects of the attraction of the mass of matter between the elevation of the point of measurement and the reference elevation. A "free-air map" is made from data with latitude and free-air corrections.

ATTRACTION OF NEAR-SURFACE MATERIAL (BOUGUER EFFECT)

Suppose that two gravity stations are at different elevations, such as points A and B (Fig. 2-1), and we wish to calculate what their gravity difference would be if they were made at the same level (say the level of A). If we simply correct station B to the elevation of A by the free-air correction discussed above, we have taken no account of the attraction at B of the mass of material below B which would not be present if B were at the same level as A. The correction for the attraction of this material is commonly called the "Bouguer* correction."

If the topography is fairly flat, the attraction of material under the point B is given quite closely as that of an infinite slab of thickness h and density σ. The

---

* Bouguer, the previously mentioned French mathematician and geodesist, made pendulum measurements in the high mountains of Peru in 1735 to 1743. He was the first to make corrections for the effect that now bears his name in connection with the reduction of these measurements.

FIGURE 2-1
The Bouguer gravity effect.

attraction of such a slab is $g = 2\pi G\sigma h$, which for $G = 6.6732 \times 10^{-8}$ (page 13) gives

$$g = \begin{cases} 0.04193\sigma h & \text{mgals/m} \\ 0.01278\sigma h & \text{mgals/ft} \end{cases}$$

The Bouguer effect under station B tends to increase the gravity and therefore is always opposite to the free-air effect (which would decrease the gravity at B relative to that at the lower point A). Thus the free-air and Bouguer corrections are always opposite in sign.

In calculating corrections to gravity stations the free-air and Bouguer corrections commonly are combined into a single factor (which depends on the density σ of the surface rocks within the range of the topography). The combined elevation correction is made to sea level or some other reference level.

If the topography is irregular, the correction for the attraction of the material is much more complicated, and the attraction of all excess masses (such as M, Fig. 2-1) above the plane of the station (or mass deficiencies from voids below the plane of the station) must be taken into account. Such corrections are usually called "terrain corrections." A more detailed discussion of them and of methods for their calculation is given below (page 92).

3

GRAVITY-MEASURING INSTRUMENTS

HISTORICAL BACKGROUND

Measurements of the magnitude of the earth's gravitational acceleration have been made since the time of Newton. The early measurements were accurate enough to show that gravity varies with latitude and with elevation. However, the demands of geophysical prospecting require that differences in gravity be measured to a precision several orders of magnitude higher than any of those early measurements. In fact, simple calculations show that the gravity effects which may be expected from moderate geologic structure are of the order of a few milligals to as little as a few tenths of a milligal. Thus, to be effective as a means of geophysical prospecting, gravity differences must be measured to 1 ppm or better, and a precision of less than 0.1 mgal, or 1 part in 10^7, is required for accurate measurement of anomalies from small geologic features. Several field gravity meters make measurements to a precision of 2 to 5 parts in 10^8 in routine operations.

Three quite different physical properties have been applied in instruments used for gravity measurement in geophysical prospecting. Historically, the first gravity instrument used by the oil industry was the torsion balance, which measures components of the space rate of change of gravity rather than gravity itself (see page 66). First used in the United States late in 1922, it was very effective in finding salt domes in the Gulf Coast but is now quite obsolete. The next was the pendulum, which had been used in Europe, America, and elsewhere in the scientific and geodetic investigations of gravity for two centuries. The precision and speed of operation required for geophysical prospecting with a pendulum, however, were first reached in about 1929 (see page 58). Its most extensive use was by the Gulf Oil Corporation, which made many thousand pendulum stations from 1930 to 1935; it has not been used for prospecting since that time, although the same pendulums are currently in use for geodetic gravity measurements (Woollard and Rose, 1963). Gravity prospecting is now carried out with the static gravity meter, or gravimeter, which was first used for geophysical prospecting in the Texas Gulf Coast in about 1932. By about 1940 gravity meters of high sensitivity, precision, stability, and convenience had reached such a stage of development that they had completely replaced the torsion balance and pendulum as field instruments for gravity prospecting.

REQUIREMENTS OF GRAVITY METERS

In principle a gravity meter is simply an extremely sensitive weighing device. The weight or force of gravitational attraction on a constant mass changes with any variation in the gravitational field. Therefore, if means are provided for detecting small enough differences in weight, the corresponding small differences in gravity can be measured.

Several different types of apparatus have been suggested and built for gravity exploration. However, with one or two exceptions, the only ones that have been reasonably successfully employed in the field in geophysical prospecting so far are essentially a mass supported by a coiled spring, a horizontal torsion fiber, or a vibrating string.

It may seem simple to hang a mass on a spring and measure the changes in weight produced by changes in gravity, but let us consider the sensitivity requirements. As mentioned above, geophysical prospecting requires the detection of differences in gravity at least as small as 1 mgal and preferably less than 0.1 mgal. To detect a gravity difference of 0.1 mgal requires a sensitivity of 1 part in 10 million of total gravity. Suppose that we hang a mass on a spring and the spring is stretched about 1 ft, or 30 cm. This stretch results from the total force of gravity acting on the mass. A change in gravity of 1 part in 10 million will produce a change in length or movement of the weight of 30×10^{-7} cm. One of the most sensitive methods of measuring small changes in length is an interferometer, which will measure lengths of the order of one wavelength of light, or about 5×10^{-5} cm. The changes in length in the spring that must be measured in a gravity meter are about one-tenth of a wavelength of light and therefore beyond the range of a simple interferometer.

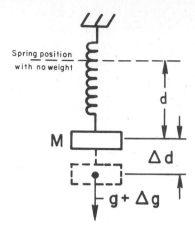

FIGURE 3-1
A simple weight on a spring.

Types of Gravity Meters

On the basis of the means used for the measurement of these very small changes in the length of a spring, gravity meters may be divided into two types: (1) the stable type uses a system which gives a high order of optical or mechanical magnification so that the change in position of a weight or associated property resulting from a change in gravity is measured directly; (2) the unstable type uses a moving system which approaches a point of instability so that small changes in gravity produce relatively large motions. Usually, stable-type instruments give readings that vary linearly with gravity over a wide range. Unstable types have a more limited range and a nonlinear response; therefore they are always read by measuring the variation of a null, or balancing, force required to bring the moving beam back to a fixed reference position. Some instruments of the first type also use nulling devices.

A null reading method keeps the moving system between very closely spaced stops. This greatly reduces hysteresis errors, which occur if the spring length is changed. Also, the beam is kept very close to a horizontal position, which minimizes sensitivity to leveling errors.

Relation of Sensitivity to Period

Instruments of the unstable type gain their sensitivity by being adjusted to a relatively long period. The relation between sensitivity and period can be shown by considering a simple weight of mass M on a spring (Fig. 3-1). The total elongation of the spring is

$$d = \frac{MG}{c}$$

where d is the extension from the zero-load condition and c is the spring constant, i.e., the ratio of stretch to force or of strain to stress. As an oscillator, the period of the system is

$$T = 2\pi \sqrt{\frac{M}{c}}$$

and

$$\frac{M}{c} = \frac{T^2}{4\pi^2}$$

Substituting in the first equation gives

$$d = \frac{T^2}{4\pi^2} G$$

For small changes in gravity Δg, the resulting change in elongation of the spring is

$$\Delta d = \frac{T^2}{4\pi^2} \Delta g$$

and the sensitivity of the system is

$$\frac{\Delta d}{\Delta g} = \frac{T^2}{4\pi^2}$$

so that the sensitivity is proportional to the square of the period. These considerations give us some idea of what is required to make a sensitive gravity meter.

If it is decided that Δd can be measured to 0.001 mm and that we need to measure Δg to 0.1 mgal, the ratio is $\Delta g/g = 0.1/10^6 = 10^{-7}$, from which $\Delta d/d$ is also 10^{-7} and the required total elongation d is 0.001 mm $\times 10^7 = 10^4$ mm = 10 m, or 33 ft. This consideration led an early consultant on the gravity-meter problem to say that it would be possible to build a meter but it would have to be over 30 ft long.

The period of this same system with elongation of 10 m, or 1000 cm, can be calculated from

$$T^2 = \frac{4\pi^2 d}{G} = \frac{40 \times 1000}{980} \approx 40 \quad \text{and} \quad T = \sqrt{40} = 6.3 \text{ sec}$$

If ways other than making the spring very long can be devised to attain the long period, the sensitivity of displacement to gravity change will be the same as for the long spring. The unstable-type meters have a mechanical system which is adjusted to operate very near a point of instability. As that position is approached, the period becomes longer because the force of the principal spring is nearly balanced by a contrary force, which results from the geometry of the system or a counteracting spring or ligament. If the system can be adjusted to have a period of, say, 7 sec, we can still measure 0.1 mgal by measuring a deflection of 0.001 mm. In some common field meters this long period and resulting sensitivity are attained in moving systems with dimensions of only 1 or 2 in.

Elimination of Nongravitational Effects

Any instrument assembly sensitive enough to detect changes of gravity to the order of 1 part in 10^7 or less is subject to disturbances from causes other than changes in gravity. Thus a gravity meter may also be a thermometer, a level, a barometer, and a magnetometer.

Temperature effects can be avoided by enclosing the sensitive element within a constant-temperature "oven" maintained at a precisely controlled temperature (variation less than 0.001°C), which must be higher than the highest ambient temperature to which the instrument may be exposed. Some gravity instruments (see page 37) use a temperature-compensating element; in this case the temperature of the entire moving system must be uniform to a high degree, which requires very good external thermal insulation, usually a vacuum flask.

Sensitivity to level can be reduced by careful design, but all instruments require a level adjustment or automatic leveling for each reading.*

Barometric pressure effects result from variable buoyancy owing to variation of the density of the air surrounding the moving system. In some instruments these effects are avoided by enclosing the sensitive element in a sealed chamber with necessary external adjustments made through packed shafts. In others a sealed cell is mounted on the moving system in such a way that variations in buoyancy of the cell with varying air density compensate very accurately the air pressure effects on the other parts of the moving system. In at least one instrument (LaCoste and Romberg) both systems are used.

Magnetic effects, if any, usually come from magnetic forces on a steel spring. These are generally avoided by using a nonmagnetic alloy for the spring material. Some instruments (now obsolete) required orientation with respect to the compass direction to avoid disturbances from such effects.

The real success of a gravity meter depends less on the physical principle involved than on the details of design and the skill of the maker. Performance depends on the features which make for resistance to shock to withstand field operation and on the adequacy of the auxiliary devices to make the gravity-sensitive elements insensitive to temperature, barometric pressure, level, magnetic field, etc. Of the many instruments which have been constructed, only a few have been used for the many millions of gravity observations made all over the world. These few are the ones which are sensitive, rugged, and reliable and will measure gravity day after day in routine field operations to less than about 0.05 mgal.

Some 20 or more different gravity-measuring instruments have been developed and applied in varying degrees to practical gravity prospecting (see Table 3-2). Most of them are capable, in principle, of measuring gravity differences to around 0.1 mgal or less, but relatively few have proved rugged, reliable, and stable

---

* Sensitivity to level can be reduced by designing the instrument so that the axis of rotation of the beam in the nulled position is horizontal and the center of gravity of the beam is on a horizontal plane passing through the axis of rotation. Then if the meter is off level, in perpendicular directions, by angles e_1 and e_2, the error in gravity will be $1 - (\cos e_1)(\cos e_2)$, which is very nearly $(e_1 e_2)^2$ for small angles.

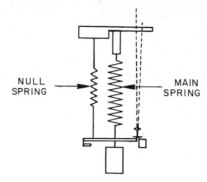

FIGURE 3-2
Principle of the Hartley gravity meter.

enough for extensive field operation. The operating principles of some of those instruments are described in the following paragraphs; those selected are either of historical interest, contributing new or basic principles, or have been used extensively in the field for gravity prospecting.

WORKING PRINCIPLES OF GRAVITY METERS

Stable-Type Gravity Meters

The Hartley gravity meter   One of the earliest published descriptions of an instrument designed for commercial geophysical exploration was by Hartley (1932). The Hartley gravity meter (Fig. 3-2) consisted of a simple mass suspended on a spring. The position of the mass with respect to the surrounding case was determined by an optical lever. In Hartley's instrument the ratio of the motion of the spot of light reflected from the optical lever to the motion of the suspended weight was about 60,000. The instrument could detect gravity differences of about 1 mgal. It was never used extensively in actual field work because of the early death of its inventor and its imperfect stage of development at the time.

This instrument is of historical interest as the first one using the null principle. The instrument was read by compensating for variation in gravity by adjusting the tension of a small, weak auxiliary spring to bring the moving system back to a fixed reference point. The reading of the gravity difference was in terms of the motion of a micrometer screw which controlled the force on the auxiliary spring. Some of the later, more successful field instruments were licensed under the Hartley patent which covered this principle.

The Gulf (Hoyt) gravity meter   This instrument does not directly measure the change of length of a spring due to changes in gravity but measures instead the rotation of the lower end, i.e., the "unwinding," of the spring associated with its elongation (Hoyt, 1938*a,b,c*). By winding the spring from a flat ribbon with a

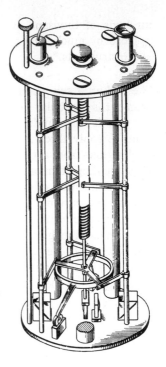

FIGURE 3-3
Working principle of the Gulf (Hoyt)
gravity meter.

width-to-thickness ratio of about 100, the amount of angular rotation associated with a given elongation is greatly increased over that which results from a spring wound with round wire. The spring carries a weight provided with electromagnetic damping vanes, and underneath the weight is a mirror which hangs close to a fixed mirror. The mirrors are partially silvered so that light from a narrow slit passing through them is reflected back and forth a number of times. Images of the slit formed from these reflections are read in a micrometer eyepiece at the top of the instrument. Thus the ultimate reading is changed from measuring a small length to measuring a small angle. With a spring length of about 30 cm and mass of about 100 g, the angular deflection is about 10 seconds of arc per milligal change in gravity. The magnitude of the angular change resulting from a given change of gravity is proportional to the order of the reflection which is read in the eyepiece. Common practice was to read the fourth reflection, which produces a sensitivity of about 0.1 mgal per scale division on the micrometer eyepiece. The drift of the instrument is quite regular, and from many repeat observations the probable error of a single observation has been determined to be about 0.04 mgal. The instrument is sensitive to temperature and therefore is provided with thermal insulation and a thermoregulator so that the temperature is kept constant to around 0.1°C. The field meter (Fig. 3-3) is rather large, with a total weight of about 80 lb. It was commonly carried in a frame suspended on elastic shock chords. This instrument was used in the field by the Gulf Oil Corporation from 1935 to about 1960 and hundreds of thousands of stations were observed with it. It has been largely replaced by much smaller, lighter, and more easily read (but not more accurate) field instruments.*

The Askania sea gravity meter Gss 3 The basic moving system of this instrument is a tubular mass T (Fig. 3-4), which is supported by a main spring S. The tube is very carefully restrained to move only on its vertical axis by five ligaments L (only three shown in the section), which prevent effects of horizontal accelerations from acting on the vertically sensitive element and thereby eliminate cross-coupling components (see page 114).

The position of the moving system is sensed by a condenser system, which controls the current in a balancing coil C through an oscillator, amplifier, and phase-sensitive rectifier. This coil, through the reaction of its variable current with the permanent magnet M, balances the changes in gravity and in vertical acceleration forces. The external electronic system measures and filters the balancing current. The separation between gravity and the forces due to motional acceleration is made on the basis of the wide difference of the rate of change of the two quantities.

The instrument is mounted on a gyrostabilized platform to maintain the measuring axis in the average vertical.

It is stated that the instrument can be read to 0.1 mgal and has measured gravity at sea to 1 mgal even under relatively heavy sea motion.

Unstable-Type Gravity Meters

The Holweck-Lejay inverted pendulum This instrument (Holweck and Lejay, 1930) is not strictly a static gravity meter since it depends on measurement of the period of a pendulum. However, the pendulum is inverted, the mass being at the upper end of a flat spring. The elastic constant of the spring and the size and shape of the mass are adjusted in such a way that the system approaches instability and the period becomes very sensitive to small changes in gravity (Hoskinson, 1936). The ratio of change in period to change in gravity is many times greater than for an

---

* In 1933 O. H. Truman, designer of the original Truman meter (page 30), who had left Humble, made a gravity meter on his own and was testing it in the field near Houston. At the same time, a Gulf pendulum party was in the field north of Houston. Truman approached the Gulf party chief (George Lamb) to arrange a test by exchanging readings at pendulum stations. This comparison demonstrated that the new gravity meter was capable of measuring gravity differences quite well. Until this time the Gulf geophysical organization had not been seriously interested in gravity-meter development but was spurred by this first knowledge that a workable gravity meter was in the field. Archer Hoyt, with a new Ph.D. in physics from Cornell, was hired and in effect told to build a gravity meter any way he could. It is a tribute to Hoyt (who died only a few years later) and to the support of the Gulf technical organization that a workable meter was in the field only 9 months later. Some 18 instruments of the same basic design were constructed and put into the field, making Gulf the most aggressive of the major oil companies in the early application of gravity-meter surveys to petroleum exploration.

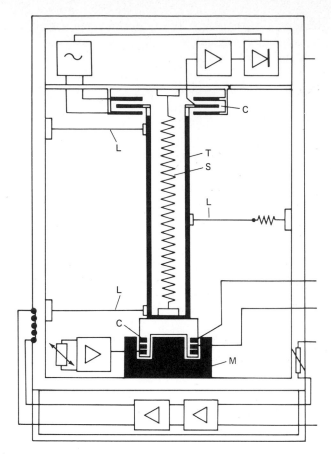

FIGURE 3-4
Diagrammatic cross section of the Askania Gss 3 gravity meter.

ordinary pendulum, so that differences in gravity can be detected by time measurements which are much less accurate and correspondingly faster than those required for an ordinary gravity pendulum. This instrument apparently has an accuracy of 1 to 2 mgal. It has been used considerably in large-scale government gravity surveys, particularly in France and China, but has had little if any commercial use in the United States.

The Humble (Truman) gravity meter   Probably the first practical field gravimeter, this instrument has been described by Bryan (1937). The mass (Fig. 3-5) is carried horizontally on a hinged triangular member connected to two springs. The right-hand spring in the figure is relatively strong and has its line of action almost through the point of support to make an unstable system. The left-hand spring is very much weaker and is used to null the deflection and measure the

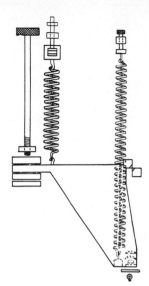

FIGURE 3-5
Working principle of the Humble (Truman) gravity meter.

gravity changes. By careful adjustment of the geometry of the system and the tensions of the springs, the entire system approaches a critical point at which the equilibrium becomes unstable. As this point is approached more and more closely, the sensitivity increases. This instrument measured gravity differences to about 0.2 mgal. Although, as described, it is no longer in use, it is of historical interest as the first practical field gravity meter used in petroleum exploration. It was put into limited use in the Texas Gulf Coast by Humble in about 1932 and by Carter Oil Company in Oklahoma shortly thereafter.

The LaCoste and Romberg gravity meter with zero-length spring As first described by LaCoste (1934, 1935), this is basically a long-period vertical seismograph using a zero-length spring. The same principle was applied to the LaCoste and Romberg gravity meter, which is one of the most stable and widely used gravity field instruments.

The zero-length spring* is made to fit the special condition that the spring force is directly proportional to the distance s between the point of support and

* The origin of the zero-length-spring gravity-meter concept is of interest. About 1932, Lucien LaCoste was a student in physics at the University of Texas under Arnold Romberg. Romberg handed out problems to his class which each student was to work at the blackboard. LaCoste's problem was to design a long-period vertical seismograph. He did not get an answer at the blackboard session but kept thinking about it, particularly in terms of the angles involved, as shown in Fig. 3-6. A week or so later, while waiting for a tennis opponent who was late, he remembered that $\sin 2x = 2 \sin x \cos x$ and that the angle at the center of a circle is measured by the subtended arc while that on the circumference is measured by half the arc. This enabled him to fit the geometry of the vertical seismograph into a circle and led to the concept of the zero-length spring. LaCoste then proceeded to make such an instrument. He wound a spring with prestressed coils (by feeding the wire back at an angle to the previously wound part, as in

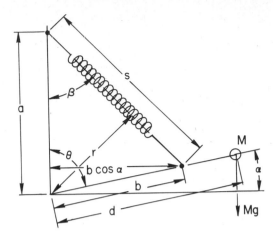

FIGURE 3-6
Diagram for theory of zero-length spring.

FIGURE 3-7
Diagrammatic sketch showing winding of prestressed spring.

the point of application of the force. This means that the stress-strain curve is a straight line passing through the origin, so that the initial length, corresponding to zero force, is zero. This condition can be met by prestressing the spring in winding so that an initial force is required before the coils begin to separate (as in an ordinary screen-door spring) and adjusting the distance between the points of support so that the zero-length condition is satisfied and $F = Ks$, where K is the spring constant.

This condition applied to a gravity meter (or vertical seismograph) permits the design of an instrument with a theoretically infinite period. This condition is also independent of the angular position of the beam carrying the weight, as shown by the following relations.

Fig. 3-7), which then requires a certain stress before the coils begin to open. The spring was terminated to fulfill the zero-length condition. A laboratory seismograph was then constructed (LaCoste, 1934) on this principle and was adjusted to periods of as long as 1 min.

A long-period vertical seismograph, of course, is equivalent to a gravity meter. The first field gravity instrument was constructed in the Physics Building at the University of Texas in 1938 but was not very successful. The second instrument, made in Romberg's basement in 1939, was usable. The partnership was formed and has been making more sensitive, more complex, smaller, and more highly developed gravity meters for various uses ever since.

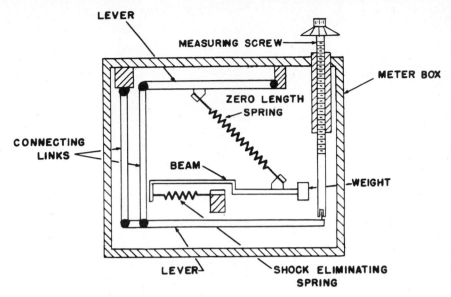

LEVER

MEASURING SCREW

ZERO LENGTH SPRING

METER BOX

CONNECTING LINKS

BEAM

WEIGHT

LEVER

SHOCK ELIMINATING SPRING

FIGURE 3-8
Diagrammatic cross section of LaCoste and Romberg gravity meter. (*LaCoste and Romberg.*)

The gravitational torque from the weight Mg is

$$T_g = Mgd \sin \theta = Mgd \cos \alpha$$

and the torque from the spring is

$$T_s = Ksr \qquad s = \frac{b \cos \alpha}{\sin \beta} \qquad r = a \sin \beta$$

so that

$$T_s = K\frac{b \cos \alpha}{\sin \beta} a \sin \beta = Kba \cos \alpha$$

At equilibrium,

$$T_g - T_s = 0$$
$$Mgd \cos \alpha = Kba \cos \alpha$$
$$Mgd = Kba$$

and the instrument is insensitive to the angles θ, α, and β. Thus it may be in equilibrium over a small range of values of the angles, there is no restoring force if it is displaced slightly from a particular equilibrium point, and it can theoretically be adjusted to infinite period.

The LaCoste and Romberg gravity meter uses a zero-length spring in the general manner shown by Fig. 3-8. The null adjustment is made by a screw operating through the linkage shown, which greatly reduces the movement of the

FIGURE 3-9
LaCoste and Romberg standard land gravity meter. (*LaCoste and Romberg.*)

upper support of the mainspring. The period is usually adjusted to about 15 sec.
A hollow buoyancy-compensating cell (not shown) is attached to the moving system
to make the instrument insensitive to changes in atmospheric pressure.

The widely used land meter (Fig. 3-9) is in a wooden case (about 12×14 in.
and about 14 in. high to the surface of the top plate), which encloses the thermal
insulating material, thermostat for temperature control, etc., and weighs about
28 lb. Power for maintaining constant temperature usually is provided by an
automobile battery. The sensitivity is approximately 0.1 mgal per scale division
(one complete turn of the 50-division reading dial is about 5 mgals). The null
indicator moves approximately one division in the microscope-eyepiece scale for
one division of the reading dial. A reset screw (not indicated in Fig. 3-8) permits a
coarse adjustment to match the range of the instrument to the general level of the
gravitational field in the area of operation. By very careful operation readings can
be made to 0.01 mgal.

The more recently developed "geodetic meter" (Fig. 3-10) has several
advantages. The meter itself is much smaller and weighs about 5 lb. The heating
battery to maintain constant temperature weighs about 6 lb. The coarse adjust--

FIGURE 3-10
LaCoste and Romberg geodetic meter with power supply. (*LaCoste and Romberg.*)

ment is eliminated by providing a single screw with worldwide range of over 7000 mgals. A carefully determined calibration table is supplied for each meter. This calibration is a property of the screw, rather than the spring or linkages, and therefore is highly stable. The moving system is pressure-compensated but also is sealed against outside pressure changes. Readings are made routinely to 0.01 mgal.

Several other gravity meters use zero-length springs and in construction and operation are similar to the LaCoste and Romberg meter, although their external appearance differs. They vary also in the details of construction, particularly in the configuration of the ligaments and hinges which support the moving system and in how the null adjustment is made. These include both the Frost and the North American gravity meters, which had some personnel or details of concept in common with the LaCoste and Romberg instrument during their development.

The Graf-Askania Seagravimeter Gss 2 Developed for measurement of gravity at sea rather than as a development from a land gravity meter (Graf, 1958), this instrument has a basic moving system (Fig. 3-11*a*) consisting of a flat horizontal bar supported by torsion springs. The gravitational moment from the weight of the bar is critically balanced by the torsion of the two springs to give a long period (about 6 sec). The bar moves within a strong field of a permanent magnet to give very strong eddy-current damping. A set of eight ligaments (Fig. 3-11*b*) stabilizes the system so that its motion is confined to rotation about the horizontal axis of the springs.

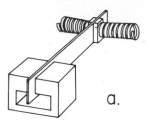

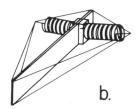

FIGURE 3-11
(*a*) Schematic diagram of moving system of Graf Gss 2 gravity meter. (*b*) Diagram of ligaments to restrict movement to rotation about the spring axis. (*From Graf, 1958.*)

A later modification (Graf, 1961) has greatly increased damping to meet operating conditions on relatively small ships with large motional acceleration (to 10 percent of gravity). It has been used quite extensively by oceanographic institutions, particularly the Lamont-Dougherty Laboratory, for long cruises in the open ocean, and also has had some application in commercial petroleum exploration. [Gss 3 (see p. 29) is a completely different instrument developed later.]

Gravity Meters with Quartz Springs and Moving Systems

Several gravity meters have been built with main support, beam, and spring elements composed largely or entirely of fused quartz. The spring element is supplied as the torsion of a quartz fiber. A compensating element or fiber supplies a counterforce to permit adjustment to a critical condition of high sensitivity.

The Mott-Smith gravity meter   This instrument (Mott-Smith, 1938) consists essentially of a horizontal quartz fiber (Fig. 3-12) carrying a perpendicular horizontal weight arm connected to a "labilizer" fiber which passes through the line of the horizontal fiber and is attached to a spring. The weight arm carries a vertical index, the position of which is measured by a microscope. The torsion of the fiber and the tension in the labilizer fiber are adjusted so that their effects on the moving system are nearly balanced; thus a small change in gravitational attraction on the weight arm produces a relatively large movement of it and of the index arm. The essential parts of the instrument are all made of quartz, fused together so that the entire moving system may be considered as continuous. The instrument is quite

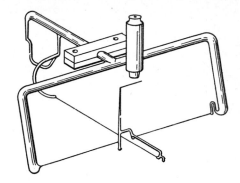

FIGURE 3-12
Moving system of the Mott-Smith gravity
meter.

small. The torsion fiber is about $1\frac{1}{2}$ in. long and 0.002 in. in diameter, the weight arm and pointer are each about 1 in. long, and the labilizing fiber is about 2 in. long and 0.005 in. in diameter. The instrument is mounted in an airtight case about 2 in. deep and 5 in. in diameter. A liquid thermostat keeps a constant temperature to about 0.001°C.

Fused quartz has very stable elastic properties, and the operating characteristics of the instrument indicate a very regular drift and stability such that gravity measurements can be made to better than 0.1 mgal. When the meter was first made, the difference in gravity was read from the motion of the index in the microscope field; later the instrument was modified to use a very weak adjusting spring and read as a null instrument.

The Worden gravity meter The Worden meter is a development from the Atlas meter, a temperature-compensated version of the Mott-Smith instrument. Figure 3-13 diagrams its essential parts. The primary moving system is carried by a horizontal bar B, which is supported by the two hinges, h and h, where the small quartz rod is drawn down to a short, very fine fiber. The zero-length main spring S is attached to a connecting arm CA, which slants downward at a critical angle. This angle and the strength of the spring S are such that the rotational period of the system is relatively long (5 to 8 sec). A vertical arm carries a pointer P, which is in the field of the microscope I. The connecting arm also carries the weight arm WA, which is looped back between stops to carry the moving mass M.

On the other side of the assembly is the bar AB which carries the adjusting and compensating system. The mainspring S is supported on this bar through the A frame AF. The temperature compensation is accomplished by the tungsten fiber T in the center of the A frame but offset at its lower end. The null-adjusting spring NS is attached to the system through the adjusting arm AA. The upper end of this spring is attached to the reading screw, which operates through a bellows to the reading head on the top plate. The range-adjustment spring RS is controlled at the top plate through another bellows.

The entire assembly of frame, springs, and connecting arms is made of fused quartz and mounted on the instrument base by a metal post P. The dimensions

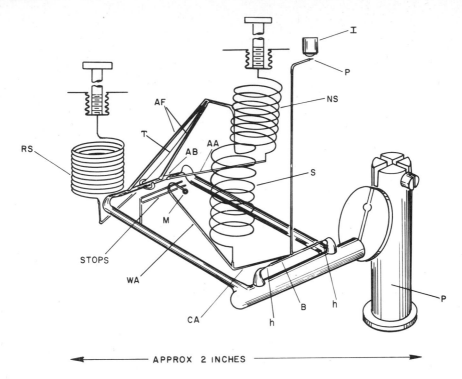

Drawn, with modification and additions,from
photograph of model furnished by Texas Instruments.

FIGURE 3-13
Moving system of Worden gravity meter.

are quite small, as indicated. There is no clamping and no mechanical damping. The moving system is so small and light that sufficient damping is obtained within the partial vacuum at which the system is sealed off.

The temperature compensation requires all parts of the moving system to be very accurately at the same temperature. Rapid temperature fluctuations are avoided by mounting the essential moving parts within a vacuum flask with a long insulating plug through which the control rods and windows for adjustment and observation are brought out to the top plate. The vacuum flask is mounted in a stainless-steel or aluminum outer case. The external dimensions are about 5 by 13 in., and the total weight is about 6 lb (Fig. 3-14).

The great advantage of this instrument is its small size and weight and the elimination of temperature control and the consequent thermostat, batteries, connecting wires, etc. This makes it particularly attractive when it must be carried, as the entire load with carrying case and pack to fit on one's back is only about 10 lb. A very large number of Worden instruments have been built and used all over the world.

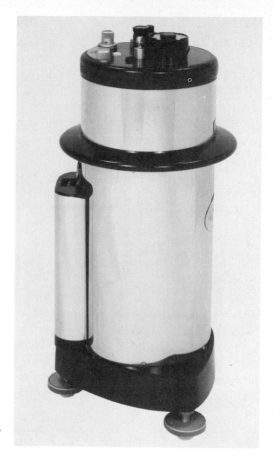

FIGURE 3-14
External view of Worden gravity meter.
(*Texas Instruments.*)

The Norgaard gravity meter Certain unique principles of this instrument, particularly a purely geometric calibration and an ingenious system of temperature compensation, make it worthy of brief description although it has not had extensive field application. Figure 3-15 is a schematic diagram of the working principles. The parts shown are all made of fused quartz, and the assembly is quite similar in its essential parts to the Mott-Smith and Worden instruments. An approximately horizontal beam A is supported on horizontal quartz torsion fibers T, which act as torsion springs. A microscope and optical system compare the relative position of mirror M_1, carried on the moving beam, with mirror M_2, which is rigidly fixed to the frame supporting the fibers. Readings are made by tilting the case about an axis parallel to the fibers T-T. At two positions, one with the beam tilted upward and one with it tilted downward, mirrors M_1 and M_2 are parallel. At these two symmetrical positions, when the case is tilted at an angle θ from the position where the beam would be horizontal,

$$g_1 l \cos \theta_1 = g_2 l \cos \theta_2 = g_3 l \cos \theta_3 = g_0 l$$

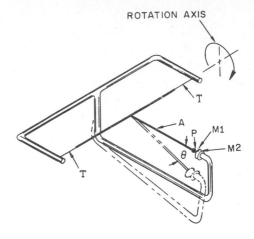

FIGURE 3-15
Schematic diagram of moving system of
Norgaard gravity meter. (*The Electrical
Prospecting Company, Stockholm.*)

where g_1, g_2, etc., are values of gravity at different places where the balance is obtained with tilt angles θ_1, θ_2, etc., and g_0 is the theoretical value of gravity such that the beam would be horizontal ($\theta = 0$). The length l from the axis of rotation to the center of gravity of the beam is constant. Since the balance is always made at the same position of the moving mirror with respect to the fixed mirror, the degree of twist and therefore the torque moment of the fibers is constant. Thus, gravity at a given point is measured by

$$g_1 = \frac{g_0}{\cos \theta_1} = g_0\sqrt{1 + \tan^2 \theta_1} = g_0(1 + \tfrac{1}{2}\tan^2 \theta_1)$$

for small values of θ, and the "calibration factor" is purely geometrical.

In practice, the inner case is rotated between two micrometer screws which measure the linear distance between the two positions of balance; this distance is a measure of twice the angle θ. In the instrument as described by the maker, $g - g_0 = 7.00m^2$ (in milligals when m, the motion measured by the micrometers, is in millimeters); it is claimed that by reading the micrometer to 0.005 mm, gravity measurements can be made with an accuracy of 0.02 to 0.06 mgal.

The Norgaard gravimeter is compensated for temperature effects by immersing the moving system in a liquid which changes its density with temperature and thereby changes the buoyancy torque on the moving system. By proper choice of liquid and the buoyancy moment of the beam (modified by the small weight P, Fig. 3-15) the effect of temperature on the elasticity of the quartz fibers is compensated exactly at one temperature and to first order for changes from that temperature. A second-order temperature correction must be made, which is approximately proportional to the square of the difference from the temperature of primary compensation and which amounts to about 7.5 mgals for a temperature difference of 20°C.

Vibrating-String Gravity Meters

The natural frequency of a vibrating string depends on its tension and on the mass per unit length of the string. If the tension is provided by the weight of a suspended mass, the frequency of the string will change with variation of gravity. With modern electronic technology frequency can be measured to a high degree of precision, which makes the vibrating-string gravity meter attractive.

The fundamental frequency F of a vertical vibrating string held in tension by the weight of a mass M is (Gilbert, 1949)

$$F = \frac{1}{2L} \sqrt{\frac{MG}{m}}$$

where m is the mass per unit length and L is the length of the string. From this,

$$F = K\sqrt{g}$$

where K is a constant, and

$$\frac{\Delta F}{F} = \frac{1}{2} \frac{\Delta g}{G}$$

The frequency also depends to much smaller degrees on the shape of the string cross section, the exact nature of its termination or fastening at the ends, temperature, air pressure, etc. The "string" may be a thin flat ribbon to minimize dependence of its frequency on the elastic constants of the material.

To make a gravity meter, the string is kept in oscillation at its natural frequency by being driven in a magnetic field by suitable electric circuitry. The frequency of the resulting alternating current is measured to very high precision, usually of the order of one part in 10^8. One method measures the beat between the string frequency and that of a suitable crystal oscillator. Another method of measurement is to count the number of oscillations of a crystal of much higher frequency, the count being initiated and stopped at the beginning and end of a selected whole number of oscillations of the string.

Although vibrating-string gravity meters have not been used for routine gravity surveys, probably because of the complications of the necessary electronic accessories and the simplicity, small size, and weight of alternative instruments, they have been used in two special applications. Their linear nature and potentially small diameter make it possible to use these meters for gravity measurements in drill holes of moderate size. The inherent averaging obtained when the number of oscillations is counted over a discrete time interval is a very useful property in measuring gravity at sea in a field of strong environmental accelerations. Vibrating-string instruments which have been developed as working instruments are described in the following paragraphs.

The Gilbert gravity meter This vibrating-string gravity meter (Gilbert, 1949) was originally designed to measure gravity in a submarine. The vibrating "string" is a beryllium-copper strip about 5 cm long. The suspended mass consists of an inverted copper cup in a magnetic field to provide electromagnetic damping.

The resonant frequency is about 1000 Hz. The string, in an evacuated constant-temperature chamber, is sharply resonant with a Q of the order of 20,000.

The frequency of the string is compared with a crystal frequency standard, and the resulting beat frequency is measured to a very small fraction of a cycle. Initial tests at sea, where comparisons were made with pendulum observations in a submarine, indicated a probable error of about 1.5 mgals.

A modified instrument operating on the same general principles was made for measuring gravity in a borehole for the determination of density (Gilbert, 1952). The external diameter of the borehole instrument is 5 in., and the total length is about 8 ft. The suspended instrument is connected to the surface equipment by a four-conductor cable. From tests made in a drill hole the probable error of a single observation derived from repeated observations at several levels is about ± 0.7 mgal.

The AMBAC Industries (Wing) vibrating-string accelerometer   This device was developed originally as an accelerometer for guided-missile applications. An instrument with the same basic components has been mounted on a gyro-stabilized platform adapted from a gyro compass and tested as a sea gravity meter (Wing, 1969). The basic sensor consists of a combination of two vibrating strings connected by a relatively soft spring (Fig. 3-16). With an increase of gravity the tension in the upper string is increased and its frequency increases, while the opposite effect occurs in the lower string.

The frequencies can be expanded in Maclaurin's series (Wing, 1967, p. 1251) as

$$F_1 = K_{01} + K_{11}g + K_{21}g^2 + K_{31}g^3 + \cdots$$
$$F_2 = K_{02} - K_{12}g + K_{22}g^2 - K_{32}g^3 + \cdots$$

where F_1 and F_2 are the frequencies of the two strings and the K's are instrumental constants, the K_0 terms being dependent on the fixed tension on the string system.

The difference frequency is

$$\Delta F = F_1 - F_2$$
$$= (K_{01} - K_{02}) + (K_{11} + K_{12})g + (K_{21} - K_{22})g^2 + \cdots$$

Thus if the pairs of K coefficients can be made identical, the odd terms cancel out and the difference frequency is a measure of g.

The vibrating "strings" actually are thin beryllium ribbons about $\frac{3}{8}$ in. long, 0.010 in. wide, and 0.002 in. thick. The overall tension is provided by the "soft" central spring, and the ends of the ribbons are firmly clamped or welded to the instrument frame. In one model, the natural frequency is about 4900 Hz. In another, it is about 9800 Hz. The permanent magnets, shown as cylinders in the figure, are divided by slots to give a magnetic field across the strings. The two strings are arranged to carry electric currents which are in oscillating circuits so that they vibrate at their natural frequency at a very low amplitude of 70 to 80 microinches (μin.). The system is very sharply resonant ($Q \approx 2000$).

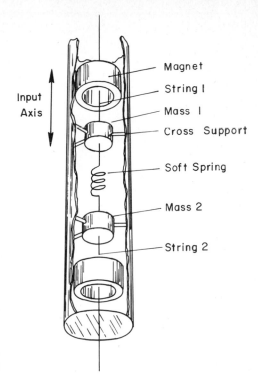

Input
Axis

Magnet

String I

Mass I

Cross Support

Soft Spring

Mass 2

String 2

FIGURE 3-16
Diagram of AMBAC double vibrating-
string accelerometer. (*From Wing, 1967.*)

The difference-frequency variation in one model is about 64 Hz for a change
of 1 g (980 gals). This is multiplied 3200 times and compared with a signal generated by a crystal oscillator. The entire system is capable of measuring gravity
changes to about 0.1 mgal practically speaking and to 0.01 mgal in the laboratory.

When the instrument is mounted on a gyrostabilized platform in a ship, the
sensitive axis is kept in the average vertical to measure $g + a$, where a represents
the vertical component of any motional accelerations. Cross-coupling effects from
horizontal components of motional acceleration are completely eliminated by
careful design and adjustment of the cross-support ligaments which keep the
center of mass of the two proof masses accurately aligned in the sensitive axis.

Recent test data indicate that this may make a successful instrument for
measurement of gravity at sea. Tests by Woods Hole Oceanographic Institution
(Bowin, Aldrich, and Folinsbee, 1972) over 1 year of operation time indicate a
reliability of 0.7 to 1.0 mgal in moderately rough seas (to state 5). Similar results
were obtained by the U.S. Navy Applied Science Laboratory and by Bedford
Institute (Canada).

The Shell vibrating-string gravity meter A description of this instrument is
given by Goodell and Fay (1964). It uses a vibrating round wire (as distinguished
from a flat ribbon) 0.001 in. in diameter and 9 in. long carrying a mass of 2 g. The

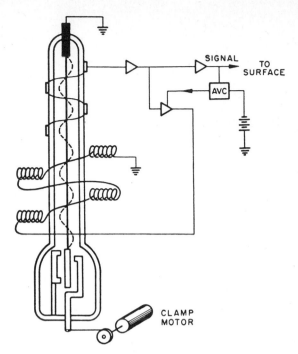

FIGURE 3-17
The Shell vibrating-string gravity meter. (*From Goodell and Fay, 1964.*)

wire is driven in its ninth mode of oscillation by four suitably placed coils and has a set of four electrostatic pick-off points to feed the drive and amplifier system (Fig. 3-17). The enclosing tube is highly evacuated to give very low damping ($Q \approx 400,000$). The normal frequency is about 1400 Hz, and the frequency is measured by timing 70,000 cycles, which requires about 50 sec.

This instrument was designed and developed for use as a borehole meter for measuring densities in sedimentary rocks. For such use it has a precision of about 1 mgal, which gives an average density value over a 1000-ft section to about 0.02 g/cm^3. Operating temperatures can be as high as 148°C, which, with commonly experienced geothermal gradients, will permit operation to depths of the order of 15,000 ft.

The Esso vibrating-string gravity meter   The basic element of this instrument (Howell, Heintz, and Barry, 1966) is a $\frac{1}{2}$-mil tungsten wire about 5 cm long which suspends a platinum mass of about 1 g (Fig. 3-18). The suspended mass is stabilized by a crossbar supported by fine tungsten torsion filaments, as indicated in the diagram. The central part of the wire is in the field of a permanent magnet. The glass insulating section of the supporting frame permits an electrical connection of each end of the fiber to the external electronic system, which drives the string at its natural frequency of about 625 Hz. To attain low damping and high resonance

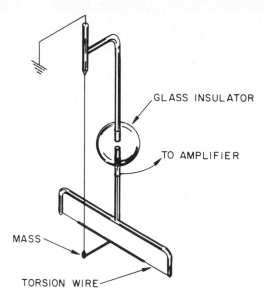

GLASS INSULATOR

TO AMPLIFIER

FIGURE 3-18
Schematic diagram of Esso vibrating-
string gravity meter. (*From Howell,
Heintz, and Berry, 1964.*)

MASS

TORSION WIRE

the moving system is in a rather high vacuum (10^{-5} to 10^{-6} mm Hg), which gives a Q of several hundred thousand.

The vibrating-string assembly is suspended by a wire to act as a plumb bob to remain vertical in the external housing. The clearance will permit deviations of up to about 4° from the vertical.

The electronic system includes a sharply resonant filter so that the oscillation is very stable. The frequency of vibration is determined by an electronic internal timer, which measures the elapsed time for a fixed number of oscillations. The basic measuring element of the timer is a crystal oscillator, and the precision of the time measurement is ± 1 μsec. If the number of oscillations counted is 100,000, which requires approximately 160 sec, the 1-μsec precision is equivalent to about 1 in 10^8, or approximately 0.01 mgal.

The application of the instrument has been to the measurement of gravity in boreholes. In a test measurement the average deviation of a single measurement from the mean value is about 1.7 μsec for the timing of 10^5 cycles. Since the usual practice is to take four readings at each station, this gives a theoretical precision of around 0.01 mgal.

In routine operation in a borehole the total time of reading, including at least four repeat timing measurements, is approximately 20 min.

The Bell Accelerometer

The basic sensor used in this instrument was originally developed for inertial navigation systems. Application to measurement of gravity at sea was made under a development contract with the U.S. Naval Oceanographic Office (90 percent)

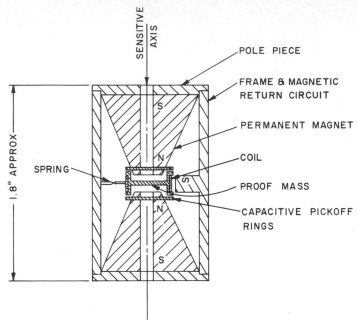

FIGURE 3-19
Schematic cross section of Bell gravity meter.

and the U.S. Coast and Geodetic Survey (10 percent). With some modifications of the associated electronics the instrument, mounted on a gyrostabilized platform, made satisfactory measurements in acceptance tests and has been in routine operation by the Gravity Division of the U.S. Naval Oceanographic Office. A somewhat more compact and simplified model has been acquired and put into routine application by some oil companies and government institutions.

The sensitive element is a coil in the field of a pair of permanent magnets. The variation in weight of the coil and accompanying "proof mass" due to changes in gravity and to motional acceleration are balanced by the force between the varying current in the coil and the constant field from the permanent magnets.

The general arrangement is indicated schematically in Fig. 3-19 and Fig. 3-20 shows a cutaway view of the actual instrument. The dimensions are very small, as the entire sensor is mounted in a flat, cylindrical case about 1.8 in. in diameter and $\frac{3}{4}$ in. thick. The moving coil is within the radial magnetic field from the conical permanent magnets. The return magnetic circuit is through radial members which are connected to the central part of the mounting frame. The proof mass serves as a form for the coil and is aligned within the gap between the magnet poles by three guide springs, only one of which is shown.

The position of the coil within the gap is sensed by capacitative pickoff rings, which are in a capacity bridge. This bridge is balanced at the reference position, and any unbalance of the bridge produces a variation in the current in the coil in the direction to restore the balance.

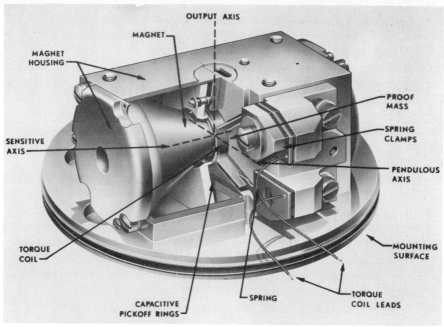

FIGURE 3-20
Cutaway view of Bell gravity meter. (*Bell Aerosystems, 1970.*)

The sensor is contained within a very carefully controlled constant-temperature oven, as the field of the permanent magnets is strongly temperature-sensitive. To make the assembly sensitive only to vertical accelerations it is mounted on a stabilized platform. Cross-coupling effects from horizontal accelerations are negligible.

The coil carries a biasing current, which is controlled very accurately by the external electronics system, one unit of which includes a precision diode. This current, in effect, takes the place of a mainspring in most gravity meters. The electronics, together with the capacitance bridge, permits a measuring range of 966 to 994 gals, which is sufficient to cover the entire range of the earth's gravity field (978 to 983 gals) with much to spare.

The variations in balancing current correspond to all changes in acceleration to which the sensor is subjected. It therefore measures the vertical component of all motional accelerations as well as changes in gravity. The changes in the balancing current are converted into digital form. The filtering to separate the desired changes of gravity from the motional accelerations is done electronically by means of a small computer which operates on the digital signal. The filter characteristics can be modified to meet varying "noise" accelerations which result from changes in sea conditions.

The Bell meter has been used in commercial geophysical exploration by several oil companies. Comparison with the gravity from bottom-meter values by the first prototype model showed most variations within 1 mgal or less. Comparison with a LaCoste and Romberg meter on the U.S.N.S. *Sgt. Curtis F. Shoup* (Coons and Smalet, 1967) indicated similar performance, discrepancies between the two instruments being mostly less than 1 mgal and dependent to some degree on the filter system used. In a stationary laboratory test, the instrument followed the tidal variation of gravity to within 0.04 mgal or less.

UNDERWATER (BOTTOM) GRAVITY METERS

Most underwater gravity meters are basically land gravity meters in a waterproof shell and with mechanical, optical, and electronic accessories arranged so that the operations which would be done by hand on shore are done remotely through an electric cable to a control box on the ship. The gravity sensor is mounted on gimbals within the waterproof case, and small motors move the instrument on these gimbals until remote indicators at the control box show that it is level. In some instruments this function is automatic, and the leveling motors are operated through a servo device to keep the instrument level.

All the operations of leveling, releasing the clamps to free the moving system, balancing the meter, obtaining the gravity value, and clamping the instrument at the end of the observation are carried out from the control box on the surface in a few minutes. The ship is not anchored but is held within a short horizontal distance from the meter by manipulation of engines and rudders.

One complication arises because the motion of the overlying water produces small movements of the instrument, particularly in shallow water with a soft bottom. It is suggested that the upper part of the soft, muddy bottom is slightly elastic, possibly from included gas bubbles caused by bacterial action on the organic material in the mud, and that this elastic bottom responds to variation in loading from the passing of surface waves. The motion obviously is caused by the surface waves as the movements and period of the beam position indicator correspond closely with that of the waves. These motions probably have both horizontal and vertical components, but the meter is sensitive only to the vertical part. Since the bottom apparently moves over a wide area and probably also to an appreciable depth, no form of wide or heavy base or stakes in the bottom has been effective in reducing the vertical motion of the instrument. Therefore, to meet this condition, and to be capable of operating in moderately rough water, the underwater gravity meter must be readable even with a limited amount of vertical motion.

One approach to this problem is that of using the instrument itself as a seismograph and containing the sensitive part within a vertically moving "elevator" inside the outer case (see Fig. 3-21 and page 49). In a later development very heavy damping is used, and the rate of change of displacement of the beam rather than the displacement itself is used as a measure of the gravity change (this is an adaptation from the shipborne gravity meter mentioned below, page 111).

Underwater gravity meters have now been developed until they are very reliable and can make observations under quite adverse conditions of sea surface and bottom motion. These motional problems are most severe in shallow water

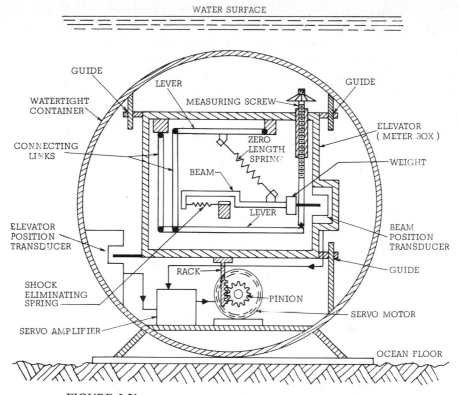

FIGURE 3-21
Schematic diagram of LaCoste and Romberg underwater gravity meter with elevator. (*From LaCoste, 1967.*)

(less than about 25 ft) with soft bottom, and they decrease in deep water. The operations of lowering the meter over the side of the ship, making the gravity reading, and returning the meter to the surface are quite routine. With good meter-handling equipment and good cooperation of the boat crew and instrument crew, underwater observations can be made in a few minutes, so that the rate of station production and the precision of the results are comparable to those obtained with land surveys, although somewhat less.

Adaptations for Underwater Use

A brief outline of the adaptations for the better-known and more extensively used underwater gravity meters is given in the following paragraphs.

The "diving bell" In this early approach to making gravity measurements under water the gravity meter and operator were simply placed in a watertight compartment which was let down on the water bottom from the side or stern of a ship or barge. In one form (Frowe, 1947) used extensively by the Robert H. Ray Company, the diving bell was conical with an opening in the side through which the operator could climb into the interior. A telephone connection through the cable and an emergency air supply were provided for safety. Fairly extensive surveys, a total of some 30,000 stations, were made in offshore Louisiana before the diving bell was superseded by the remote-control instruments.

With its large size the diving bell was susceptible to disturbances from motion of the water, and its operation was rather largely confined to the summer periods of quiet water in the Gulf of Mexico. Measurements were actually made in water as deep as 125 ft.

The operation also used the diving bell in its conventional form, i.e., with the bottom open and the water kept out by air pressure. With lead weights and with tripod legs which would sink into the mud this form was particularly advantageous for operation in areas with very soft bottom.

The Gulf (Hoyt) meter   This meter as modified by Pepper (1941) used small internal electric motors operating on gimbals to level the instrument, photocells controlled by light through the level bubbles to indicate when the instrument was level, a modified Leica camera to photograph the multiple reflected images of the light slit, and motors to clamp and unclamp the moving system. The photographic record of the day's work was removed from the instrument and measured under a traveling microscope to make the readings. When the instrument was unsteady, due to motion of the bottom, an average value was determined by reading the record at several places and averaging the results.

The LaCoste and Romberg meter   Early versions of this underwater meter (Fig. 3-21) were leveled by motor-operated jacks in the external support feet (not shown in this figure). Levels were indicated by a reflected grid and photocell arrangement operating on a pair of pendulums. A similar arrangement indicated the position of the moving beam. A selsyn motor operated the null-adjusting dial in response to the motion of a similar dial on the control box at the surface. The response to bottom motion was counteracted by an "averager" or elevator. A servomotor was arranged to move the elevator vertically, and this motor was controlled through a servomechanism which was responsive to the position of the beam between its stops. When the outer case moved downward, the beam, acting as a stationary weight in a seismograph, tended to move upward; this motion decreased the light on one photocell and increased that on another. These changes act through the beam-position transducer to control the elevator motor so that when the outer case moved downward in response to an external disturbance the elevator moved the gravity meter upward; if the outer case moved upward, the sensitive element was moved downward. In this way the meter itself stayed relatively stable in space while the outer case moved up or down; thus the external accelerations were largely prevented from affecting the gravity-sensitive system. This device was developed to such a degree that vertical motions of approximately $\frac{1}{4}$ in. were effectively compensated. This permitted readings comparable in precision with those on solid ground to be made in spite of motional acceleration of the outer case equivalent to several tens of milligals.

This version of the underwater meter is now largely replaced by adaptations of the principles of the shipborne gravity meter (see p. 112). The most important change is that the moving system now uses very heavy damping, and the rate of change of the position of the moving system is measured rather than its deflection. This adaptation has resulted in somewhat simpler and faster (but not more accurate) results.*

ABSOLUTE GRAVITY MEASUREMENT BY A FALLING BODY

In principle, a very simple way of measuring gravity is by determining the distance over which a body falls in a given time (or the time for the body to fall through a known distance). The basic relation is

$$S = \tfrac{1}{2}Gt^2 \quad \text{or} \quad G = \frac{2S}{t^2}$$

where g is the acceleration of gravity and S is the distance traversed by the falling body in time t.

There have been a number of attempts in the past to measure gravity by this principle. Both distance and time must be measured to a precision of better than 1 ppm to achieve accuracy of the order of 1 mgal. Three developments of recent years have made new approaches to the falling-body technique feasible:

1   The development of the laser, which has made it possible to devise interferometer systems with long light paths (1 m or more) because of the coherent, single-frequency light beam which it produces

2   The development of highly precise electronic gating and timing systems which can measure time intervals to a precision of 1 part in 10^8 or better

3   The application of photomultiplier tubes to sense very short light flashes to operate the timing devices

The following paragraphs describe two such devices which have been able to measure absolute gravity to a precision of a few parts in 100 million. Only the general principles of these systems are indicated without attempting to describe the many mechanical, optical, and electronic details involved.

The Falling-Corner-Cube Instrument

Hammond (1970) and Faller (1965) have described a device which makes use of an interferometer formed between one beam, which is reflected from a falling corner

---

* A comparison of two independent but overlapping surveys made with elevator meters showed an average difference between station values of one survey and values interpolated from contours of the other of 0.14 mgal.

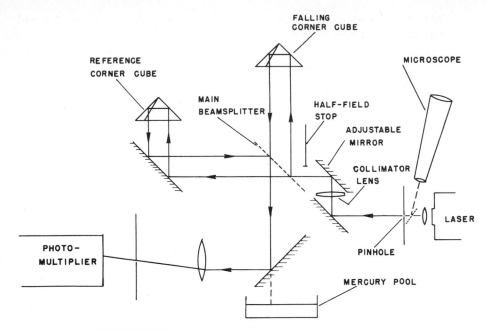

FIGURE 3-22
Schematic diagram of JILA falling-body absolute gravity meter. (*From Hammond, 1970.*)

cube, and a second beam which is reflected from a stationary corner cube. The basic principles are shown in Fig. 3-22. The corner cube is a prism from which a light beam is returned after internal reflection from three mutually perpendicular surfaces. This cube has the property that a reflected beam is exactly parallel with the incident beam. (This is a three-dimensional analog of a billiard ball being shot toward the corner of the table which, on reflection by both rails, returns on a path which is displaced from its initial course but parallel to it.)

The collimated laser beam is divided by the "main beam splitter" into one part that goes upward to the falling corner cube and another part that is reflected to the reference corner cube. The interference of the two beams reflected from the corner cubes makes moving fringes which are seen by the photomultiplier tube. The output of the tube is an alternating current, which is the input to the scaler (not shown). Since the wavelength of the laser light is accurately known, the distance traveled by the falling cube is accurately determined by the number of fringes passing during the interval from a starting time t_1 to an intermediate time t_2 and from t_2 to a final time t_3.

The electronic scaler determines the time between whole numbers of fringes to within 10^{-9} sec = 1 nanosecond (nsec). The scaler is calibrated by reference to the standard frequencies broadcast by the National Bureau of Standards from Station WWVD in Fort Collins, Colorado.

With the time intervals between an initial and intermediate point (T_1) and the intermediate and final points (T_2) measured by the electronic scaler, and with the distances S_1 and S_2 between these intervals known from the fringe counts, we have

$$S_1 = V_0 T_1 + \tfrac{1}{2}g T_1{}^2 \quad \text{and} \quad S_2 = V_0 T_2 + \tfrac{1}{2}g T_2{}^2$$

from which
$$g = 2 \frac{S_2 - (T_2/T_1)S_1}{T_2{}^2 - T_1 T_2}$$

or, since each fringe corresponds to a distance of one-half wavelength of the laser light,

$$g = \lambda \frac{M_2 - (T_2/T_1)M_1}{T_2{}^2 - T_1 T_2}$$

where M_1 and M_2 are the numbers of fringes counted in the time intervals T_1 and T_2, respectively.

The achievement of a precision of better than 0.1 mgal in the gravity measurement has involved a great deal of careful attention to many details. The falling corner cube, of course, is in a high vacuum with provisions for elimination of electrostatic and magnetic effects. It is essential that the light beam to the falling cube be accurately vertical; this is achieved by a test with the beam reflected back on itself from a mercury pool at the bottom of the system. An important element in achieving accurate results is the suspension of the reference corner cube from the mass of a long-period vertical seismograph to eliminate microseismic motion.

The system is arranged so that successive drops are carried out automatically, the essential data being recorded digitally on tape; a single measurement is usually the average of 50 drops. The mechanical system finally devised automatically lifts the moving unit, drops it, and measures and records the essential data on magnetic tape at a rate of 50 drops each half hour or 36 sec/cycle. The calculations including statistical data and average gravity value for each series of drops are carried out by electronic computer programs suitable for use on several different computers.

Up-and-Down Corner Cube

This gravity-measuring system, developed at the International Bureau of Weights and Measures (Sèvres, France) (Sakuma, 1963; Sakuma, Chartier, and Duhamel, 1967), also uses a freely falling corner reflector but differs in many fundamental details from that discussed above. The reflector is catapulted upward, and measurements are made both of the deceleration going up and the acceleration coming back down. The fundamental length interval is that between a pair of fixed mirrors, their separation being determined in terms of a standard light wavelength by interference methods. The fundamental time measurements are made by an electronic gating and counting system. The very short light flashes between which time measurements are made depend on the fact that in a Michelson interferometer reinforcement of white light occurs only when the lengths of the light paths in the two arms are exactly equal.

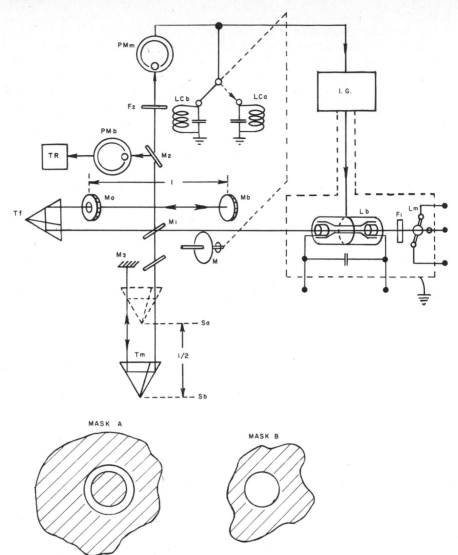

FIGURE 3-23
Schematic diagram of catapult falling-body apparatus: F_1, F_2, monochromatic filters; IG, impulse generator; L_b, white-light flash lamp; L_m, mercury-vapor lamp; LC_a, LC_b, tuned circuits; M, motor; M_1, M_2, half-silvered mirrors; M_3, opaque mirror; M_a, M_b, interferometer mirrors separated by distance l; PM_a, PM_b, photomultiplier tubes; S_a, S_b, measuring positions; T_f, fixed corner reflector; T_m, moving corner reflector; TR, electronic timer. (*From Sakuma, 1963.*)

The essential components of the system are shown by Fig. 3-23. The basic Michelson interferometer light beam is from the white-light flash source L_b or the monochromatic source L_m to the beam-splitting mirror M_1. The light passing through M_1 is reflected by the fixed corner reflector T_f to either mirror M_a or M_b, then returned over the same path to mirror M_1, where it is reflected upward to M_2 and to photomultiplier tube PM_b, which controls an electronic gating and timing unit TR. The portion of the initial beam reflected downward by M_1 is reflected back by the moving corner reflector T_m to the fixed mirror M_3 and then back over the same path to mirror M_2, photomultiplier PM_b, and timer TR. There are two positions of T_m at which the two light paths are identical, i.e., at the lower position, where the light path is matched by the other beam being reflected at M_b, and at the upper position, where the other beam is reflected at M_a. The distance between these two positions is exactly half the distance l between mirrors M_a and M_b. Thus, the high-intensity source provides high-intensity flashes of white light at two positions of the moving-corner-cube reflector.

The triggering of the flash tube L_b is controlled by light from the monochromatic source L_m, which passes over the same Michelson-interferometer light path as the flash, except that it passes through the partially silvered mirror M_2 and filter F_2 to photomultiplier tube PM_m. The output of this tube controls the instant of flash. As the corner reflector approaches either of the two positions of equal path length, monochromatic fringes formed at F_2 cause a sinusoidal output of PM_m at a frequency depending on the velocity of the moving reflector. This signal is detected and amplified by the tuned circuits LC_a and LC_b, which are resonant to the two different fringe frequencies. The monochromatic fringe intensity increases as the identical path distance is approached and triggers the impulse generator IG to start the flash very slightly before maximum so the flash is initiated at the time when it has maximum intensity at the instant of white-light interference.

The optical-path lengths to mirrors M_a and M_b are separated by use of the hole in M_a and the two masks, A and B, shown in the lower part of the diagram. The hole in M_a may be covered by either of the two masks. With mask A in place, reflection is from the exposed ring around the hole in M_a. With mask B in place, the light passes through the hole and is reflected from M_b. A mechanical system driven by the motor M alternates the masks and also switches the electrical filters, LC_a and LC_b, to the proper tuned circuit for control of the impulse generator for the white flash. With mask A in place, there is a flash when the corner cube passes the upper station, and with mask B in place, there is another flash when the corner cube passes the lower station. Thus there are four successive flash times, that is, t_1 at the lower station on the way up with mask A in place; t_2 and t_3 as the corner cube passes the upper station, once on the way up and once on the way down, with mask B in place; and finally, t_4, with mask A in place again, as the corner cube passes the lower station on the way down. With a distance l of 1 m the time available for the switching of masks and circuits is about 0.2 sec.

The interferometric measurement of the distance between M_a and M_b is made to a precision of about 1 part in 10^8 and is done before and after each measurement. The scaling circuitry TR, activated by the white-light impulses from the photomultiplier PM_b, determines the time intervals to about 5 nsec.

With the time intervals and the distance interval between the two positions precisely known, gravity can be calculated by the relation

$$g = \frac{8H}{T_1^{\,2} - T_2^{\,2}}$$

where H = distance between two levels of measurement

$\quad\quad T_1$ = time interval between upward and downward passings of lower level

$\quad\quad T_2$ = time interval between upward and downward passings of upper level

It can be shown (Cook, 1969) that if air pressure is low (10^{-4} atm), the gravity value is independent of variations in residual air pressure.

With time and distance measurements to the order of 1 part in 10^8, the precision of actual gravity measurements is a few hundredths of a milligal.

Some mechanical details of the system are of interest. The moving corner reflector consists of three mutually perpendicular mirrors on a metallic frame. This frame carries a second identical mirror assembly so arranged that the optical apex is at the center of gravity of the moving system. This is done to eliminate effects of any small rotational movement of the falling system.

The mirrors on the frame are mounted around a relatively large central vertical hole. The catapult is a lifting base, fitting into the frame and attached to an elastic fiber stretched through the hole. This lifting base is normally latched in the lower position. When the latch is released, the corner-cube assembly is tossed upward and is caught by the base and fiber when it returns.

As in other falling-body gravity-measuring systems, the effect of microseisms causes a problem. The equipment described here is mounted on a seismically stabilized table supported at each corner by two piezoelectric units, one above the other. The upper unit gives out an electric signal from any force due to seismic motion; this signal is integrated, reversed in phase, and fed back to the lower unit, which then moves up or down by the proper amount to compensate for the seismic movement of the ground. (This is similar in concept, although entirely different in instrumentation, to the elevator of the earlier form of the LaCoste and Romberg underwater gravity meter, page 50).

The final value A, based on a series of 50 measurements for the gravity at Sèvres, is given as 980.95675 gals with an uncertainty of 0.03 mgal.

Other Falling-Body Gravity-Measuring Systems

At least two other falling-body systems that have measured gravity to about 1 mgal or less deserve brief mention.

A free-falling scale   Developed earlier at Sèvres (Cook, 1969, p. 7), this system makes observations by photographing a freely falling scale, essentially a very refined meterstick with very sharply etched measuring marks. The scale is illuminated by extremely short-period high-intensity flashes at precisely controlled time intervals. Images of the scale are formed on a rapidly moving photographic film. In

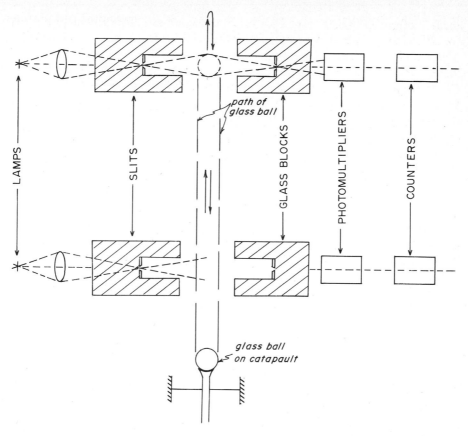

FIGURE 3-24
Principle of National Physical Laboratory free-fall gravity measurement.
(*After Cook, 1969.*)

principle, a value of gravity can be determined from a set of three measurements at known time intervals. Actually, some 50 pairs of images and times are determined for each drop. The reported accuracy is somewhat better than 1 mgal.

An up-and-down-glass-ball system   A glass ball (Fig. 3-24) is tossed upward by a catapult and passes through holes in two glass blocks (Cook, 1967; 1969, p. 14). Each block has two horizontal slits. Light from a bright source is focused on the first slit. The glass ball acts as a lens to focus the light on the slit on the other side of the block. This makes a very sharply defined and very brief flash of light through the exit slit when the ball passes the plane of the slits. This light falls on a photomultiplier which activates an electronic counter. Thus the time of upward and downward passing of the planes of the slits is precisely determined.

The distance between slits is determined by interferometric measurements between the optically flat surfaces of the slit blocks. Two sets of measurements are made, both blocks being turned over for the second set. This eliminates the need for knowing the position of the slits within the blocks accurately as any slight asymmetry disappears when two sets are averaged. The derivation of a gravity value from the measurements is the same as that of the Sèvres measurements, as the fundamental data, i.e., the time intervals and the distance between the levels of measurement, are the same.

The principal source of error is from ground motion, and no provision was made for its elimination or compensation. By averaging about 100 measurements, the reported precision is about 0.2 mgal.

THE POTSDAM BASE FOR ABSOLUTE GRAVITY

Of particular interest are the results of recent absolute gravity measurements which are relevant to the worldwide gravity system and the international gravity formula. The first term of this formula, that is, 978.049, the value of gravity at the equator, is based on absolute gravity measurements at Potsdam made in 1906 (page 17n). In recent years a number of new measurements of gravity have been made, by free-fall or by rise-and-fall methods at primary base stations. In earlier years several of these bases were tied to Potsdam by relative gravity measurements, by pendulum or gravity meter.* Thus a new absolute measurement at a base tied to Potsdam gives a new value there. Measurements made with the falling-body instrument described above in Washington; Teddington, England; and Sèvres, France, have indicated corrections to the Potsdam values of −13.76, −13.60, and −13.80, respectively. The spread of these values probably reflects errors in the earlier ties from Potsdam to the points of observation. It is quite clear that the Potsdam value and the first term of the international gravity formula need correction (reduction) by a little less than 14 mgals.

* Access to Potsdam for later direct measurements has been denied.

Table 3-1*   REFERENCE VALUES AT POTSDAM

Reference	Place	Standard deviation, mgals	Correction to Potsdam value, mgals
Free-fall:			
Thulin (1961)	Sèvres	0.7	−11.8
Preston-Thomas et al. (1960)	Ottawa	1.5	−13.1
Tate (1966)	Gaithersburg	0.3	−12.8
Faller (1965)	Princeton	0.7	−13.7
Rise-and-fall:			
Cook (1967)	Teddington	0.13	−13.7
Sakuma (Cook, 1968)	Sèvres	0.1	−13.8

* From Garland (1971, p. 144).

Table 3-1 lists revisions of the Potsdam value based on a number of new measurements at several primary base stations previously tied to Potsdam.

Changes may be made in future in the official value at Potsdam or in the first term of the gravity formula. In a review of the SI units, Markowitz (1973, p. 235) gives a value of 9.78032 m/sec², which is 17 mgal lower than the old value. From Table 3-1 this seems to be about 3 mgals too low. Changes will have no effect on the use of gravity in petroleum prospecting, as all such uses depend on relative and not on absolute values.

THE GRAVITY PENDULUM

Measurements of gravity with pendulums, which have been used to only a limited extent for commercial geophysical prospecting, have been used quite extensively for geodetic and other scientific purposes. Gravity measurement with a pendulum depends on the fact that the period of a freely swinging pendulum is inversely proportional to the square root of the gravitational acceleration. If the physical dimensions of a pendulum are held constant and the period can be measured sufficiently accurately, small changes in the period will indicate small changes in gravity. The period can be measured by comparing the pendulum with an accurate chronometer, the rate of which is independent of small variations in gravity.* The period can also be measured by simultaneously comparing the period of a "base pendulum," kept at a fixed reference point, with that of a "field pendulum," which is set up at the field station, the two pendulums being connected by telegraphy or radio. This method was used for geophysical prospecting.

Comparison of pendulums has been done in two ways. (1) In the coincidence method† the individual swings of two pendulums or of a pendulum and a chronometer are compared continuously and the time measured in which one pendulum gains or loses an integral number of half swings. (2) In the phase-position method‡ the phase relations of the two pendulums are compared with a simultaneous time mark at the beginning and at the end of a run. Either method, of course, requires telegraph or radio connection between the base and field pendulums.

More recently, the crystal clock has been used to time the pendulum. The timing element of such a clock is an oscillating quartz crystal and associated electrical circuits. When electric power and temperature are carefully controlled, the period of a crystal is constant to the order of 1 part in 100 million and is independent of gravity. A pendulum and clock together form a complete gravity-measuring unit and avoid the necessity of a communication link with a stationary reference pendulum.

---

* This method is discussed in detail by Haalck (1934, pp. 21–27).

† This is described briefly by Haalck (1934, pp. 27–28) and in detail by Berroth (1927).

‡ This method is described in detail by Bullard (1933). In the measurements described in Horsfield and Bullard (1937), the reference pendulum was in England and the field pendulum in Africa, the time marks being supplied by recording commercial radio-telegraph signals.

Table 3-2 GRAVITY METERS

Many different gravity meters have been built over the years, mostly from about 1932 to 1945. A few have already been described. This table lists the instruments known through publications or patents or through application in the field. It also lists their operating features and descriptive references. The listing is according to the general type to which the instruments belong.

Name	Null system	Reading system	Sealed (S) or buoyancy-compensated (C)	Field use	Reference	Remarks
			Stable type			
Hartley	Separate spring	Optical lever	S	Nil	Hartley (1932)	First use of null spring
Boliden	Electrical capacitor	Beat oscillator	S	Some in mining exploration	Hedstrom (1938)	
Gulf (Hoyt)	None; reads angular deflections	Multiple reflections	S	Extensive	Hoyt (1938a), Wyckoff (1941)	Uses flat ribbon spring
Gulf underwater meter	None; reads angular deflections	Photographic images of multiple reflections	S	Extensive	Pepper (1941)	Basically the same as land meter
Haalk	None; reads total deflection	Scale on capillary tube	S	Nil	Haalck (1938)	Uses compressed gas for spring
Askania Gss 2	Separate spring	Electrical; by grid and photoelectric cells	S	Mostly European	Graf (1938)	
Askania Gss 3	Magnetic	Electronic	S	Testing only	Askania (1970)	Position detected by condenser system and maintained by current in magnetic field
Bell	Coil in magnetic field with balancing servomechanism	Records coil current on magnetic tape		Limited; at sea	Bell Aerosystems (1970)	Coil continuously balanced; variation of current measures changes in gravity and in ambient accelerations

Table 3-2—continued

Name	Null system	Reading system	Sealed (S) or buoyancy-compensated (C)	Field use	Reference	Remarks
				Unstable; Quartz systems		
Ising	Tilting case	Microscope	S	Some in Scandinavia	Ising (1937, 1940)	Variation of inverted pendulum; made trans-Atlantic traverse
Norgaard	None; read relation of two inverted pendulums	Microscope	S	Some in Denmark and at sea	Norgaard (1939)	
Norgaard	Tilting case	Microscope and mirror	S	Extensive	Norgaard (1945)	Temperature-compensated system; no thermostat
Mott-Smith	Main spring support	Microscope	S	Extensive	Mott-Smith (1938)	First meter with very small quartz system
Worden	Main spring support	Microscope	S	Very extensive	Heiskanen and Vening Meinesz (1958, p. 109)	Temperature-compensated system; no thermostat
Atlas	Main spring support	Microscope	S	Limited		Temperature-compensated

Table 3-2—continued

Name	Null system	Reading system	Sealed (S) or buoyancy-compensated (C)	Field use	Reference	Remarks
			Unstable; steel spring and lever systems			
Humble (Truman)	Separate spring	Microscope	C	Extensive	Bryan (1937)	First commerical instrument in field
LaCoste-Romberg	Main support	Microscope	C	Extensive	LaCoste (1934, 1935)	Zero-length spring
LaCoste-Romberg underwater meter	Main support	Reflected grid and photocell	C	Extensive		Acceleration compensation
North American	Beam support	Microscope	C	Extensive		Zero-length spring
Frost	Separate spring	Microscope	C	Extensive		Zero-length spring
Western	Separate spring or adjustable main spring	Microscope	C	Extensive		Zero-length spring
Humble (Boucher)	Auxiliary spring	Microscope	C	Extensive	Boucher (1943)	
Brown	Torsion head	Microscope	S	Extensive	Brown (1938)	Gravity change produces angular rotation
Thyssen	None; reads deflections of one scale over another	Microscope		Limited; mostly European	Heiland (1939), Thyssen-Bornemisza (1938b, c)	Buoyancy effect computed as a correction
Wright	Horizontal torsion head	Microscope	S	Nil	Wright and England (1938)	First instrument used for geodetic applications
Humble (Zenor)	Electrostatic	Beat oscillator	S	Limited; underwater meter	Zenor (1943)	Uses flat beam spring

Table 3-2—continued

Vibrating-string systems

Name	String	Reading system	Approximate dimension			Field use	Reference	Remarks
			Length	Mass, g	Frequency, Hz			
Gilbert	Flat ribbon 0.010 by 0.002 in.	Beat with crystal oscillator	5 cm	65	1000	Limited	Gilbert (1949, 1952)	Used at sea and in boreholes
Shell	0.001 in.	Electronic timer	9 in.	2	1400	Internal only	Goodell and Fay (1964)	Borehole only, vibrates in ninth harmonic
Esso	0.0005 in. tungsten	Electronic timer	5 cm	1	625	Extensive in boreholes	Howell, Heintz, and Barry (1966)	Application in borehole only, for density measurement
AMBAC (Wing)	Two flat ribbons 0.101 by 0.002 in.; tension spring	Frequency multiplier and crystal oscillator	$\frac{3}{8}$ in.	Large	4800 (9600)	Extensive in tests at sea	Wing (1967)	Measures difference frequency

Instruments for absolute gravity; free-fall systems

Falling body	Distance measurement	Time measurement	Precision, mgals	Field use	Reference	Remarks
Corner cube	Interference fringes	Counting interference fringes	0.2	None*	Faller (1965) Hammond (1970)	One-way motion
Corner cube	Interferometer	Electronic counter	0.2	None*	Sakuma (1963) Sakuma, Chartier, and Duhamel (1967)	Up-and-down motion
Falling scale	Flash photos of scale	Timing of flash	1+	None*	Cook (1969)	One-way motion
Glass ball	Interferometer	Electronic counter	0.2†	None*	Cook (1967, 1969)	

* Laboratory only. † Average of 100 drops.

Theory of Pendulum Measurements

Either the comparison with a fixed pendulum or with a crystal clock gives the variations in period caused by variations in gravity between the different locations at which the measuring pendulum is observed. The period of a physical pendulum may be written

$$T = 2\pi \sqrt{\frac{I}{mgh}}$$

where I = moment of inertia about knife-edge

m = total mass of pendulum

h = distance from knife-edge supports to center of gravity

g = acceleration of gravity

The equation can be written

$$T = \frac{K}{\sqrt{g}}$$

where $K = 2\pi\sqrt{I/mh}$, which will be constant if the physical dimensions of the pendulum are constant. By differentiation, it is evident that

$$\frac{dT}{T} = -\frac{1}{2}\frac{dg}{g}$$

so that, for a given fractional increase in G there will be a fractional decrease of the period T which is half as large. This means that for a pendulum to measure a change of gravity of, say, 0.2 mgal the change in period must be measured to 1 part in 10 million. Also, this means that the physical dimensions of the pendulum must be held constant to the same degree.

To achieve this precision, great care must be taken in the timing or comparison measurements. The pendulum case must be held at nearly constant temperature (to avoid changes in I, the moment of inertia, due to expansion or contraction of the pendulum). The low air pressure in the case must be kept constant (to avoid any change in the effective mass of the pendulum because of air carried with it in its motion). Any spurious forces on the pendulum must be avoided; a small amount of radioactive material in the case ionizes the residual air to make it conducting, preventing the accumulation of electrostatic charges and resulting forces on the moving pendulums. Any sway of the case and pendulum supports must be avoided or corrected for because such motion would affect the period.

Pendulum Field Apparatus

Pendulum apparatus was developed by the Gulf Oil Corporation for commercial oil exploration (Fig. 3-25) and used in the field for several years (1929 to 1935) until it was rendered obsolete by the development of much smaller, lighter, faster, and more accurate field gravity meters. The same pendulums, with crystal clocks for timing, have been used very extensively to establish a worldwide network of primary gravity base stations (Woollard and Rose, 1963).

FIGURE 3-25
Field pendulum apparatus. (*Gulf Res. & Dev. Co.*)

Each original field unit consists of two pendulums in a single case. The periods of the two pendulums are carefully adjusted so that throughout a normal run of up to 1 hr the total number of swings is the same within a small fraction of one swing. The pendulums are started swinging in opposite phases, and because of the equality of the periods this phase relation is maintained throughout the run, which eliminates sway of the case. Amplitude corrections are eliminated by always starting the pendulums with the same amplitude. The decrement is quite constant, so that the final amplitude also is the same from one run to another. Also, it was the practice in field operations to use two complete sets of apparatus so that at each station two independent values of gravity were determined.

The pendulum itself consists of a fused-quartz bar which is carefully designed as a "minimum pendulum."* The bar is carried by a quartz knife-edge resting on

---

* A minimum pendulum is proportioned so that its period is a minimum. For a physical pendulum oscillating about a knife-edge at E at a distance S from the center of gravity G_0 the reduced length l, or length of the equivalent simple pendulum (Fig. 3-26), is

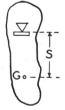

$$l = \frac{K_0 + S^2 M}{SM}$$

where K_0 is the moment of inertia about the center of gravity and M is the total mass.

The period of the pendulum is a minimum when

$$\frac{dl}{dS} = 0 \qquad \text{or when} \qquad S_0{}^2 = \frac{K_0}{M}$$

FIGURE 3-26

which gives

$$l = 2S_0$$

Thus, a minimum pendulum is so proportioned that the distance from the knife-edge to the center of gravity is half the length of the equivalent simple pendulum. The advantage of a minimum a solid Purex glass flat. The pendulum period (complete cycle) is about 0.9 sec. The motion of the pendulum is recorded by a beam of light reflected from a mirror surface polished on the upper end of the quartz bar. The pendulums are mounted in a vacuum-tight cast-aluminum case. Clamping, starting, and stopping of the pendulums are all done from the outside through vacuum-tight stuffing boxes without disturbing the vacuum. The cases are tight enough for a vacuum of 0.1 mm Hg to be maintained for months without pumping out. The pendulum case is surrounded by a net of electric heating wires and covered on the outside with a thick eiderdown blanket. This permits maintaining the case at a temperature which is constant within about 0.1°C, the heating power being supplied by automobile storage batteries and controlled by a thermostat mounted on the metal case.

A beam of light passes into the pendulum case and is reflected from the mirror surfaces on the faces of both pendulums and by other mirrors in the case, so that two reflected beams are returned to the recorder, where they fall on a moving photographic tape. By means of radio signals from the base stations, recording tapes are started simultaneously at the base and field stations and the same time signal is recorded on each at the beginning of the run. Transits of the pendulums are recorded for a few swings, and by measurements on the tapes the phase position of base and field pendulums relative to the time signal is determined very accurately, using the record from a 500-cycle tuning fork as a time scale. The same process is repeated at the end of the run after an interval of 30 or 60 min. The total number of whole swings in the unrecorded interval is calculated from the interval between the starting and ending time signals. The measurements then give the total number of swings and fractions of a swing between the beginning and ending time signals, to high precision. The period of the field pendulum is compared with that of the base pendulum, first with the two side by side at the base station and then with the field pendulum at a field station. The gravity difference between base and field stations is calculated from this change in period.

With time measurements made to about one 1/10,000 sec and with a total run usually of $\frac{1}{2}$ hr (1800 sec), the theoretical precision is better than 1 part in 10 million. Actually, in routine field pendulum operations, gravity determinations were made regularly to a precision of about 1 part in 4 million.*

THE EÖTVÖS TORSION BALANCE

The torsion balance is now obsolete, but from its first extensive application to petroleum prospecting, about 1920, to its complete replacement by gravity meters,

---

pendulum for an invariable pendulum is that small changes in the distance S will not change the period. Thus, a minimum pendulum is much less sensitive to wear of the knife-edge or slight movement of the knife-edge with respect to the pendulum itself than the more common simple pendulum with a heavy weight on a relatively long staff. For more complete theory and design of minimum pendulums, see Meisser (1930).

* Gay (1940) gives details of Gulf pendulum performance during the establishment of a precise gravity difference for gravity-meter calibration. The many possible sources of error, their determination or elimination, and details of data reduction are included. The final value for the gravity difference was determined as 27.53 ± 0.05 mgals.

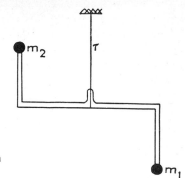

FIGURE 3-27
Moving system of Z-beam torsion
balance.

about 1936 to 1940, it was the principal instrument for gravitational exploration. Well over 100 instruments were in the field at the time of maximum application. A very brief explanation of its principles of operation, the physical meaning of the quantities measured, and their geological significance is included because of its physically interesting principles, the historical importance of the method, and the recent reappearance of gravity-gradient measuring devices. Much more complete accounts are given by Nettleton (1940), Heiland (1940), Jakosky (1940), and Dobrin (1952).

Physical Principles and Basic Theory

The torsion balance measures the distortion, or warping, of the gravitational field rather than the intensity of the field, measured by the pendulum or gravity meter. It was invented by Baron Roland von Eötvös, a Hungarian physicist, about 1880. Eötvös (1896, 1909) was interested in the space variations of gravity as a measure of the departure of the shape of the earth from a spherical form. The instrument was first used in geophysical prospecting in 1915 and was introduced into the United States for oil prospecting in 1922.

The essential elements of a torsion balance are a pair of masses m_1 and m_2 (Fig. 3-27) suspended by a sensitive torsion fiber τ and supported so that they are displaced both horizontally and vertically from each other. If the gravitational field is not perfectly uniform, there will be a very slight difference in the direction of the gravitational force on the two weights. This slight difference in direction causes slight horizontal force components on the two weights, which tend to rotate the suspended system about the suspending fiber. By measuring the rotation produced, it is possible to calculate the amount of distortion of the gravitational field and to infer something about the nature of the mass irregularities that have produced the gravitational disturbance.

If the gravitational field is not perfectly uniform, the lines of the vertical are slightly curved. Horizontal forces acting on the torsion-balance masses result from this curvature of the vertical, as indicated in Fig. 3-28. The gravitational forces acting on the two masses are F_1 and F_2. Because the vertical is curved, these two

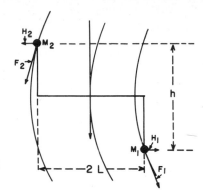

FIGURE 3-28
Torsion-balance beam in a curved field
of force.

forces are not strictly parallel and hence have horizontal components such as H_1 and H_2, which may be in any direction with respect to the beam and therefore can exert a torque on the suspension.

The forces tending to rotate the torsion-balance beam depend on the differences in the force components on the two weights and therefore on the rate of change of these forces within the dimensions of the balance. Since the forces are first derivatives of gravity, their rates of change are second derivatives of the potential; it is these second derivatives which produce the torque on the torsion-balance beam. It can be shown (Nettleton, 1940, pp. 67–69) that the torque on the suspension fiber is

$$T = mhl \left(\frac{\partial^2 U}{\partial y\,\partial z} \cos \alpha - \frac{\partial^2 U}{\partial x\,\partial z} \sin \alpha \right)$$

$$+ ml^2 \left[2 \frac{\partial^2 U}{\partial x\,\partial y} \cos 2\alpha + \left(\frac{\partial^2 U}{\partial y^2} - \frac{\partial^2 U}{\partial x^2} \right) \sin 2\alpha \right]$$

where m = mass of weights on beam
   h = vertical distance between weights
   $2l$ = horizontal distance between weights
   α = angle between beam and x axis (usually north)

$\dfrac{\partial^2 U}{\partial y\,\partial z} \ldots$ = second partial derivatives of gravitational potential U

The torque produces a rotation of the suspending fiber, and the resulting angle of twist is measured by the movement of a small multiply reflected spot of light on a photographic plate. If the deflected position is n and its undisturbed position is n_0, the working equation for the torsion balance becomes

$$n - n_0 = P(U_{yz} \cos \alpha - U_{xz} \sin \alpha) + Q(2U_{xy} \cos 2\alpha + U_\Delta \sin 2\alpha)$$

where $$P = \frac{2NDmhl}{\tau} \quad \text{and} \quad Q = \frac{2NDml^2}{\tau}$$

P and Q are fixed constants depending on the physical parameters of the instruments, for

N = number of reflections of light beam
D = length of optical path
τ = torsion constant of the suspending fiber

$$U_{yz} = \frac{\partial^2 U}{\partial x\, \partial z}, \quad U_{xy} = \frac{\partial^2 U}{\partial x\, \partial y}, \quad \text{etc.} \quad \text{and} \quad U_\Delta = U_{yy} - U_{xx}$$

The equation contains four unknown quantities, i.e., the four second derivatives U_{yz}, U_{xz}, U_{xy}, and U_Δ. Also the undisturbed position of the beam n_0 is unknown. By making readings with the beam in five different azimuths (different values of α) it is possible to solve the five simultaneous equations to determine the five unknowns. Actually, the instrument was always made with two parallel beams side by side but with the lower weights on opposite ends so that each position gave two values for α, differing by 180°, but another unknown is added, i.e., the n_0 for the second beam. With the double instrument, readings in three azimuths, differing by 120°, gave the required six equations from which the six unknowns, i.e., the four second-derivative quantities and the two n_0's, could be determined.

Torsion-Balance Instruments and Field Operations

The actual forces represented by the differences in the second-derivative quantities are extremely small. Torsion balances were designed to be sensitive enough to determine the derivative values to about 1 Eötvös unit, or 10^{-9} gal/cm. One unit would produce a deflection $(n - n_0)$ of about 0.1 mm of the spot of light on a photographic plate. To achieve this sensitivity requires an extremely small torsion constant for the long suspending wire compared with the moment of inertia of the beam and therefore a very long torsional period. A commonly used instrument had a torsional period of about 25 min. Such an instrument required about 1 hr to come to rest after being set up and thus a minimum of 3 hr per reading. Usually more than the minimum three positions were used, so that each reading required about 5 hr and one instrument usually made three stations per day.

Although its fundamental parts are simple, the practical torsion-balance field instrument is a quite precise and moderately complex piece of apparatus. The extreme sensitivity of the torsion fiber requires special protection from temperature gradients, which would produce convection currents in the surrounding air of sufficient magnitude to disturb the instrument. For this reason, the instrument was commonly built with three separate thermally insulated metal cases, one inside the other, to ensure temperature uniformity and absence of air convections in the inner case holding the moving system itself. The instrument was nearly always placed in a protecting house, or "hut," during the time required for an observation. Once set up, the operation was automatic for most instruments with a clockworks to rotate the balance to its different recording positions, turn on the recording light, move the photographic plate, etc.

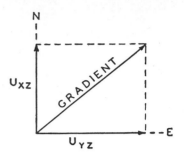

FIGURE 3-29
Gradient components.

The Meaning of the Torsion-Balance Quantities

We have seen that a torsion balance measures two groups of second derivatives. The first group, U_{xz} and U_{yz}, measures the gradient; the second, U_Δ and $2U_{xy}$, measures the curvature.

Gradient The gradient quantities U_{xz} and U_{yz} are, respectively, the north and east components of the horizontal rate of change of gravity. Their vector sum is the total rate of change, or gradient, which is represented on a map by an arrow with its base at the station location, pointing in the direction of the maximum gradient and with its length proportional to its magnitude (Fig. 3-29).

Gravity differences along a line of torsion-balance stations can be determined by integrating the average gradient components along the line multiplied by the distances between stations. Thus a gravity contour map can be made from a torsion-balance survey which ideally would be the same as a Bouguer map from a gravity-meter survey. Actually, the torsion-balance map is nearly always more irregular because the instrument is very sensitive to local density inhomogeneities of terrain near the station.

From their relation to the gravity field it is evident that gradients pointing in the direction of increasing gravity will be directed toward an area of high-density material and away from one of low density. Thus for a simple anticline, gradients on each side are directed toward the axis.

Curvature The curvature quantities measured by the torsion balance are more difficult to appreciate, and reference may be made to Nettleton (1940) or to other early texts (e.g., Heiland, 1940; Jakosky, 1940) for a more complete and mathematical explanation. We shall point out, however, the general nature of the effects measured and their geophysical application.

The quantities U_{xy} and U_Δ (or $U_{yy} - U_{xx}$) are closely related to the curvature of the equipotential or level surface. When the gravitational field is distorted, the level surfaces are curved. Over small distances (of the order of magnitude of the dimensions of the torsion balance) the level surface may be considered as a second-order surface. There is a theorem of differential geometry which shows that, in general, through any point on a curved surface two planes can be passed, in one

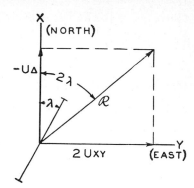

FIGURE 3-30
Map representation of the torsion-balance curvature quantities.

of which the radius of curvature is a maximum and in the other of which the radius of curvature is a minimum and that these two planes are mutually perpendicular. The curvature of the potential surface in each of these two planes is $1/\rho_1$ and $1/\rho_2$, where ρ_1 and ρ_2 are the radii of curvature. The greater the curvature (or the smaller the radius of curvature) the greater the horizontal forces experienced by the torsion balance. If the curvature were equal in all directions (corresponding to a spherical surface), the horizontal forces would be uniform and there would be no differential force and therefore no torque on the beam. For any other surface, the horizontal forces are not equal, and there is a torque depending on the difference in curvature in the two principal planes.

A mathematical analysis shows that the magnitude of this force is related to the second derivatives of the potential as

$$g\left(\frac{1}{\rho_1} - \frac{1}{\rho_2}\right) = [(U_{yy} - U_{xx})^2 + (2U_{xy})^2]^{1/2} = \sqrt{U_\Delta^2 + 4U_{xy}^2} \equiv \mathcal{R}$$

Since the effect depends on the difference in curvature, it is often referred to as the "differential curvature."

The azimuth angle λ between the x axis and the principal plane in which the curvature is algebraically minimum is given by

$$\tan 2\lambda = \frac{2U_{xy}}{-U_\Delta}$$

Since the terms for the curvature forces are multiplied by functions of twice the azimuth angle (sin 2α and cos 2α, page 68), the quantity is double-valued in azimuth and repeats after 180°. The convention used to represent the curvature is a bar or line (Fig. 3-30) the orientation of which corresponds to the azimuth of the plane in which the curvature is a minimum and the length of which is proportional to the magnitude $\mathcal{R}$. The curvature has the same dimensions as the gradient and is measured also in Eötvös units of 10^{-9} gal/cm.

Corrections to the Measured Quantities

Two corrections to the torsion-balance quantities as measured by the field instrument are required before final mapping.

Normal corrections The shape of the earth makes these corrections necessary. They are, in Eötvös units,

> $U_{xz} = 8.122 \sin 2\varphi$ for the gradient because of the increase of gravity toward the poles
>
> $U_\Delta = 10.36 \cos^2 \varphi$ because of the flattening of the spheroid, where φ is the latitude

There are no corrections for the east-west component of the gradient (U_{yz}) or for the other curvature component (U_{xy}).

Terrain corrections The sensitivity of the instrument to nearby inhomogeneities makes these corrections necessary. It was usual field practice actually to level the ground off in a circle about 3 m in radius around each station location to eliminate the nearby irregularities. Elevation differences beyond this circle were measured in a system of sectors commonly to a distance of 100 m. These elevations were entered into a calculation form from which corrections were determined.

In any but quite regular topography terrain effects may become larger than the desired corrected components and at best are not determined very accurately. The effects for gradients decrease approximately as the square of the distance, while the effects for curvatures decrease only as the first power. Therefore curvature values tend to be uncertain or even useless in moderate to rugged topography.

In final mapping it was nearly always the practice to derive and contour gravity by integration of the gradients. Since this integration is often uncertain, particularly when gradients are erratic, there were nearly always substantial closure errors in any network of stations. Some system of adjustment therefore was required. Adjusting and contouring a torsion-balance survey was an art which is now almost unknown to geophysical map makers and interpreters.

The problems of terrain corrections and the inherently slow operation of the torsion balance led to its rapid obsolescence when reliable field gravity meters became available. With 30 to 50 stations per day readily observable versus 3 stations per day per instrument for the torsion balance, the advantage of the gravity meter is obvious. The requirement of leveling to determine accurate elevations of gravity-meter stations, which is not needed for torsion-balance mapping, is more than compensated for by the much greater field progress and the greater certainty of measured gravity differences over those obtained by integrating gradients.*

---

* A special instrument, called a "gradiometer," designed to measure gradients but not curvatures, has been used in some British geophysical work. It was designed to give gradient measurements as accurate as those by an ordinary torsion balance but to have a much shorter period and therefore to be capable of making a measurement in a much shorter time. For a very complete and mathematically elegant treatment of the general theory of torsion balances, as developed for the design of the gradiometer, see Lancaster-Jones (1932).

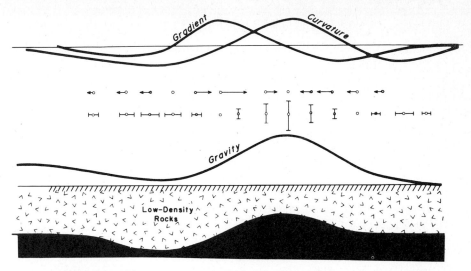

FIGURE 3-31
Profile and cross section showing gravity, gradient, and curvature changes across anticlinal structure.

Mapping and Geological Meaning of Torsion-Balance Results

The four torsion-balance quantities are shown on a map (Fig. 3-30) by a vector representing the resultant of the two gradient components and a bar of length $\mathcal{R}$ representing the magnitude of the resultant of the two curvature components plotted at the angle λ, which is half that of the vector $\mathcal{R}$. A common plotting scale was 1 mm = 1 Eötvös unit (1 $E°$). The quantities were usually in the range of zero to around 30 $E°$ but occasionally ran to 100 $E°$ or more. A gravity change of 1 mgal/mile is about 6 $E°$; 1 mgal/km is 10 $E°$.

The geological significance of the curvature is more obscure than that of the gradient but may be appreciated in a general way by reference to a simple anticline and syncline. The form of the potential surface is roughly similar to that of the density increases and decreases. For an anticline, the direction of minimum curvature is parallel with its axis. By definition, the orientation of the curvature symbol is parallel with the direction of minimum curvature, and therefore the bar on a map is parallel with an anticlinal axis. For a syncline the curvature is negative in the direction perpendicular to the axis, and this is algebraically less than that along the axis; therefore the minimum curvature is perpendicular to the axis.

The approximate relative relations of gravity, gradient, and curvature over a simple anticline-syncline combination are indicated in profile and cross section in Fig. 3-31.

OTHER GRAVITY INSTRUMENTS

In recent years there have been some instrumental developments which are related to the quantities measured by the torsion balance. A rotating "gradiometer" carries two symmetrical horizontal bars, at right angles, with masses on their ends. Forces acting on the weights, corresponding to those of the torsion-balance curvature components, cause very small torque differences between the two bars. These torque effects are measured by an elastic torque sensor between the bars and have a frequency double that of the frequency of rotation (corresponding to the 2λ term of the torsion-balance curvature relations, page 71). The sensitivity is greatly magnified by making the two-bar system sharply resonant and controlling the rotation frequency to be exactly half that of the torsional vibrations. The torques and hence the horizontal gradients, i.e., curvature components, are derived from the resonance amplitude measured by the torque sensors.

Some experiments have been carried out with a similar system on a horizontal axis which would measure the torsion-balance gradient quantities, but this appears to be a much more difficult instrumental problem.

These systems were proposed as a means of measuring gravity in the air, as (to a first degree at least) they would not be directly sensitive to motional accelerations.

4

FIELD OPERATIONS, DATA REDUCTION, AND MAPPING

INTRODUCTION

This chapter describes the basic features of gravity field operations as carried out in different environments, the necessary corrections to these measurements, and the mapping procedures. Details of field operations vary somewhat with policies of the operating company and the gravity meter used. The wide range of field conditions which may be encountered—topography, vegetation, wetness, and the availability of roads or trails—determines the vehicles used, the mode of travel, and corresponding variations of field procedures. The geological purpose of the survey controls, to some extent, the degree of detail and the precision of the gravity and elevation measurements.

To produce the necessary information for the preparation of the final Bouguer gravity map, the following essential data must be determined at each observation point:

1 Relative gravity difference from one or more reference, or "base," stations
2 Relative elevation for making the elevation correction
3 Relative position for making latitude corrections and for mapping the final results

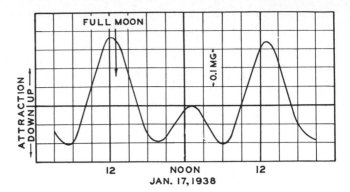

FIGURE 4-1
Theoretical tide-producing gravitational attraction at full moon. Calculated on tide-predicting machine 2. (*Courtesy of the U.S. Coast and Geodetic Survey.*)

DRIFT, OR TIME VARIATIONS, OF GRAVITY METERS

All gravity instruments sensitive enough for commercial geophysical prospecting have a certain amount of drift, or time variation, because any scheme used to measure the very small elongation or torsion of a spring system due to small changes in gravity will also measure small mechanical changes from other sources. Even under a well-controlled environment, at constant temperature, the spring and the associated mountings and connections are not perfectly stable, and slow or abrupt changes may occur. These may cause spurious changes in meter reading which are larger than those due to the small gravity differences being measured. These instrumental changes are called drift, and the field work must be conducted in such a way that this drift can be determined and corresponding corrections made.

Gravity Tidal Effects

In addition to the drift from internal causes there are external time variations of gravity which result from the variation in the gravitational attraction of the sun and the moon as their positions change with respect to the earth. For certain configurations of the sun and moon, the rotation of the earth produces changes that have a maximum amplitude of about 0.3 mgal and occur in a period as short as about 6 hr. The details of the tidal gravity change vary widely with the different phases of the moon, as indicated by Figs. 4-1 and 4-2. At full moon or new moon the tidal amplitude is a maximum, and, as indicated by the figures, the gradients can be as much as 0.05 mgal/hr. At the quarter phases the amplitudes are much smaller.

    The theoretical tidal gravity effects can be calculated for any time and location from readily available tables. An expression for the tidal gravitational effects, modified from the form given by Heiskanen and Vening Meinesz (1958, p. 119) is

$$\Delta g = \frac{3GrM_m}{2D_m^{\ 3}}\left(\cos 2\alpha_m + \tfrac{1}{3}\right) + \frac{3GrM_s}{2D_s^{\ 3}}\left(\cos 2\alpha_s + \tfrac{1}{3}\right)$$

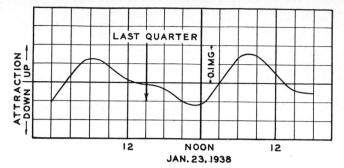

FIGURE 4-2
Same as Fig. 4-1 but calculated when the moon is in the last quarter. (*Courtesy of the U.S. Coast and Geodetic Survey.*)

where r = radius of rigid earth
M = mass
D = distance from earth
α = geocentric angle

The subscripts m and s designate the quantities for the moon and the sun, respectively. This expression gives the vertical attraction at the surface of a rigid earth. When the angles are expressed in terms of the latitude and time at the observation point and of the positions of the sun and moon, the tides consist of a component that peaks approximately twice a day, a component that peaks approximately daily, and longer-period components which peak every 14 days as a result of the moon and 6 months as a result of the sun. Each component varies differently with latitude, the daily component being maximum at middle latitudes and antisymmetric about the equator, the twice-daily component being maximum at the equator, and only the very long-period components existing at the poles.*

The calculations can be reduced to a routine operation with the necessary data taken from an astronomical ephemeris (Adler, 1942). The European Association of Exploration Geophysicists for years has published annually a set of tables reducing the tide calculation to a simple operation which can be done in the field (Goguel, 1954). Also, the tidal calculations can be made by electronic computer, using any one of a number of programs based on the closed-form solution of Longman (1959) or the programs described in Melchior (1966, chap. 8).

Observations of tidal effects with fixed gravity meters capable of measuring gravity changes to a few microgals have shown that the actual variation in gravity at the earth's surface is greater than that calculated from the theoretical attraction of the sun and moon; this results from slight tidal deformation of the earth's surface. For most purposes this variation can be approximated by increasing the

---

* A readable treatment of the theory, effects, measurement, and calculation of earth tides appears in Melchior (1966). The description of the three types of tides discussed above appears in chap. 1 of part 1 of that book.

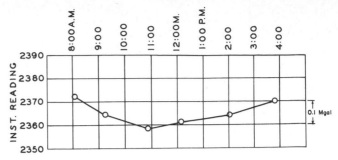

FIGURE 4-3

A gravimeter drift curve (Gulf gravimeter). One unit of the "instrument reading" scale is approximately 0.1 mgal.

calculated effects by 20 percent, although there is some variation in this factor, depending particularly on the distance from ocean shores and very little if any on the geology around the point of observation (Kuo et al., 1970). This factor, usually a uniform 20 percent, is added automatically in some tide-effect-calculation charts and tables.

Field Observations for Drift Corrections

To provide data for the drift and tide corrections, the gravity meter is returned to a reference station at intervals depending on the stability of the meter and the precision expected of the survey. It is not necessary for drift readings to be made at the same station if other points with previously established gravity values are more convenient. The readings can be reduced to what they would have been at a single reference point or base station by properly taking into account the gravity differences between bases which have been established by the procedures described below.

These drift readings impose conditions on the field operations which are often onerous but are necessary. If drift observations are made at intervals of about 2 hr or less and meter performance is good, there is little loss in precision by simple linear interpolation between drift readings.

A curve (Fig. 4-3) is plotted of the readings at the base station including those at other bases, referred back to the first by the established differences. This curve presumably is that which the instrument would have if continuous readings were made at the same point. Readings at other stations are then referred to this curve; i.e., the difference of the reading at a station from the value interpolated along the drift curve at the corresponding time gives the number of instrument scale divisions due to the gravity difference between the reference point and the station. This scale-division difference is then converted into a gravity difference from the calibration data of the instrument.

It is a frequent field practice to determine the drift rate (as milligals per minute) between base readings and use this rate to calculate drift from the reading

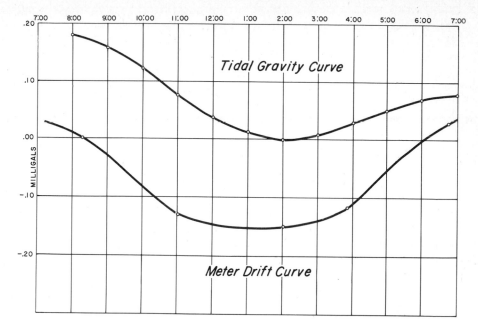

FIGURE 4-4
Drift curve with calculated tidal curve.

time at each intermediate station. This corresponds to interpolating along the straight-line segments of Fig. 4-3 but avoids plotting such a curve. However, as noted below, there are other advantages in plotting a daily drift curve. This simple interpolation procedure corrects for the total drift as a combination of the internal or instrumental effects and the external or tidal effects.

A better, although somewhat more laborious, field practice is to precalculate, by tables or computer, the tidal variation at the general locality of the field work and plot the varying tidal curves on daily work sheets (Fig. 4-4). The drift observations of the instrument are then plotted on the same sheets. Then a curve can be interpolated between points of drift control, using the calculated tidal curve as a guide. Also, this gives a very simple and effective indication of day-to-day meter performance. With a very stable gravity meter it may be possible, by using the tidal drift curve, to interpolate between considerably longer times of meter-drift observations and thereby extend the time interval between such observations. Where travel is rapid the time saving may be small and not worthwhile, but under difficult field conditions, where returning to base stations for drift observations becomes a serious burden, the use of calculated tidal curves for interpolation can materially increase station production. It must be reiterated that a very stable meter is required to give a linear long-time drift.

A long-range picture of meter performance can be gained by plotting a continuous drift curve day after day, with all base stations reduced to a common datum (Fig. 5-9). This curve is from a stabilized-platform ship meter, but the same system can be used for ordinary land meter observations if the principal base stations are tied into the absolute gravity network. Such a curve will show at a glance when meter performance is good or bad or when sudden discontinuities, or "tares," occur and is useful in evaluating a map of results when certain stations appear erratic or out of line with nearby stations made on different days.

ELEVATION REQUIREMENTS

We have seen (page 21) that the combination of free-air and Bouguer effects causes a change of gravity with elevation. Within the range of usual surface densities, of 2.6 to 1.8, the factor varies from about 0.06 to 0.07 mgal/ft. This means that if reduced gravity measurements are to be accurate to 0.1 mgal or less, the relative elevations must be determined to 1 ft or less. If the ultimate precision of modern gravity meters, about 0.01 mgal, is to be retained in the final values, relative elevations must be reliable to about 0.2 ft.

POSITION REQUIREMENTS

The required relative precision of station locations (horizontal control) is dictated by the latitude corrections. The horizontal gradient K, or rate of change of gravity with the north-south component of distance, can be derived from the gravity formula of the form

$$g_0 = A(1 + B \sin^2 \varphi - C \sin^2 2\varphi)$$

as

$$K = \frac{dg_0}{dx} = \frac{dg_0}{r \, d\varphi}$$

Table 4-1

Latitude φ, deg	K	
	mgals/mi	mgals/km
0	0	0
10	0.447	0.278
20	0.840	0.522
30	1.132	0.703
40	1.287	0.800
45	1.307	0.812
50	1.287	0.800
60	1.132	0.703
70	0.840	0.522
80	0.447	0.278
90	0	0

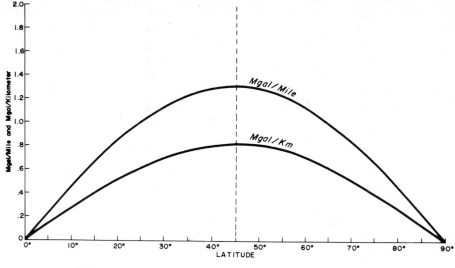

FIGURE 4-5
Latitude correction gradient.

for $dx = r\, d\varphi$, where r is the radius of the earth. Then

$$K = \frac{1}{r}\frac{dg_0}{d\varphi} = \frac{A}{r}(B \sin 2\varphi - 2C \sin 4\varphi)$$

From this formula, we can compute the value of K for any given latitude by inserting the values of A, B, and C from the gravity formula. For the international gravity formula (page 17) the values are

$$A = 978.049 \qquad B = 0.0052884 \qquad C = 0.0000059$$

For the radius r we can use the mean radius of the earth; that is, $(a + b)/2 = 6{,}368{,}000$ m. Inserting these numerical values in the equation above for K and neglecting C, which is too small to have an appreciable effect, gives

$$K = \begin{cases} 0.8122 \sin 2\varphi & \text{mgals/km} \\ 1.307 \sin 2\varphi & \text{mgals/mi} \end{cases}$$

Values of K for several latitudes are given in Table 4-1 and are shown by the curve in Fig. 4-5. To retain a precision of 0.1 mgal in the final values in middle latitudes, relative north-south distances must be accurate to roughly 100 m, or 400 ft, and to only one-tenth of these figures if mapping to 0.01 mgal is attempted.

SURVEYING METHODS AND PRACTICES

The required precision in elevation is not particularly difficult for engineering-type leveling, but this is a comparatively slow process. In routine gravity field operations elevations are commonly determined by level transit sights along with transit and stadia surveying of positions. A shortcut procedure of reading vertical angles or of "stepping" (adding stadia units by repeated vertical steps) is sometimes used when the topography permits only short level sights, but it needs skill, practice, and care to be successful and generally is not recommended. The transit-survey system is adequate if very carefully done and much faster than using levels, but it requires a practiced operator to make the observations accurately at the rate required for economical field work. A skillful surveyor can make the necessary elevation and position observations quite rapidly. The speed of surveying depends very much on the field conditions. In flat country foresights and backsights of $\frac{1}{4}$ mile can be made to stations $\frac{1}{2}$ mile apart, which is about the maximum distance for satisfactory precision and is a very common station interval for routine surveys. In more rugged country more and shorter sightings must be made, and field work is slower.

Gravity surveys also can be made by alidade and plane table. This, of course, gives a location map in the field rather than one computed later from surveyors' notes.

The balance between field personnel depends largely on the nature of the country covered. Commonly, there are two survey crews (surveyor, rodman, and vehicle) for each gravity-meter operator. In hilly, brushy country, a meter operator can keep up with three survey crews. Depending on field conditions, the cost of the surveying for elevations and positions for a gravity field operation is twice to several times the cost of the gravity measurements. For very precise work elevations must be determined by a separate survey operation using levels. In routine field practice, the source of errors in final map values is much more commonly traceable to errors in elevation than to errors in the gravity observations.

MAPPING GRAVITY-SURVEY RESULTS

There must be a map to show the station locations and contours of the gravity field. The scale and precision of the map may vary widely, depending on the purpose of the survey and the map source. The map may be entirely new, based on the horizontal control from the surveying of the gravity operation, but nearly always it is tied to some prior control such as triangulation stations and elevation bench marks.

In commercial operations, a base map commonly is provided or specified by the company for which the survey is made. This may be the same base map as used for other geophysical work or for land and leasing operations.

Almost the first step in planning a gravity survey in a new area is to find whatever published or other maps are available. There is, of course, a very wide range in the quality, scale, and detail of such maps. For reconnaissance surveys it may be possible to avoid any field surveying by making station observations at points where elevations are given on the maps. In areas in the United States where the $7\frac{1}{2}$-minute quadrangles at scale 1:24,000 (1 in. = 2000 ft) are available, they can serve quite well as a base for reconnaissance gravity maps, but the elevations are not accurate enough (only to 1 ft) for high-precision surveys. Also, care must be taken to be sure that the original sites for elevation points have not been disturbed by subsequent construction, particularly roadwork or road relocations.

Station Gravity

The field observations produce gravity values which are relative to some starting point or reference. This is very closely analogous to leveling operations, which produce elevations referred to some starting point or bench mark. If the bench mark is in a standard system, such as sea level determined by earlier surveys, the new measurements will be on a sea-level datum. Similarly, station gravity values may be established on a standard system by reference to base stations which are tied to a previously established or reference system. The U.S. Coast and Geodetic Survey and the Army Map Service call this value "observed gravity." If no prior gravity work has been done or no reference to other gravity work is anticipated, the datum must be arbitrary.

It is very good practice, where feasible, to tie the entire gravity net to reference points at which absolute gravity has been determined.* The absolute gravity value at such a reference station is the measured value without any reduction and corresponds to "station gravity." Such values are in numbers of the order of 978 to 983 gals, depending on latitude and (to a lesser extent) on elevation and correspond roughly to the values in the tables of normal gravity. When the latitude correction† is made, a number of similar magnitude is subtracted and gives the sea-level anomaly. If a free-air correction is added, i.e., the elevation (in feet) multiplied by 0.09406, the result will be the free-air anomaly. When the elevation correction is made in the usual way, i.e., with the elevation factor (free-air and Bouguer combined) multiplied by the sea-level elevation of the station, the result is the Bouguer anomaly. The corrected values thus determined then are on the same basis as the commonly published gravity maps of the various public agencies such as the U.S. Coast and Geodetic Survey, the Dominion Observatory of Canada, etc. These published Bouguer maps are usually reduced for a density of 2.67 g/cm³, which has been accepted as an average density for the surface rocks of the earth. When

---

* The data needed for such a reference station are the latitude, longitude, elevation, and absolute gravity value. Such values for many points in the United States and over the world, largely at airports, are given by Woollard and Rose (1963). Many values in Canada are available from the publications of the Dominion Observatory. Values from the older pendulum surveys in the United States are given by the "principal facts" tables published by the U.S. Coast and Geodetic Survey. A very large volume of base stations in the United States and Europe with latitude, longitude, elevation, and gravity value, together with a location plot of the point of observation, to permit exact reoccupation, is available from Aeronautical Chart and Information Service, St. Louis, Mo. Such values are still based on the old Potsdam gravity value, which is known to be about 13 mgals too high (see page 59).

† It must be remembered that the latitude "correction," as usually carried out, includes the total normal gravity of the spheroid and is not limited to the *change* of gravity with latitude.

FIGURE 4-6
Sequence of "looping" gravity-meter observations.

local densities differing from this value are used, there will be a substantial difference in the Bouguer anomaly, which, of course, will increase with elevation. For example, if a survey is corrected to sea level for density of, say, 2.40 g/cm³, the elevation factor will be 0.06339 mgal/ft while the factor for a density of 2.67 is 0.05994. For an elevation of 2000 ft, these differences in factor would make a difference of nearly 7 mgals in Bouguer gravity anomaly values.

Base Stations and Base Ties

A gravity survey nearly always consists of a series, or net, of field stations. The base-station net is the frame which supports the field stations. As outlined in the previous section, the gravity datum for this frame should be established by a tie to one or more previously established absolute gravity stations.

The ties between base stations commonly are made by looping (Fig. 4-6), i.e., by an observation, say, at station 1, ahead to station 2, looping back to 1, then on to 2 again, then on to 3, back to 2, ahead to 3, on ahead to 4, etc. Thus the successive observations at the same station are, in effect, short segments of a drift curve to which the next station is referred (Fig. 4-7). The average differences between these curves give the scale division differences between the successive stations. Closed traverses of such loops should be set out so that the accuracy of the work can be checked by the closure errors (Fig. 4-8). Modern gravimeters can carry closures of many miles or several tens of miles with closure errors of only a few tenths of a milligal.

Base-Loop Maps

It is good practice to make up base-loop maps in the field. These maps, usually at a much smaller scale than the station location maps, show base stations only. Gravity differences are shown for each "leg" between base stations (Fig. 4-8). It is convenient to show, for each such difference, an arrow pointing in the direction of gravity increase or positive gradient. Then for each closed loop, these differences are added around the loop with values in the direction of the arrows positive and those against the arrows negative. The sum of these values is the error of closure, which can be shown by a partial circle with an arrow indicating the direction of positive error (see Fig. 4-8). The originally observed gravity difference is shown by the lower figure on each leg; the upper figure shows the adjustment. Around each loop these adjustments add up to the closure error. Base-loop maps can be made separately for gravity differences and for elevation differences or for both together (in the example shown the elevation differences were on a separate similar map).

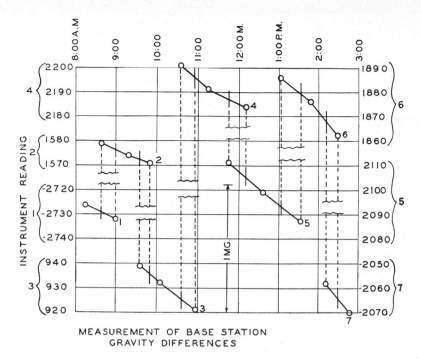

FIGURE 4-7
Segments of drift curves for looping base-tie observations.

The magnitude of the closure errors is a direct measure of the quality of the gravity and elevation data for the survey.

Before assigning final values to station gravity and to elevation at base stations, the errors of closure must be adjusted out. There are well-known statistical least-square methods developed for adjustment of level networks which can be used. Also, there are computer systems for making the adjustments. These more elaborate systems usually are not applied to gravity surveys because, with good field work, the closure errors are small enough for the manner of adjustment to make little difference in the practical application of the results. Careful inspection of the closure nets can indicate those legs most probably in error. For instance, if adjacent loops have closure errors in opposite directions, it is evident that adjustment of the common leg will improve both closures. In this way, empirical application of adjustments to the various legs of the survey can lead to satisfactory final values at each base station, and these values are then used to establish the values for ordinary field stations tied to these bases.

A useful, less empirical procedure is to use averages along different paths to establish base values. Starting at a primary base, differences to another base, preferably on the far side of the survey area, are determined along several different connecting paths. These differences are averaged and added to the base value to

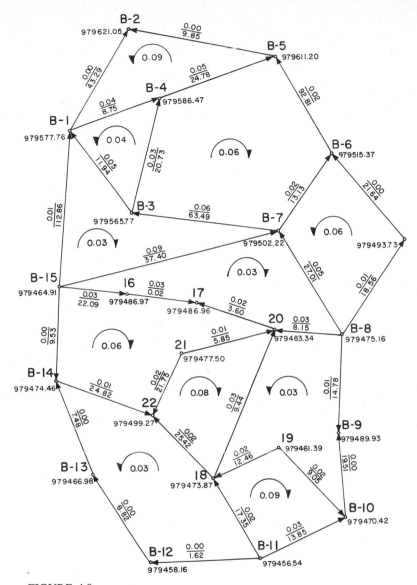

FIGURE 4-8
Gravity-difference observations, closure errors, adjustments, and adjusted base-station gravity values.

give an adjusted value for that station. Then the value at an intermediate station can be determined as the average of the values obtained from these two stations over different paths and the process repeated until values are assigned to all base stations.

For large programs, there is a definite advantage in setting out base stations and adjusting the base net far in advance of the observation of field stations. This permits the determination of base-station values which are not subject to change by later extensions of the survey. If small extensions of the base net are made piecemeal, there is a tendency to push adjustments into the new loops and errors may accumulate to a degree that requires readjusting the base net in previously completed areas and recalculating the stations observed from the corresponding base stations. There is a natural reluctance to make such recalculations.

REDUCTION OF FIELD OBSERVATIONS

The field observations themselves produce a series of meter readings. The usual final map is of Bouguer gravity, which is in gravity (milligal) units with meter readings multiplied by the instrument scale factor and with observations corrected for the shape of the earth, elevation, and (sometimes) surface irregularities.

Instrument Scale Factor

Modern gravity meters are carefully calibrated by their makers. For most land meters a simple scale constant is supplied (usually engraved on the instrument nameplate), and conversion to milligals is made simply by multiplying scale readings by this number. For some meters with very long range, the so-called "geodetic" meters and some shipborne meters, there is a very slight deviation from perfect linearity and a table rather than a simple constant is used. This calibration is determined by small weights temporarily added to the moving system, in various parts of its total range and then removed (LaCoste and Romberg meters). However established, the calibration is quite permanent as long as the spring is not modified in any way.*

In routine land surveys the meter scale factor is incorporated in the calculation of the observer's field notes, along with latitude and elevation corrections.

Latitude Corrections

The latitude correction takes account of the increase of gravity from equator to pole, as discussed in detail above (page 16). Latitude corrections are nearly always based on gravity differences derived directly from the gravity formula. This can be done by use of a table of gravity values, such as that computed from the international gravity formula (Appendix to Chap. 4). If the latitudes of stations are expressed in degrees and minutes, the correction tables can be made accordingly and the latitude correction can be read off from such tables directly.

* For any meter which has a continuous calibration over a worldwide gravity range (7000 mgals), the system of carrying a meter-zero curve is very useful (see page 122). If the meter drifts or has discontinuities, the curve gives a continuous and dramatic picture of meter performance. Furthermore, when gravity values at any station (in milligals) are added to the meter-zero value for that date, the data are automatically on the worldwide absolute gravity datum.

If the base-station net is tied to absolute gravity, the station gravity values will be (for middle latitudes, say 45°) near 980.6000 gals. The latitude corrections will be of similar magnitude (980.6294 for 45°00′). If these particular values were encountered, the corrected station gravity (with latitude correction only) would be 980.6000 − 980.6294 = −0.0294 = −29.4 mgals and the elevation correction (including Bouguer correction) would be applied to this value.

An efficient reduction practice is to use values directly from the table to calculate the latitude correction.* It is convenient to make up scales from these tables, corresponding to the scale of the map used and with the scale extending for a range of $\frac{1}{2}$ degree of latitude (for maps at scale of around 1 in. = 1 mile) or $\frac{1}{4}$ degree (for maps at large scale such as 2 in. = 1 mi. or 1 in. = 2000 ft or equivalent). Reference marks corresponding to 15 or 30 minutes of latitude are made on the scale and the intervals corresponding to whole milligals in latitude effect, taken from the table, are inserted between these marks. The scale is then laid across the map in a north-south direction with the reference marks matched to the corresponding latitudes, and the correction is read off from the position of the station with respect to the scale. Alternatively, the scale can be used to draw east-west lines on the work map corresponding to whole milligals of latitude correction. A small auxiliary scale is used to interpolate the position of a station between these lines and read off its latitude correction. Since modern instruments can be read to a few hundredths of a milligal, it is desirable to attempt to keep the corrections accurate to 0.01 mgal, which corresponds to around 50 ft in north-south direction.

Elevation Corrections

The corrections to gravity values which must be made on account of differences in elevation take care of two effects: (1) the free-air effect and (2) the Bouguer effect.

The free-air correction   The vertical decrease of gravity with increase of elevation, considered above (page 20), is taken care of by the free-air correction. As pointed out there, this variation has a magnitude of

$$0.3086 \text{ mgal/m} \quad \text{or} \quad 0.09406 \text{ mgal/ft†}$$

The correction can be made to any arbitrary reference or datum level that is convenient, e.g., the elevation of a base station for a survey, or it may be made to sea level. Since a station at a relatively higher elevation has a lower gravity (because it is farther from the center of the earth), the correction must be *added* to stations

---

* The first three numbers (whole gals) may be dropped and the corrections listed starting with the first number to the right of the decimal point. If base ties to absolute gravity values are used, as in the preceding paragraph, the same number in whole gals can be dropped there also to give the difference values correctly.

† There is a small additional term which depends on the slight change of the vertical gradient with elevation (Heiskanen and Vening Meinesz, 1958, p. 149) which is usually negligible. Its effect is 1.7 mgals for an elevation of 5000 m, 0.3 mgal for 2000 m, and 0.067 mgal for 1000 m.

at a *higher* elevation and must be *subtracted* from stations at a *lower* elevation than the reference level.

The Bouguer correction   The attraction of the material between a reference elevation and that of the individual station is taken care of by the Bouguer correction. The term is used here in a restricted sense to designate the correction for the attraction as approximated by considering the material as an infinite horizontal slab. The gravity attraction for a point on the surface of a slab of thickness h and density σ is $g = 2\pi G\sigma h$, which, for $G = 6.6732 \times 10^{-8}$,* gives

$$g = \begin{cases} 0.04193\sigma h & \text{mgals/m} \\ 0.01278\sigma h & \text{mgals/ft} \end{cases}$$

If a given station is *higher* than the reference elevation, its gravity value is *increased* because of the attraction of the slab of material between it and the reference level and the correction is negative. If the station is lower than the reference elevation, its gravity value is decreased because of the lack of attraction of the absent material between it and the reference level and the correction is positive. Therefore, the Bouguer correction is always opposite in sign to the free-air correction.

Combined Elevation-Correction Factor

Since both the free-air and Bouguer correction are simple constants multiplied by the elevation, they are nearly always combined into a single "elevation factor" f. From the values given above, it is evident that

$$f = \begin{cases} 0.3086 - 0.04193\sigma & \text{mgals/m} \\ 0.09406 - 0.01278\sigma & \text{mgals/ft} \end{cases}$$

where σ is the density of surface material within the range of the topography. Since the net *effect* is negative, i.e., a higher station has lower gravity, the *correction* is positive, and the higher the station, the greater the elevation correction to be added.

Values of the factor f for different densities are given in Table 4-2 and Fig. 4-9.

The choice of density for the elevation factor for a survey in a new area is often a troublesome problem. It is quite common to assume a value based on the general nature and age of the surface rocks and make preliminary maps using this factor. After the party is in the field, "density profiles" can be run over selected topographic features to give a measure of the average density within the elevation range of the topography.

The density profile (Nettleton, 1942) consists of a line of stations at close spacing observed over a topographic feature. The measurements are reduced with different elevation factors to find the one which minimizes the correlation of

---

* The dimensional units of G are $L^3 T^{-2} M$.

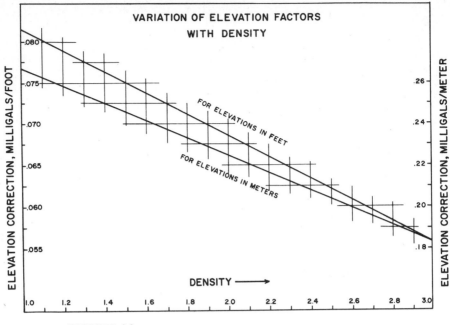

FIGURE 4-9
Variation of elevation correction factor with density.

gravity with topography. The corresponding density is an average value for the topographic feature sampled. An example is given by Fig. 4-10.

In choosing a topographic feature it is preferable to select one over a hill rather than a valley because the latter may have nontypical floodplain sediments in the bottom. Also, it is desirable for the elevations beyond the feature to be roughly the same to give a definite maximum elevation, that is, a one-sided rise or fall is not suitable.

Table 4-2 COMBINED ELEVATION CORRECTION FACTOR

Density	mgal/m	mgal/ft	Density	mgal/m	mgal/ft
0.0	0.3086	0.09406	2.3	0.2122	0.06467
1.0	0.2667	0.08128	2.4	0.2080	0.06339
1.6	0.2415	0.07361	2.5	0.2038	0.06211
1.7	0.2373	0.07233	2.6	0.1996	0.06083
1.8	0.2331	0.07106	2.67	0.1966	0.05994
1.9	0.2289	0.06978	2.7	0.1954	0.05955
2.0	0.2247	0.06850	2.8	0.1912	0.05828
2.1	0.2205	0.06722	2.9	0.1870	0.05700
2.2	0.2164	0.06594	3.0	0.1828	0.05572

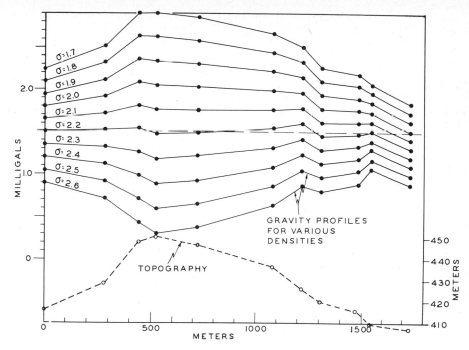

FIGURE 4-10
Density profile over local topographic feature. The several gravity profiles are reduced with elevation factors corresponding to the densities shown. The indicated density is 2.2, as that profile has minimum correlation with topography.

After an area has been mapped and a correlation of gravity with topography is suspected, a check can be made by the method of triplets. Profiles of elevation and gravity are plotted along the more rugged lines of the survey. For each group of three stations, the differences of elevations and of gravity of the middle station from a straight line between the two outside stations are determined graphically, as indicated by Fig. 4-11, and an average value of $\Delta g/\Delta h$ for the line or area is calculated. This value is a correction to be applied to the elevation factor used for the original reduction of the field observations. The individual values commonly are very ragged, but if they are mostly of the same sign, they should be averaged and the elevation factor should be corrected; i.e., the value of the elevation factor is decreased if $\Delta g/\Delta h$ is positive and increased if $\Delta g/\Delta h$ is negative. Then the area is remapped. A statistical system (unpublished) which accomplishes the same result by a more elegant procedure has had limited use, but the simple graphical procedure described above is usually adequate and is more convenient and much more readily understood by field personnel.

The use of elevation-factor profiles or the triplet or statistical methods implies that surface-elevation changes do not correlate with structures. It is quite common for subsurface structure to influence surface topography; if the structure has a

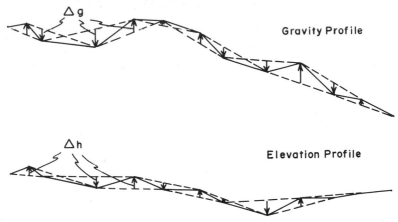

Gravity Profile

Elevation Profile

FIGURE 4-11
Use of triplets to determine error in elevation factor.

gravity effect, its anomaly will correspond with that produced by the topography and modify it. Attributing the entire anomaly to the elevation effect may lead to ridiculous, even negative, densities for the topography. This possibility must be kept in mind in using correlations of gravity with topography as a measure of the elevation correction factor.

The density measured by elevation-gravity correlation is that of rock within the local range of the topography. It does not apply to the geologic section below the minimum elevation.

The elevation factor used for a Bouguer gravity map should always be shown on the map, a practice which is not followed as meticulously as it should be.

As pointed out above (page 80), elevations of gravity stations must be known to less than 0.2 ft if the elevation correction is to be accurate to 0.01 mgal. For such precision it is necessary to survey the elevation of each station, as elevation contour maps of sufficient accuracy are seldom if ever available. If errors of $\frac{1}{2}$ mgal or more are tolerable and topography is not too rugged, a good topographic map may give adequate vertical control. Attempts have been made to use aneroid barometers for determining elevations, but the accuracy attainable by such methods is not sufficient to make corrections to 0.1 mgal since good surveying aneroids (even if very carefully used) have probable errors of 1 ft or more. Entirely aside from the precision of the barometer itself, it has been shown (Vacquier, 1937) that horizontal variations in the air pressure make uncertainties of elevation of about 1 ft over distances of only a few miles. Experience has shown that gravity surveys depending on barometric elevations are very seldom reliable to better than about 1 mgal.

Terrain Corrections

The simple Bouguer correction described above is calculated as if the surface were horizontal at the elevation of the station corresponding to the dotted horizontal

FIGURE 4-12
Topographic irregularities which produce effects requiring terrain corrections.

line through the station (Fig. 4-12). The value is calculated as the effect of an in-finite horizontal slab, that is, $g = 2\pi G\sigma h = 0.01278\sigma h$, where h is the elevation in feet. There may be voids below the plane of calculation, such as at a (Fig. 4-12), or higher topography above the plane, as at b. The void makes a deficiency in gravity which is not taken into account by the horizontal-plane calculation. The hill at b has an upward component of attraction, tending to decrease the gravity, which also is not taken into account. Thus both the voids below and the hills above the plane at the elevation of the station produce negative effects, and the terrain correction which compensates for these effects is always positive. The magnitude, of course, is proportional to the density of the rock, which (except for very special circumstances) is the same as that used for the Bouguer correction.

For many years, terrain corrections have been made by charts having com-partments of increasing size and with corresponding tables or curves (Hammer, 1939; Bible, 1962) which give the effect of each compartment as a function of its elevation above or below the station. The chart is laid over a topographic map with its center at the station, and the average elevation is estimated and listed for each compartment. The gravity effect for each compartment is then read from the table or curve for the zone in which it occurs. Finally all the effects are added and the sum multiplied by a factor for the density to give the total terrain correction for the station. The chart is then moved to the next station and the process repeated. This is, at best, a laborious process, requiring many minutes to 1 hr or more per station, depending on the severity of the topography, the distance to which the calculation is carried, and its precision. Even with fairly detailed maps the defi-nition of the topography in the immediate vicinity of the station will not be ade-quate in moderate to severe topography. This is particularly true for stations run along roads in steep valleys or canyons. To fill in the topography of the inner zones [through zones D or E of the Hammer (1939) tables, i.e., out to distances of 558 or 1280 ft] it is common practice to estimate elevation or slopes in the field, usually for zones B through E. A small zone chart, on a sheet to fit the engineer's notebook, is provided, and the surveyor writes the estimated elevation in each compartment directly on the chart. These inner-zone corrections are added to the outer-zone corrections determined from a topographic map or from a computer operation.

A number of systems for calculating terrain corrections by digital computer have been published (e.g., Kane, 1962; Bott, 1959). These generally use a scheme of digitizing the topography by writing the elevations at points on a uniform grid so that the same grid points can be used for all stations. Then, from the coordinates

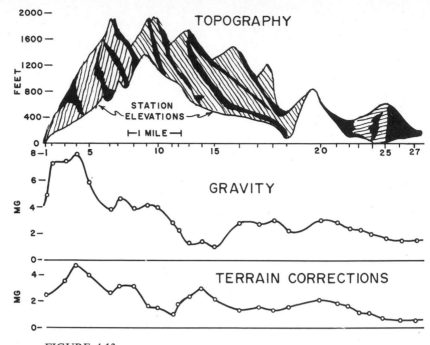

FIGURE 4-13

Sample traverse in Alberta foothills, showing relative station elevations, topography laterally adjacent to the line of stations (drawn to scale), and the reduced (Bouguer) gravity and terrain corrections. Corrections were calculated through Hammer zone *E* in the field and from 1 in. = 1 mile topographic maps through zone *J* (21,826 ft) and occasionally through zone *M* (71,996 ft). (*From Hastings, 1945.*)

of the point and its elevation the distance and elevation difference to each digital unit of topography are determined and the effects calculated and summed. A system of this kind is quite effective for a large survey, where the considerable labor of digitizing the elevations over a large area is applied to many stations. For widely spaced stations in a large area it may be more economical to make the computation by hand using charts and tables.

Some shortcut schemes are applicable in certain situations. For instance, if there are high distant mountains, their effects can be contoured from calculations at a small number of stations and values at individual stations interpolated from the contours if the range of station elevations is relatively small. A system for distant corrections is given by Winkler (1962). A small example from a survey with severe topography and large terrain corrections is given by Fig. 4-13 (Hastings, 1945).

In calculations from topography digitized on a regular grid there is always a compromise between adequate detail near the station and the increase in digitizing and computer costs when the grid interval is reduced. This can be met, in part, by

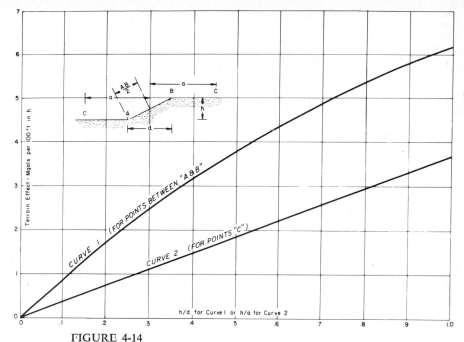

FIGURE 4-14
Terrain correction chart for two-dimensional topography with a density of 2.0.
Line 1 is for points on slope between *A* and *B*; line 2 is for points below the slope
(left of point *A*) or above the slope (right of point *B*). (*Compiled from charts by
Hubbert, 1948b.*)

omitting the area immediately around the station from the digital operation and
calculating inner-zone corrections by conventional chart methods (Kane, 1962)
or by estimating effects near the station in the field (page 93). Then the local
effect at each station is added (in the computer program) to the more distant effects
derived from the digital operation.

Hubbert (1948*b*) gives a series of charts for calculating terrain corrections for
two-dimensional topography which is applicable to stations in long valleys or at
some distance from essentially linear mountain ranges. The several charts are for
either (1) a point along a uniform slope or (2) a point at a distance from the foot or
the top of a uniform slope. The two conditions represented by all the charts can
be approximated by the two curves of Fig. 4-14, which is for a density of 2.0.
While the representation is not quite exact, it is as close as the two-dimensional
approximation of the topography. The system is useful for the special circumstances for which it is applicable and particularly in estimating the magnitude of
terrain corrections to determine whether more elaborate methods should be
carried out.

Terrain corrections are omitted in many field operations where they should
be included simply because of the labor and cost involved. This is especially true

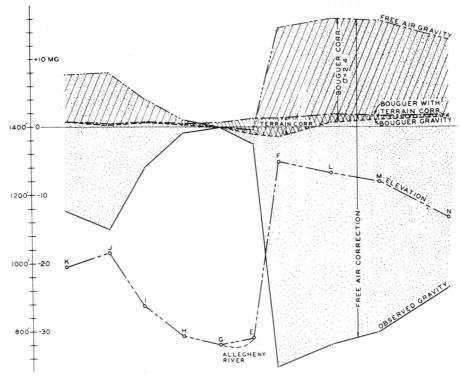

FIGURE 4-15
Profile of experimental gravity traverse showing station elevations, observed gravity, and gravity values after successive reductions for free-air, Bouguer, and terrain effects. (*From Nettleton, 1940.*)

if no topographic map is available. In such circumstances, the larger effects of the nearby terrain may be determined approximately by special observations in the field (an extension of the inner-zone corrections mentioned above). These can be quite effective if skillfully and carefully done, especially in reducing erratic differences between nearby stations.

Example of Reduction of Gravity Measurements

An example of the reduction of a few gravity stations in a rugged area is shown by Table 4-3 and Fig. 4-15. This was an experimental line crossing the Allegheny River valley near Pittsburgh with some 500 ft of sharp relief. Station *G* was used as a base to which gravity and elevation differences were referred. The free-air and Bouguer corrections are listed and plotted separately, rather than being combined (as is common practice) in order to show the relative magnitudes of the different corrections.

Table 4-3

Station	Elevation, ft	Observed gravity re station G, mgals	Latitude correction, re station G, mgals	Observed gravity + latitude correction, mgals	Free-air correction, mgals	Free-air gravity, mgals	Bouguer correction, (σ = 2.4) mgals	Bouguer gravity, mgals	Terrain correction, mgals	Final corrected, gravity, mgals
K	991	−12.38	−1.49	−13.87	+21.63	+7.76	−7.04	+0.72	+0.07	+0.79
J	1017	−14.97	−1.04	−16.03	+24.08	+8.05	−7.84	+0.21	+0.23	+0.44
I	877	−5.97	−0.74	−6.71	+10.91	+4.20	−3.55	+0.65	+0.13	+0.78
H	785	−0.92	−1.30	−1.22	+2.26	+1.04	−0.73	+0.31	+0.20	+0.51
G	761	0	0	0	0	0	0	0	+0.77	+0.77
E	781	−2.50	+0.22	−2.28	+1.88	−0.40	−0.61	−1.01	+2.33	+1.32
F	1289	−35.37	+0.44	−34.93	+49.66	+14.73	−16.17	−1.44	+2.78	+1.34
L	1266	−31.98	+0.57	−31.41	+47.50	+16.09	−15.47	+0.62	+1.24	+1.86
M	1240	−30.14	+0.89	−29.25	+45.05	+15.80	−14.67	+1.13	+0.53	+1.66
N	1134	−23.58	+1.48	−22.10	+35.08	+12.98	−11.42	+1.56	+0.24	+1.80

This particular example shows some stations that would be considered poorly located for a geophysical survey. This applies particularly to stations E and F, which are at the bottom and top of the high bluff at the river bank and have very large terrain corrections. Terrain corrections of this magnitude are difficult to calculate accurately from a zone chart, and the corrections used were calculated in part with the aid of a dot chart (see page 206). However, by selecting a location back from the bluff and with relatively flat topography near the site, such as station H, the terrain correction can be greatly reduced and can be calculated quite accurately with a zone chart. Thus, if some freedom in choice of station locations is allowed, fairly good gravity stations can be made in moderately rough topography.

Reduction of Combined Land and Water Stations

In routine gravity surveying, the field conditions and nature of the observations are the same, and all stations are reduced in the same way. Problems may arise when different surveys with different field techniques, often made at widely different times, or surveys made on the surface or bottom of adjacent water-covered areas are combined to make a uniform and continuous map. Then it may be desirable to modify parts of maps or change the data to bring them together without hiatus at survey boundaries. Similarly, if the surveys overlap, it is essential that both sets of stations have similar corrections applied before attempting to contour the combined data. Otherwise, one set of station values may be high or low with respect to the other, the map will be erratic, and nonexistent anomalies will be generated.

The system indicated below will give consistent reductions and continuous mapping of stations in the several situations illustrated provided the original gravity observations are corrected for consistent datum differences. For simplicity and uniformity, the station elevation is taken as the elevation of the *ground* (except for surface ship meters), as indicated by h_1, h_2, etc., of Fig. 4-16. Then, for the various situations of the diagram, the corrections are as outlined in the following paragraphs. Values are in milligals for elevations in feet (depths are in fathoms for ship meter observations) and give Bouguer gravity.

CASE 1: LAND STATION ON GROUND   For an instrument on a low pinboard resting on the ground the correction $= fh_1$, where f is the ordinary Bouguer elevation factor, depending on the density of the ground.

CASE 2: LAND STATION WITH RAISED INSTRUMENT   Correction $= fh_2 + 0.0941e$. The second term is the free-air correction for the additional height of the instrument above the ground. If all stations are at the same elevation, as when the instrument is used on a tripod, this term drops out in taking differences. It must be considered when some stations are on a pinboard on the ground and some on a tripod.

CASE 3: WATER STATION WITH INSTRUMENT ON A TRIPOD ABOVE WATER   Correction $= fh_3 + 0.0813d + 0.0941e$. The second term is the ordinary elevation factor for a density of 1, which is the density of fresh water.

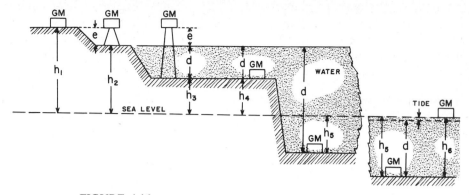

FIGURE 4-16
Schematic diagram and notation for consistent reduction of gravity stations under various field situations.

CASE 4: UNDERWATER STATION   Correction $= fh_4 + 0.01278d$. The second term is to correct for the upward attraction of the water above the instrument.

CASE 5: UNDERWATER STATION BELOW DATUM LEVEL   Correction $= -fh_5 + 0.01278d$. This is the same as case 4 except for change of sign for station below elevation datum.

For offshore underwater meter operations the meter elevation (h_5') is the sea-level elevation of the bottom. There may be tidal variations of water depth which are taken into account, and therefore the water depth d and the meter elevation h_5 may not be the same and the water-depth correction may be carried as a separate item in the data-reduction calculations.

CASE 6: SHIP METER ON SURFACE SHIP   Usually the elevation is taken as sea level, and there is no elevation of free-air correction because at offshore locations there is no readily available means to determine tidal changes in sea level. The correction becomes the Bouguer correction only, which is Bh_6, where $B = 0.01278(\sigma - 1.03)$ mgal/ft $= 0.07656(\sigma - 1.03)$ mgal/fathom, where σ is the density of the rock material below water bottom and 1.03 is the density of seawater. This correction reduces the gravity value to what it would be if the water were replaced by the rock below sea bottom. A chart for the variations of this water-depth correction with density is given by Fig. 5-9 (page 125).

Terrain effects from irregular bottom topography under gravity measurements at sea are quite different from those discussed above, where the meter is on the surface; they are discussed in Chap. 5.

A few unusual or special cases, e.g., gravity meter in a submarine or datum plane between meter and water bottom, are not included in the examples above. The proper combinations of corrections should be evident from the six cases given.

APPENDIX TO CHAPTER 4

TABLES OF NORMAL GRAVITY

The following tables give the value of theoretical gravity on the international ellipsoid, calculated from the international gravity formula,

$$\gamma = 978.0490 \, (1 + 0.0052884 \sin^2 \varphi - 0.0000059 \sin^2 2\varphi)$$

for each 10 minutes of latitude.

Since the tables are given to the sixth decimal place and the second differences are almost constant for a range of a few degrees, it is possible to interpolate accurately to single minutes or tenths of minutes by using a suitable second-order interpolation formula. However, values accurate to the fifth decimal place are obtained by simple linear interpolation.

If a survey of a general area is to be reduced with respect to a local latitude, tables of latitude effects from that reference latitude can be prepared by subtracting the normal gravity for the reference latitude from the values interpolated from the tables over the latitude range of the area being covered.

VALUES OF THEORETICAL GRAVITY ON THE INTERNATIONAL ELLIPSOID[1]
$$\gamma = 978.0490(1 + 0.0052884 \sin^2 \varphi - 0.0000059 \sin^2 2\varphi)$$

Geographic latitude φ	Gravity, cm /sec ^{2}	Geographic latitude φ	Gravity, cm /sec ^{2}	Geographic latitude φ	Gravity, cm /sec ^{2}
0° 0′	978.049000	6° 0′	978.105265	12° 0′	978.271635
10	.049044	10	.108422	10	.277770
20	.049174	20	.111665	20	.283983
30	.049392	30	.114993	30	.290277
40	.049697	40	.118405	40	.296649
50	.050089	50	.121903	50	.303100
1 0	978.050568	7 0	978.125484	13 0	973.309630
10	.051134	10	.129150	10	.316238
20	.051789	20	.132901	20	.322926
30	.052528	30	.136737	30	.329690
40	.053356	40	.140656	40	.336531
50	.054270	50	.144659	50	.343453
2 0	978.055272	8 0	978.148747	14 0	978.350450
10	.056360	10	.152918	10	.357524
20	.057535	20	.157173	20	.364675
30	.058797	30	.161511	30	.371903
40	.060146	40	.165934	40	.379207
50	.061582	50	.170439	50	.386587
3 0	978.063104	9 0	978.175027	15 0	978.394043
10	.064714	10	.179699	10	.401574
20	.066408	20	.184452	20	.409182
30	.068192	30	.189289	30	.416863
40	.070060	40	.194208	40	.424620
50	.072016	50	.199209	50	.432452
4 0	978.074057	10 0	978.204293	16 0	978.440358
10	.076185	10	.209458	10	.448337
20	.078398	20	.214706	20	.456392
30	.080699	30	.220034	30	.464519
40	.083086	40	.225443	40	.472720
50	.085559	50	.230935	50	.480993
5 0	978.088116	11 0	978.236507	17 0	978.489339
10	.090761	10	.242161	10	.497757
20	.093491	20	.247896	20	.506248
30	.096306	30	.253710	30	.514810
40	.099208	40	.259605	40	.523445
50	.102194	50	.265580	50	.532150

[1] W. D. Lambert and F. W. Darling. *Bull. Géodesique*, No. 32, October, November, December, 1931.

VALUES OF THEORETICAL GRAVITY ON THE INTERNATIONAL ELLIPSOID
(*Continued*)

Geographic latitude φ	Gravity, cm /sec ^{2}	Geographic latitude φ	Gravity, cm /sec ^{2}	Geographic latitude φ	Gravity, cm /sec ^{2}
18° 0′	978.540926	24° 0′	978.901505	30° 0′	979.337764
10	.549773	10	.912682	10	.350787
20	.558690	20	.923917	20	.363853
30	.567678	30	.935210	30	.376963
40	.576736	40	.946560	40	.390115
50	.585864	50	.957968	50	.403309
19 0	978.595059	25 0	978.969432	31 0	979.416545
10	.604324	10	.980953	10	.429822
20	.613658	20	978.992530	20	.443140
30	.623059	30	979.004164	30	.456498
40	.632530	40	.015851	40	.469897
50	.642067	50	.027592	50	.483337
20 0	978.651671	26 0	979.039389	32 0	979.496812
10	.661343	10	.051240	10	.510328
20	.671081	20	.063144	20	.523882
30	.680886	30	.075102	30	.537473
40	.690756	40	.087112	40	.551103
50	.700692	50	.099174	50	.564768
21 0	978.710694	27 0	979.111288	33 0	979.578470
10	.720761	10	.123454	10	.592207
20	.730892	20	.135671	20	.605981
30	.741089	30	.147940	30	.619789
40	.751348	40	.160257	40	.633632
50	.761671	50	.172626	50	.647508
22 0	978.772057	28 0	979.185044	34 0	979.661419
10	.782507	10	.197510	10	.675362
20	.793019	20	.210025	20	.689338
30	.803594	30	.222589	30	.703345
40	.814230	40	.235201	40	.717385
50	.824928	50	.247860	50	.731455
23 0	978.835687	29 0	979.260565	35 0	979.745556
10	.846506	10	.273318	10	.759688
20	.857387	20	.286117	20	.773850
30	.868327	30	.298961	30	.788040
40	.879327	40	.311850	40	.802260
50	.890386	50	.324785	50	.816507

VALUES OF THEORETICAL GRAVITY ON THE INTERNATIONAL ELLIPSOID
(Continued)

Geographic latitude φ	Gravity, cm /sec ^{2}	Geographic latitude φ	Gravity, cm /sec ^{2}	Geographic latitude φ	Gravity, cm /sec ^{2}
36° 0′	979.830784	42° 0′	980.359132	48° 0′	980.899782
10	.845087	10	.374093	10	.914747
20	.859417	20	.389063	20	.929704
30	.873773	30	.404042	30	.944650
40	.888155	40	.419028	40	.959587
50	.902563	50	.434022	50	.974511
37 0	979.916995	43 0	980.449023	49 0	980.989425
10	.931453	10	.464031	10	981.004328
20	.945934	20	.479043	20	.019218
30	.960438	30	.494062	30	.034095
40	.974966	40	.509086	40	.048959
50	979.989517	50	.524114	50	.063809
38 0	980.004089	44 0	980.539146	50 0	981.078646
10	.018682	10	.554182	10	.093466
20	.033297	20	.569220	20	.108271
30	.047931	30	.584262	30	.123062
40	.062587	40	.599304	40	.137835
50	.077261	50	.614349	50	.152591
39 0	980.091955	45 0	980.629394	51 0	981.167331
10	.106667	10	.644439	10	.182053
20	.121397	20	.659486	20	.196755
30	.136146	30	.674530	30	.211438
40	.150911	40	.689574	40	.226103
50	.165692	50	.704616	50	.240748
40 0	980.180490	46 0	980.719656	52 0	981.255373
10	.195303	10	.734694	10	.269977
20	.210131	20	.749726	20	.284560
30	.224975	30	.764758	30	.299120
40	.239832	40	.779783	40	.313658
50	.254702	50	.794805	50	.328175
41 0	980.269585	47 0	980.809821	53 0	981.342667
10	.284481	10	.824832	10	.357135
20	.299389	20	.839836	20	.371579
30	.314308	30	.854834	30	.385999
40	.329240	40	.869825	40	.400393
50	.344181	50	.884807	50	.414761

VALUES OF THEORETICAL GRAVITY ON THE INTERNATIONAL ELLIPSOID
(*Continued*)

Geographic latitude φ	Gravity, cm /sec ^{2}	Geographic latitude φ	Gravity, cm /sec ^{2}	Geographic latitude φ	Gravity, cm /sec ^{2}
54° 0′	981.429104	60° 0′	981.923902	66 °0′	982.362437
10	.443419	10	.936939	10	.373622
20	.457708	20	.949932	20	.384749
30	.471968	30	.962881	30	.395815
40	.486200	40	.975785	40	.406823
50	.500404	50	981.988642	50	.417770
55 0	981.514578	61 0	982.001455	67 0	982.428657
10	.528723	10	.014222	10	.439482
20	.542837	20	.026941	20	.450248
30	.556921	30	.039613	30	.460950
40	.570974	40	.052239	40	.471591
50	.584996	50	.064816	50	.482171
56 0	981.598985	62 0	982.077344	68 0	982.492687
10	.612942	10	.089824	10	.503141
20	.626866	20	.102255	20	.513530
30	.640757	30	.114636	30	.523857
40	.654613	40	.126967	40	.534120
50	.668435	50	.139248	50	.544319
57 0	981.682222	64 0	982.151478	69 0	982.554452
10	.695974	10	.163656	10	.564520
20	.709691	20	.175782	20	.574522
30	.723371	30	.187856	30	.584460
40	.737014	40	.199878	40	.594331
50	.750620	50	.211846	50	.604135
58 0	981.764188	64 0	982.223763	70 0	982.613873
10	.777718	10	.235624	10	.623543
20	.791209	20	.247431	20	.633146
30	.804662	30	.259184	30	.642681
40	.818076	40	.270882	40	.652148
50	.831448	50	.282525	50	.661548
59 0	981.844781	65 0	982.294112	71 0	982.670877
10	.858073	10	.305642	10	.680138
20	.871323	20	.317116	20	.689330
30	.884531	30	.328532	30	.698452
40	.897697	40	.339893	40	.707504
50	.910821	50	.351194	50	.716485

VALUES OF THEORETICAL GRAVITY ON THE INTERNATIONAL ELLIPSOID
(*Continued*)

Geographic latitude φ	Gravity, cm /sec ^{2}	Geographic latitude φ	Gravity, cm /sec ^{2}	Geographic latitude φ	Gravity, cm /sec ^{2}
72° 0′	982.725396	78° 0′	982.996761	84° 0′	983.164537
10	.734236	10	983.002866	10	.167636
20	.743005	20	.008889	20	.170648
30	.751702	30	.014832	30	.173574
40	.760328	40	.020694	40	.176415
50	.768881	50	.026477	50	.179169
73 0	982.777361	79 0	983.032177	85 0	983.181836
10	.785769	10	.037795	10	.184417
20	.794104	20	.043333	20	.186912
30	.802367	30	.048788	30	.189319
40	.810554	40	.054161	40	.191640
50	.818669	50	.059452	50	.193873
74 0	982.826710	80 0	983.064661	86 0	983.196021
10	.834676	10	.069787	10	.198080
20	.842566	20	.074830	20	.200052
30	.850383	30	.079791	30	.201938
40	.858124	40	.084668	40	.203736
50	.865790	50	.089463	50	.205446
75 0	982.873379	81 0	983.094173	87 0	983.207070
10	.880893	10	.098801	10	.208606
20	.888331	20	.103344	20	.210054
30	.895691	30	.107805	30	.211415
40	.902975	40	.112181	40	.212689
50	.910182	50	.116472	50	.213874
76 0	982.917312	82 0	983.120679	88 0	983.214972
10	.924365	10	.124801	10	.215982
20	.931341	20	.128840	20	.216906
30	.938238	30	.132793	30	.217740
40	.945056	40	.136661	40	.218487
50	.951796	50	.140444	50	.219146
77 0	982.958458	83 0	983.144142	89 0	983.219718
10	.965040	10	.147755	10	.220201
20	.971543	20	.151283	20	.220597
30	.977967	30	.154725	30	.220904
40	.984311	40	.158081	40	.221124
50	.990576	50	.161352	50	.221256
				90 0	983.221300

5

THE MEASUREMENT OF GRAVITY AT SEA
AND IN THE AIR

INTRODUCTION

The desirability of measuring gravity in water-covered areas for oil exploration has been mentioned in the section on underwater gravity meters (page 48). Very extensive areas of the continental shelves have been covered by such measurements made with bottom meters since about 1946.

Difficulties increase as measurements are extended to deeper water, and, for the most part, underwater meters have been used only to the depths of the continental shelf, i.e., around 600 ft. A few measurements in special circumstances have been made at depths as great as 3000 ft (Beyer et al., 1966), but the rate of observation becomes very slow as depths increase.

The difficulties of bottom-meter observations in deep water, the much greater speed of measurement, and the interest in petroleum prospects beyond the continental shelves make it very desirable to make gravity measurements continuously on surface ships while under way. In addition to their application to commercial exploration, measurements over the deep ocean are useful for geodetic purposes. The earlier studies of the shape of the earth from its gravity field were hampered by the fact that most of the data available were taken over the one-fourth of the earth's surface covered by the continents, leaving great blank areas over the oceans. The relatively rapid and now very extensive gravity measurements at sea have made possible much broader and more detailed studies of the earth's crust, its deformation, and its probable history.

The first extensive gravity measurements over water-covered areas were made by Vening Meinesz, a Dutch geodesist, who first tried to measure gravity in the unstable areas of Holland. This led to the development of a pendulum apparatus which could be used on an unstable base. By swinging two identical pendulums in opposite phase the forces on their supporting frame are canceled out (Vening Meinesz, 1941). As adapted to measurement of gravity at sea the pendulum apparatus consists of three pendulums with the outer two swinging in opposite phase and the middle one taking up any motion imparted by the external movement. An optical system uses a light beam reflected from all three pendulums in such a way that the motions imparted to the system by the motion of the base on which it is mounted are very largely eliminated.

This equipment was used extensively by Vening Meinesz and his associates in Dutch submarines in most of the oceans of the world and in American and British submarines. The results of these measurements gave the first knowledge of the nature of gravity at sea, including the spectacular anomaly patterns of the island arcs of the East and West Indies.

A few early attempts were made to measure gravity at sea with gravity meters, including one trans-Atlantic traverse, but these measurements were crude and of much lower accuracy than the pendulum measurements in submarines and did not contribute much usable information.

ADAPTATION OF GRAVITY METERS TO SHIPBORNE USE

The problem of making gravity observations on a moving ship is primarily that of observing a very small quantity (the desired gravity variation) within an extremely "noisy" environment (the motional accelerations from movement of the ship).*

Vertical Accelerations

The vertical and horizontal accelerations due to the movement of a ship, even in rather moderate seas, can be a tenth or more of total gravity. This means that to make a measurement comparable with that made with land gravity meters we must detect changes of the order of 1 mgal or less within a background noise (the vertical

---

* When magnetic measurements began to be made quite accurately from airplanes, the question was frequently asked: Why not make gravity measurements in the air? The answer, of course, is that the magnetic field measurement is not affected by motion. On the other hand, gravity has the physical and mathematical dimensions of an acceleration ($f = ma$), and there is no physical principle by which motional and gravitational accelerations can be separated.

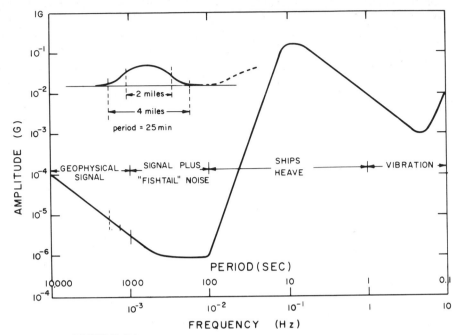

FIGURE 5-1

Frequency distribution of the magnitude of motional acceleration field within which a shipborne gravity meter must operate and minimum measurable anomaly. (*Modified from Wing, 1969.*)

component of the motional accelerations) which may reach the order of 100,000 mgals.

The problem is basically one of filtering a low-frequency signal from high-frequency noise with a maximum noise-to-signal ratio of the order of 10^5. Figure 5-1 is an approximation of the frequency spectrum of the vertical component of ship motion.

The "frequency" (or conversely, the "wavelength") of the desired gravity effect is a function of both the horizontal extent of the gravity disturbance and the speed of the ship. These factors determine the time required to pass over the real gravity anomaly to be measured. The frequency of the noise is primarily that of the motion of the ship and is usually in the range of about 1 to 0.1 Hz, depending on the size of the ship and the sea conditions in which it is operating. High frequencies from ship vibrations may be present, but their influence can be eliminated by suitable shock-absorbent mounting of the instrument frame. Relatively large long-period accelerations may be present because of the fishtailing motion of the ship when the steering (either automatic or manual) tends to hunt about an average course. These result primarily in horizontal acceleration components but can have very disturbing vertical components as well.

The order of magnitude of the vertical accelerations from ship's motion can be approximated by considering the ship as moving in simple harmonic motion. Then, the acceleration a is given by

$$a = r\omega^2 = r\left(\frac{2\pi}{T}\right)^2 \approx r\frac{40}{T^2}$$

where r is the radius of the motion in centimeters and T is the period of the motion in seconds. As an example, let us take the heave amplitude r as 6 ft (200 cm) and the period as 10 sec, from which we calculate

$$a = \frac{200 \times 40}{10^2} = 80 \text{ gals} \approx 0.08 \; g$$

This corresponds approximately to the peak acceleration of about 0.1 g at a frequency of 0.1 Hz (period = 10 sec), as shown by Fig. 5-1.

In Fig. 5-1 there is added a hypothetical anomaly with a half-amplitude width of 2 miles; this is about as narrow an anomaly as one would hope to define with a shipborne gravity meter. The equivalent wavelength of this anomaly would be about 4 miles, as indicated. With a ship speed of, say, 10 miles per hour, the 4 miles would be traversed in 24 minutes, giving a period of 1440 sec, which is within the geophysical-signal zone of the frequency spectrum. In considering the filtering it would be desirable to include a somewhat shorter period to give a chance for some detail to be shown within a minimal anomaly, which would then correspond, approximately, with the beginning of the geophysical-signal zone of the figure.

The necessary filtering for elimination of motional acceleration effects and retention of the desirable signal in this extremely noisy background are accomplished in different ways by different shipborne gravity meters and are mentioned briefly in discussions of those instruments. A discussion of measured motional accelerations on the Lamont-Doherty ship *Vema*, their relation to sea conditions, and the effectiveness of various meter-filter combinations is given by Neuman and Talwani (1972).

The separation of gravity effects to be measured from disturbing motional accelerations is somewhat like the Heisenberg uncertainty principle, which states that the position and momentum of an electron cannot be measured simultaneously with unlimited accuracy. If a gravity value is to be determined at a closely defined point, its precision must be low because the short-time value near that point is strongly influenced by short-period motional accelerations. If a value is to be determined precisely, it cannot apply to a point but must be related to the time and therefore to the distance over which the observations, including the disturbances, are averaged or filtered to determine the gravity value. The higher the precision the greater the time over which the average must be taken and, depending on the speed of the ship, the greater the distance over which the average is applicable.

Horizontal Accelerations

The above discussion has considered the sensitivity of the meter to vertical accelerations, which are in the same direction as the gravity to be measured and must be
eliminated by the damping and filtering operations. Horizontal accelerations result
from the roll or pitch of the ship and from its horizontal motion, or sway. They
cannot be eliminated by averaging or smoothing as their effects add a horizontal
force vector. The resultant vector is always greater than the average vertical component. Thus any cross coupling of horizontal forces to the highly sensitive vertical
response must be eliminated or measured and removed.

The cross-coupling effects can be eliminated, to a high degree, by mounting
the gravity sensor on a gyroscopically stable platform which is kept level, i.e.,
perpendicular to the average vertical. All shipborne gravity meters now use
gyrostabilized platforms, which have become highly practical with the development
by the aerospace industry of small, reliable, long-lived, and relatively inexpensive
gyroscopes.

THE GIMBAL-MOUNTED SHIPBORNE GRAVITY METER

Before sufficiently accurate long-lived gyroscopes were available at a reasonable
price, a gimbal-mounted version of the LaCoste and Romberg meter was used
quite widely. In this instrument the entire meter is suspended on gimbals so that
it can swing in any direction (LaCoste, 1967, p. 486). The horizontal accelerations
cause the meter assembly to swing off the vertical, and the acceleration quantity
measured is the vector sum of the vertical and horizontal components. The excess
acceleration can be determined if the angle of deflection from the vertical is con-
tinually measured. This was accomplished by the use of two mutually perpendicular horizontal accelerometers (HAMs) mounted in the directions along and across
the axis of the meter beam.

Each HAM contains a horizontal bar which is free to rotate on a perpendicular horizontal axis and which is very sensitively adjusted to have a rotational
period of about 2 min. These serve as long-period levels and give a reference from
which the angle of deflection from the vertical of the suspended meter assembly
can be measured and recorded. An automatic computer within the mechanical
and electronic accessories accepts a signal from each of the two HAMs, squares it,
and adds the two together to give the total excess gravity measured and then sub-
tracts this excess quantity from the total to give the true vertical gravity.

Gimbal-suspended meters were used for some 10 years until the gyrostabilized platform mounting was developed. They gave quite acceptable results in
smooth to moderate seas with HAM corrections up to the order of 200 mgals or
even larger. The gimbal meter is not subject to cross-coupling effects (page 114)
because as it swings on its gimbals, the meter is suspended in the direction of total
acceleration. The meters are more difficult to maintain in operation than those on
stabilized platforms, largely because the HAM is an extremely delicate and sensitive
device.

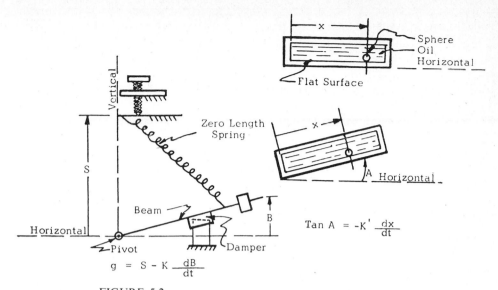

FIGURE 5-2
(a) Moving system of damped gravity-meter beam and (b) analog of slope
measurement to determine gravity change. (*Nettleton, LaCoste, and Glicken, 1962.*)

THE GYROSTABILIZED SHIPBORNE GRAVITY METER

This meter, as developed by LaCoste and Romberg, is described in some detail because the modifications made on the original land meter to adapt it to measurement
of gravity at sea illustrate the difficulties to be overcome and the methods of meeting
them which have evolved over a period of some 10 years or more.

The basic moving system or gravity-sensitive element of the ship gravity
meter is the same zero-length spring assembly which has been used for many years
on land meters, as described on page 31 and illustrated here by Fig. 5-2a. The
basic modifications are that the damping is very greatly increased and the instrument is adjusted to a very long period so that the net restoring force is near zero.
Also, important mechanical modifications such as stiffening of the beam and
changes in the ligaments have been made to reduce cross-coupling effects.

The high damping and low restoring force change the basic measurement
from that of displacement of the beam to that of its rate of displacement. This can
be illustrated in terms of the basic differential equation (LaCoste, 1967, p. 483),
which is

$$b\ddot{B} + f\dot{B} + kB - cS = g + a$$

where B = displacement of beam (Fig. 5-2a)
  $\dot{B}, \ddot{B}$ = first and second time derivatives
  cS = vertical force on mass exerted by spring acting through various
    mechanical links when meter is nulled
  g = gravity to be measured
  a = vertical acceleration of meter case

For a simplified, but close, first approximation we assume that the coefficients b, f, and k are constant. Somewhat complicated second-order effects are present and are treated in detail by LaCoste (1967).

Making the damping very large* gives a large value to the coefficient f; making the restoring force very small gives a low value to the coefficient k. Because of the heavy damping and resulting slow relative motion, acceleration forces are small, giving a small value to the first term, $b\ddot{B}$. For a given position of the spring tension adjustment (the screw shown in Fig. 5-2a) the coefficient c is constant. Thus the principal result of a change in gravity is a change in $\dot{B}$, that is, the rate of change of the displacement or slope of the curve of the position of the moving system.

The physical meaning of the system can be approximated as indicated by Fig. 5-2b. The upper part of the diagram indicates a flat dish filled with a viscous fluid, such as oil, within which there is a heavy sphere (ball bearing). If the dish is level, the ball will not move and the distance x is constant; this corresponds to a balanced or null position of the meter beam when B is constant. If the dish is tilted from the horizontal, as shown by the lower part of Fig. 5-2b, the ball will immediately begin to move to the left at a rate dx/dt determined by the equation

$$\tan A = -k' \frac{dx}{dt}$$

This corresponds to a change of g, so that B begins to change at a rate determined by the unbalance of the spring and

$$g = cS - k \frac{dB}{dt} = cS - k\dot{B}$$

Thus, if the spring adjustment is not changed and the ship passes over an area of changing gravity, there will be a corresponding variation in the rate of change of the displacement $\dot{B}$ like that illustrated by Fig. 5-3, and the changes in gravity can be derived from the value of the spring tension S and the slope of the curve.

The actual instrument contains numerous internal and external complications over the simplified basic features considered above. These are indicated in part, schematically, in Fig. 5-4.

The position of the beam is determined by photocells and external electronic circuitry. The balancing, or spring-tension, screw is controlled by a feedback circuit so that the beam is kept constantly in a near-balanced position.

A pair of gyroscopes (3, Fig. 5-4) counters the tilt of the ship in transverse and longitudinal directions. The gyroscopes are referenced to horizontal accelerometers (4, Fig. 5-4), which act as long-period levels to keep the gyros from precessing off of the vertical, so that the sensitive axis of the gravity meter is maintained in the direction of the average vertical. The period of response of the gyros can be controlled by the tightness of coupling between the gyros and the horizontal accelerometers.

---

* The damping is so large that for a 1-mgal unbalance it would take 2 weeks for the beam to move from one stop to the other (about 0.05 in.).

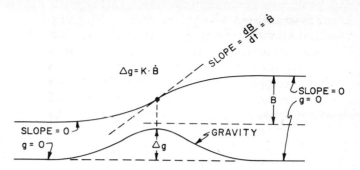

FIGURE 5-3
Diagram of gravity anomaly and measurement of slope to determine gravity change.

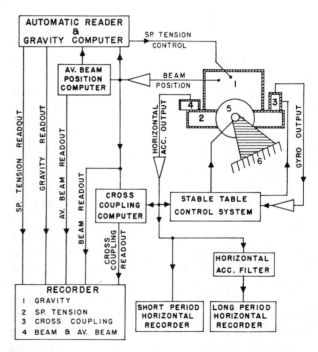

FIGURE 5-4
Schematic diagram of LaCoste and Romberg gravity meter on a gyrostabilized platform and accessories. (*From LaCoste, Clarkson, and Hamilton, 1967.*)

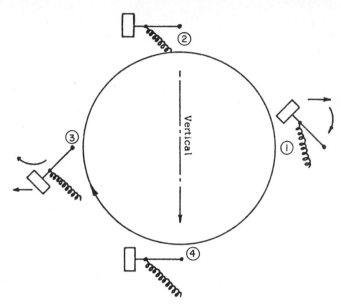

FIGURE 5-5
Inherent cross coupling. (*From LaCoste, Clarkson, and Hamilton, 1967.*)

The table (2) carrying the meter (1) is driven by torque motors (5, one only shown) on each of the gimbal axes to compensate for any tilt of the floor of the vehicle (6) on which the assembly is mounted.

A number of somewhat complex electronic components are indicated only in part by Fig. 5-4. The various outputs are recorded graphically to give analog records by which performance can be monitored. In modern installations, component values which are involved in the data reduction commonly are also recorded digitally on magnetic tape. This greatly facilitates processing the large volume of numerical data by digital computer by which values from the meter can be combined with base ties, position, and water-depth values to calculate latitude and Eötvös and Bouguer corrections to yield Bouguer gravity.

Cross Coupling

Even though it is maintained level on a gyroscopically stabilized platform, the gravity meter is still subject to horizontal accelerations resulting from the horizontal components of the motions of the ship. The mechanical parts of a gravity meter are not perfectly rigid. There can be small distortions of the ligaments, mainspring, or beam or other effects, depending on the details of construction, which are such that horizontal forces can result in small vertical effects. Also, for an instrument with a horizontal beam, even though mechanical cross coupling is completely and ideally eliminated, there can be a coupling which results from the phase relations between the horizontal and vertical components of motion. These effects are indicated in Fig. 5-5. The motion of the meter is assumed to be circular, corresponding to circular motion of the water in the waves. At time 1 on the diagram, the ship is moving downward, the meter is deflected upward, and the horizontal motion is to the left, making a force to the right; the two combined make a clockwise torque on the moving system. If, at time 3, while the meter is moving upward, the horizontal motion is to the right, the force is to the left and again the torque is clockwise.* Thus there can be cross-coupling effects which depend on the nature of the individual instrument and the magnitude and phase relations of the motions to which it is subjected. To a large extent, these cross-coupling effects can be determined and measured and proper allowance made for them in calculating final gravity values, either by subtracting them from the observations or by incorporating a cross-coupling correction in the instrument itself. The details of such a correction are determined in the laboratory by a program of testing in which the motions to which a meter may be subjected are simulated and their cross-coupling effects noted.

OTHER INSTRUMENTS FOR MEASURING GRAVITY AT SEA

The LaCoste and Romberg shipborne gravity meter has been described in some detail as an example of the developments and modifications required for a gravity meter to produce useful measurements in the background accelerations encountered on a moving ship. A number of other instruments have been used to varying degrees.

The Gilbert vibrating-string gravity meter (page 41) was originally developed to make gravity measurements on ships. Apparently only one such instrument was made.

The Askania Sea gravimeter (Gss 2) (page 35) has been used quite extensively in Europe and to a limited extent for commercial petroleum exploration. Many improvements have been made over a period of several years, particularly from experience gained by extensive use of the instrument for gravity measurement on the high seas by the Lamont-Doherty Geological Observatory of Columbia University.

The later (1970) Askania Gss 3 instrument (page 29), which is quite different in its fundamental operating principles, was designed to eliminate some of the difficulties, particularly the cross coupling, of the original Graf design.

The AMBAC vibrating-string accelerometer developed by Wing (1969) for measurement of gravity at sea (page 42) has been tested quite extensively, particularly on ships of the Woods Hole Oceanographic Institute, and is reported to have demonstrated that measurements have an rms accuracy of 0.6 mgal.

---

* These phase relations are valid for a highly damped meter. If the meter is not overdamped, the maximum deflections occur near the top and bottom of the circular motion and cross-coupling effects are greatest for ramp, i.e., linear, sloping, motions rather than for circular motions.

The first prototype Bell instrument (page 45) was used at sea in extensive acceptance tests under the development contract for its production. The more compact and somewhat simplified model is in use by several oil companies on oceanographic ships in commercial exploration and by some oceanographic institutions.

A review (Henderson, Strange, and Iverson, 1970) showed a total of 61 gravity instruments in use at that time. Of these, 4 were AMBAC, in academic research or geodetic surveys by oceanographic institutions, naval or national geodetic organizations and 1 in commercial exploration; 6 were Bell, of which 4 are in naval applications and 2 in commercial exploration; and 25 LaCoste and Romberg (including 2 gimbal-mounted instruments), of which 12 were in academic research or geodetic surveys by naval or government institutions and 13 were in petroleum exploration by oil or geophysical service companies.

THE EÖTVÖS EFFECT

A body moving over a curved, rotating earth is subject to an acceleration which is not present when it is stationary on the earth's surface. This introduces a major correction for ship- or airborne gravity measurements which is not present in land measurements. The effect was first realized by Baron von Eötvös (Eötvös, 1919) (the same Eötvös who invented the torsion balance described on page 66). He devised a rotating balance to demonstrate the existence and magnitude of the effect and pointed out that it would have to be taken into account when and if gravity measurements were made on moving vehicles.

The outward acceleration due to the earth's rotation is about 3500 mgals at the equator and decreases to zero at the poles (Fig. 5-6). On a vehicle moving on the earth's surface this acceleration is modified because the east-west component of its motion increases or decreases the rate of rotation with respect to the earth's axis and therefore changes the effective value of ω.

At a point on the surface at latitude φ (Fig. 5-6) the outward acceleration, $R\omega^2$ at the equator, decreases toward the poles as $R\omega^2 \cos \varphi$ and is directed outward perpendicular to the axis of rotation. The component perpendicular to the earth's surface, i.e., in the direction in which gravity is measured, is $R\omega^2 \cos^2 \varphi$. Any east-west component of motion over the surface produces a change $d\omega$ in angular velocity, and by differentiation the corresponding change da in acceleration is

$$da = 2R\omega \cos^2 \varphi \, d\omega$$

The change in ω in terms of the east component V_E of motion over the surface is

$$V_E = R \cos \varphi \, d\omega \quad \text{and} \quad d\omega = \frac{V_E}{R \cos \varphi}$$

so that the Eötvös effect da is

$$E = da = 2V_E\omega \cos \varphi = 2V\omega \cos \varphi \sin \alpha$$

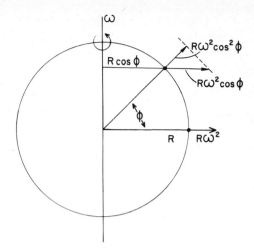

FIGURE 5-6
Schematic diagram of centrifugal force
from earth's rotation and Eötvös effects.

where V is the velocity of motion over the earth's surface and α is the heading angle
between the direction of motion and the astronomic north.*

In addition to the term resulting from the modification of the rotational
acceleration there is a V^2/R term from the simple outward acceleration in moving
over a curved surface. This term is independent of direction.

* The Eötvös effect includes a component of the Coriolis force. The Coriolis acceleration
is defined as $2(\omega \times V)$, that is, twice the vector product of the rotation vector (directed north
along the earth's axis of rotation) times the velocity vector V. The magnitude of this product is
$2\omega V \sin \theta$, where θ is the angle between the vectors. The direction of the acceleration is perpendicular to the plane containing the ω and V vectors.

The relation to the Eötvös effect can be clarified by considering the east and north components of V separately.

The plane containing ω and the east component of V is parallel with the earth's axis, and
the vertical component of this acceleration is $2\omega V_e \cos \phi$. But $r\omega \cos \phi = V_\phi$ and, therefore,

$$2\omega V_e \cos \phi = \frac{2 V_\phi V_e}{r}$$

which is the first term of the Eötvös correction equation. Thus, the vertical component of the
Coriolis acceleration is included in the Eötvös correction.

The plane containing ω and the north component of V is vertical, and therefore, the
Coriolis acceleration for this component is horizontal. For speeds of 200 mph, this component
may amount to 1 gal (1000 mgals). Nevertheless it can be shown as follows to contribute a
negligible amount to the quantity measured. If we let G represent the vertical gravity and C the
horizontal component of the Coriolis acceleration, then the measured value will be

$$G_m = \sqrt{G^2 + C^2} = G \sqrt{1 + \left(\frac{C}{G}\right)^2} = G \left[1 + \frac{1}{2}\left(\frac{C}{G}\right)^2\right]$$

Since C/G is on the order 0.001, the 1-gal horizontal effect would contribute only $\frac{1}{2}$ mgal to the
measured effect. Thus, the horizontal component of the Coriolis acceleration is negligible even
at airplane speeds.

If applied to measuring gravity in the air, at an altitude h, the effective radius of the earth is increased by that amount. The complete expression for acceleration due to motion over, or parallel with, the earth's surface becomes

$$E = \frac{R + h}{R^2} (2R\omega V \cos \varphi \sin \alpha + V^2)$$

For shipborne gravity measurements, $h = 0$ and

$$E = 2\omega V \cos \varphi \sin \alpha + \frac{V^2}{R}$$

To determine the numerical coefficients, we take the average radius of the earth as 6.371×10^8 cm, $\omega = 2\pi/T$, where T is the length of the sidereal day, or 86,164 sec, to give $\omega = 7.2921 \times 10^{-5}$ rad/sec, from which

$$E = (2 \times 7.2921 \times 10^{-5} \times \cos \varphi)V \sin \alpha + \frac{V^2}{6.371 \times 10^8}$$

$$= 14.584 \times 10^{-5}V \cos \varphi \sin \alpha + 1.569 \times 10^{-9}V^2$$

This gives values in gals for speeds in centimeters per second. To change the speed unit to knots and gravity to milligals, 1 knot $= 1.85325$ km/hr $= 51.479$ cm/sec.

For the first term, the numerical coefficient becomes $14.584 \times 10^{-5} \times 51.479 = 750.8 \times 10^{-5}$ gal/knot $= 7.508$ mgals/knot. In moving *eastward*, the normal outward acceleration is increased, the effect is to decrease the meter reading, and the correction is *positive*. In moving *westward* the correction is *negative*.

For the second term,

$$1.569 \times 10^{-9} \times 51.479^2 = 4.158 \times 10^{-6} \text{ gal/knot}^2$$

$$= 0.004158 \text{ mgal/knot}^2$$

The working expression for shipborne measurements is

$$E = 7.508V \cos \varphi \sin \alpha + 0.004154V^2$$

to give E in milligals for V in knots.

The V^2 term is usually neglected but becomes appreciable for higher ship speeds. For it to amount to 1 mgal,

$$V^2 = \frac{1}{0.00416} = 240$$

$$V = \sqrt{240} = 15.5 \text{ knots}$$

Since this effect is always outward and tends to decrease gravity, the correction is always positive.

NAVIGATION REQUIREMENTS

From the expression for Eötvös effects it is evident that corrections become quite large even for moderate ship speeds. At a speed of 10 knots on an eastward course at the equator there is a 75-mgal positive correction and on a westward course a 75-mgal negative correction, so that the difference between east and west course is 150 mgals. To make the correction to a precision of 1 mgal requires that the speed be known to about 0.1 knot. For a true north or south course the correction is zero ($\alpha = 0$ and $\sin \alpha = 0$), but it is very sensitive to the course direction. If the course changes by $1°$ ($\sin \alpha = 0.017$) and the speed is 10 knots, the change in the Eötvös effect (at the equator) is $0.017 \times 7.5 \times 10 = 1.3$ mgals. Thus, to make Eötvös corrections to 1 mgal or less, for nearly east or west courses the speed must be known to about 0.1 knot and for nearly north or south courses the azimuth must be known to about $1°$. This means that in routine shipborne operations with a meter performing well and measuring gravity changes to less than 1 mgal, uncertainties in the final results are likely to be greater because of inaccurate Eötvös corrections from lack of precision in navigation than from errors in the gravity measurement itself.

The Eötvös effects require that for accurate gravity values at sea some sort of precise navigation system be used. On continental-shelf areas, usually within 150 miles or less of shore, much work has been done with positioning by electronic navigation systems such as Shoran, Raydist, Lorac, or other range-measuring or phase-comparison systems (see page 348). For measurements in the open ocean, far from shore and out of range of accurate shore-based electronic systems, use is being made of worldwide navigation systems. The one with the most apparent promise is that depending on earth satellites, which can give locations within a fraction of a mile. Observations are limited to times when satellites are in favorable positions, which are usually separated by 1 or 2 hr. Other devices must be used to interpolate the course of the ship between satellite fixes. Within its limitations, sonar Doppler is very effective; it measures course and speed by sonic signals reflected off the bottom and is reliable to depths of around 200 fathoms (but also may give speed and direction with respect to the water, based on reflections from stratifications within the water itself). Doppler reflections are being used to interpolate between satellite fixes. By use of an electromagnetic speed recorder and a recorder of the course with respect to a gyroscopic compass it is possible to make continuous calculations of Eötvös corrections and to make a reliable gravity record even through relatively sharp maneuvers of the ship (Fig. 5-7).

A recent development by LaCoste and Romberg is to mount the stable platform of their ship gravity meter so that it can rotate on a vertical axis. Its orientation in space is held fixed by a third gyroscope, torque motor, and servosystem. Then, by integrating the two accelerometer outputs the north and east components of the ship's speed can be determined. Experiments are under way to use this inertial navigation system as a means of interpolating between position fixes by satellite. Also holding a fixed orientation gives some improvement in meter performance.

A system using a gyro compass as a platform for a vibrating-string accelerometer is described by Bowin, Aldrich, and Folinsbee (1972). This also is a platform

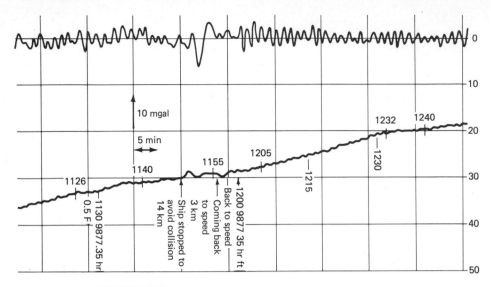

FIGURE 5-7
Sample record of real-time gravity with Eötvös corrections carried out during ship maneuvers: (*lower curve*) gravity corrected in real time for short-term Eötvös errors due to changes in course and speed; (*upper curve*) the corrections applied. (*From Wing, 1969.*)

with three-axis control. It provides means by which ship's speed (from an electromagnetic ship's log) and direction can be used for interpolation of the course between position fixes, such as by satellite. By incorporating the various items into a data-reduction system, using a relatively small on-board computer, it is possible to make the Eötvös and latitude corrections to produce a real-time free-air gravity record on the ship.

A highly sophisticated marine geophysical system including seismic, gravity, and magnetic observations, described by Brown (1970), includes satellite fixes for primary control, sonar Doppler where applicable (less than 600 ft water depth) or by Sal Log (which gives speed with respect to the water), and gyro compass. All data are recorded on magnetic tape, and preliminary real-time records of reduced gravity and magnetic data are superimposed on seismic record sections. All this is accomplished by moderately sophisticated computers on the ship.

The computer system itself for the same operation is described in some detail by Darby et al. (1973). This is probably the most sophisticated shipboard geophysical data-processing and recording system developed to that time. The data input are from

  1 Four simultaneous aquapulse sound sources
  2 A LaCoste and Romberg gravity meter
  3 A proton-precession magnetometer
  4 A satellite receiver

    5 A gyrocompass
    6 Sonar Doppler
    7 Sal Log
    8 Velocimeter
    9 Fathometer
   10 Shoran
   11 A master clock

Not all are used at all times. The outputs are

    1 A 24-stack seismic record
    2 Single-trace seismic record
    3 Magnetometer record
    4 Gravity record with latitude, Eötvös, and water-depth corrections
    5 Ship's speed
    6 Ship's heading

Items 3 to 6 are superimposed on the single-channel record section. All the processing is done in real time so that results are available as the recordings are made
and can be used to monitor and, if desirable, modify the operation immediately.

    An extensive review of precise navigation and its application to the measure-
ment of gravity at sea is given by Talwani (1970). This describes, in some detail, the
operation and precision of the U.S. Navy's satellite system and the means which can
be used to interpolate positions between fixes. A brief review of overall operations
of a navigation system and combined (gravity, magnetic, and seismic) geophysical
system is given by Kologinczak (1970).

    For all precise gravity-measuring operations at sea it is highly desirable that
the ship sail on a straight line with constant course and speed for periods of at least
1 hr. For this reason ships in gravity-measuring operations are nearly always
controlled by an autopilot, which maintains a constant ship's course so that
Eötvös effects from steering errors or fishtailing are minimized. Fishtailing errors
from manual steering are particularly disturbing because they can occur in the
same part of the frequency spectrum (see Fig. 5-1) as the gravity signal; i.e., their
period may be of the order of several minutes depending on the skill and attention
of a manual helmsman. An example of the difference in a ship gravity-meter record
between manual and automatic steering is given by LaFehr and Nettleton (1967).

THE REDUCTION OF GRAVITY OBSERVATIONS AT SEA

In a previous section (page 89) we have described the necessary reductions of land
gravity observations at specific station locations and the procedures necessary for
producing a Bouguer gravity map from such observations. The process for producing a map from shipborne gravity data is considerably more complex, particularly because of the Eötvös effects.

    The rather complex interrelations between the gravity and navigation ob-
servations and the large volume of data produced by a relatively rapidly moving

ship operating continuously, 24 hr a day often for many days, require that the data-reduction process be mechanized into operations with an electronic computer so that they can be carried out in reasonable time. For this reason the gravity, navigation, and water-depth observations are often recorded digitally on magnetic tape to facilitate electronic data processing. In some modern operations computer facilities are provided on the ship so that at least preliminary reduced gravity data are available in real time. Then the results may be used, if desirable, to control the observation program rather than waiting for the data to be sent to a processing or computing center and further program instructions being sent back to the ship.

The details of the systems of data processing vary for different operations and among different operators, but all must include the essential steps described below, which are those taken in the processing of commercial shipborne gravity data for petroleum exploration, using an intersecting network of ship's tracks. Such a network is necessary for high-precision mapping. The intersecting lines give pairs of values at crossing points from which a measure of the precision of the operation can be determined. Also the errors of closure of loops can be determined from the sums of the gravity differences between these points of intersection and systematic or least-squares adjustments made to reduce loop closures to zero. Either the line-crossing or closure errors can provide a basis for systematic adjustment into a continuous map, which, with modern techniques, can be reliably contoured at a 1-mgal interval. Such procedures, of course, cannot be used for isolated traverses on long lines across the open oceans such as have been conducted extensively by governmental and academic oceanographic institutions. Their reliability can be judged by the occasional intersections which occur, by ties to submarine pendulum stations, or by observations at ports or docks where gravity reference stations have been established. For such checks, over long time intervals, the magnitude of the discrepancy will depend very largely on the long-time drift characteristics of the instrument.

The Meter Zero

A desirable practice is to keep a long-time drift curve of the instrument in terms of the meter zero. This is a fictitious gravity value which, if it could be attained, would reduce the meter reading to zero. A meter zero can be established any time a "still reading" is taken at a dock or other place where an absolute value of gravity is known. Then, if G_a is the absolute gravity value and G_s is the still reading converted to milligals, the meter zero G_0 is simply

$$G_0 = G_a - G_s$$

It is a physical characteristic of a particular instrument. Any drift, change in the meter assembly, results of working on the meter, shock from rough handling, etc., immediately become apparent in the meter-zero curve. Also, it provides a continuous value for a base constant for calculating gravity values which automatically places the reduced values on the international datum.

Figure 5-8 is an example of a meter-zero curve for a LaCoste and Romberg ship meter over a period of some 6 months. It shows a quite regular upward meter

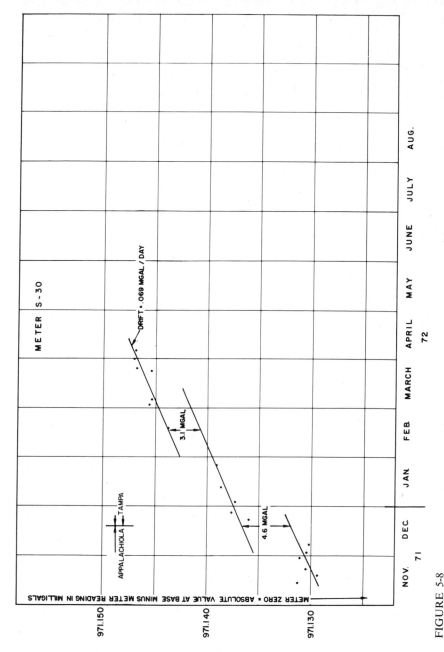

FIGURE 5-8

Meter-zero curve for LaCoste and Romberg gyrostabilized ship gravity meter. The plotted points are the absolute gravity value at the dock minus the meter reading reduced to milligals.

drift of about 0.069 mgal/day over the entire period. It also shows a jump of about 4.6 mgals in December which accompanied a base change; this may or may not be caused by an incorrect base value. Another jump of about 3.1 mgals in mid-February probably was caused by some work on the meter.

The individual still readings vary by as much as 2 mgals from the general slope. If these readings are very carefully made, and if corrections are made for tidal elevation changes at a dock, they should be somewhat less.

One possible cause for the jumps is that the LaCoste and Romberg stabilized platform meter is not equipped with a buoyancy-compensation cell (the earlier gimbal-suspended meter has such a cell). Therefore, if there are leaks, or if the valve to the inner chamber containing the moving system is opened, there may be a small change in pressure of the enclosed air with a resulting meter jump due to the change in buoyancy.

Required Corrections

The several corrections and the sources of the data for making them are given in the following paragraphs.

The latitude correction   This is the same as that made for ordinary land gravity observations. It can be calculated directly by including the geodetic formula for gravity at sea level in the computer program and calculating the correction from the latitude as given by the positioning system.

The elevation correction   Since the instrument is always at sea level, this correction is not needed. There can be small elevation adjustments when making ties to land-based stations set at dockside. Also there are elevation effects resulting from tidal changes in the water-surface elevation. These are usually ignored, except in making dockside ties, because they are unknown in open ocean observations and in surveys with intersecting traverse lines they become part of the intersection or closure adjustment.

The Bouguer correction   The correction is made from fathometer measurements of water depth. In gravity operations which accompany continuous seismic profiling the water depth can be obtained from the water-bottom reflection, but it is usually taken from a fathometer signal included in the digital recording. The calculation assumes that the water is replaced with rock at the same density as that of the earth material within the range of the ocean-bottom topography. This, theoretically, eliminates the contrast at the water bottom between the density of the water and the density of the underlying rock. Figure 5-9 shows the Bouguer factor as a function of the density used for the rock and for an assumed density of seawater of 1.03.

The terrain correction   Unlike the terrain correction applicable to land gravity observations, which is always positive, this may be either positive or negative. When the Bouguer correction is made in the usual way, considering the topography

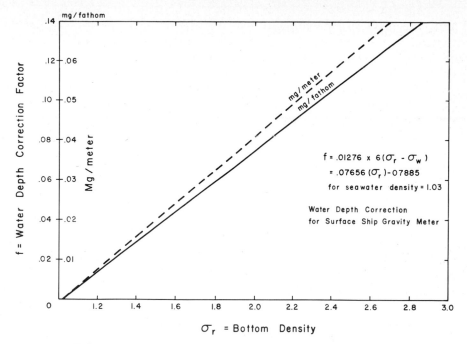

FIGURE 5-9
Bouguer correction factor for different subbottom rock densities.

as horizontal at the water depth under the point of calculation, the departures from this condition have gravity effects corresponding to terrain corrections. When measurements are made near a coast with mountains beyond, there will be terrain corrections for the topography above sea level. These are made in the same way as the distant terrain correction for land gravity observations.

The condition for a simple submerged vertical cliff is shown in Fig. 5-10. For a measurement with the ship at position 1, the Bouguer effect is calculated as if the depth h extended continuously to the right, as indicated by the dashed line, and does not take into account the deeper water to the right of the cliff. Therefore the calculated Bouguer effect is too small, and the terrain correction is positive. With the ship at point 2, the Bouguer calculation does not see the higher topography, with thickness t, to the left of P, the calculated effect is too large, and the correction is negative. For the ideal case shown, the positive correction increases to the point P and then suddenly reverses, as shown by the terrain-correction curve in the upper part of the diagram. For a vertical cliff, with a relief of 1 km and with the densities shown, the terrain correction reverses from $+35$ to -35 mgals and the meter would register a change of 70 mgals. These large effects are present because the instrument is above the topography rather than on the surface, as in land measurements. Note that the Bouguer and terrain corrections have large discontinuities at the cliff, but ideally the corrected gravity is smoothly continuous.

SUBMARINE TERRAIN CORRECTION FOR VERTICAL CLIFF

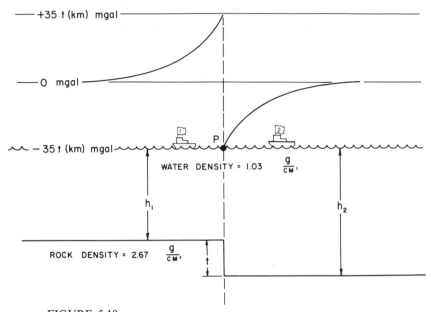

FIGURE 5-10
Idealized terrain corrections for a shipborne gravity meter. (*From Nettleton, 1971; by permission of the Society of Exploration Geophysicists.*)

In actual topography the cliff would not be vertical and the two sharp peaks are smoothed out, but the curve is otherwise generally similar. For example, a survey off southern California, over an area having bottom depth relief of several thousand feet, had terrain corrections of ± 25 mgals. Terrain corrections for a survey over the mid-Atlantic ridge had maximum values of 35 mgals (Woodside, 1972, p. 947).

The correction is made by a computer operation for which the input is a digital representation of the topography. This means that the hydrographic surface is approximated by discrete values of water depth at a regular array, or grid, of points. This is done by laying down the grid over a bottom topography or hydrographic contour map and recording the water-depth value at each grid point. These values and their grid coordinates listed on punched cards or magnetic tape become the input to a computer operation which is similar (but with entirely different coefficients) to some published systems for computer determination of terrain corrections (page 93).

Special corrections Reduction of gravity at sea requires special corrections when the Bouguer correction is not made and the result is a free-air map. This means that in the interpretation of the results the layer of water and its density become a part of the interpretation calculations.

In some special cases a more complex Bouguer correction may be made where the actual geologic section is not well represented by a single density. For instance, in the California offshore, the water depth varies by several thousand feet, and the material within this range is partly recent deposits with relatively low density and partly much older rocks with considerably higher density. An approximation to this condition can be made by representing the top of the heavier material by a smooth surface at a variable depth below the actual water bottom. A partial Bouguer correction is made for depths from the water bottom to that surface with a density appropriate for the recent material. The correction for the effect of the material below the smooth surface is made with a higher density appropriate for the older and more consolidated rocks.

Data Recording and Reduction Procedures

As mentioned above, data may be recorded on the ship in analog form, i.e., as a graphic record by a chart recorder, or in digital form, or both. The analog record is almost essential for monitoring the overall operation and to determine that all mechanical and electronic components of a complex system are functioning properly. The digital record is used as input for the electronic processing.

The details of the processing, the combinations of computer and manual operations, graphical smoothing or editing of erratic results, adjustments, etc., vary considerably between different operators and with different instruments. A description of one example of the basic process as used in a test of a LaCoste and Romberg stabilized-platform meter is given by LeFehr and Nettleton (1967). Figure 5-11 is a schematic diagram of a procedure used for that test.

The final result of the data-handling procedure is a set of maps, usually of Bouguer gravity, and quite commonly of free-air gravity also. With modern shipborne gravity meters and accurate navigation data such maps can be reliably contoured to about 1 mgal. A histogram of a comparison of shipborne with underwater gravity-meter data (Fig. 5-12) shows that 50 percent of the comparisons have differences of 0.5 mgal or less between the map from a bottom-meter gravity survey and the shipborne measurements.

As mentioned previously (page 124) the precision of shipborne gravity measurements depends upon the magnitude of the background of motional accelerations and the filtering system. Part of the filtering is inherent within the meter itself, and part is in the averaging of values or the external or digital filtering, all of which are functions of time. At high boat speeds the smaller gravity details are necessarily smoothed out by the time functions of the filtering. If attempts are made to map anomalies of small lateral extent, boat speed must be slower.

The greatest single source of uncertainty in mapping shipborne gravity results is in location data. Even with modern electronic positioning there are errors, sometimes attributable to lane jumps and sometimes to unknown sources which

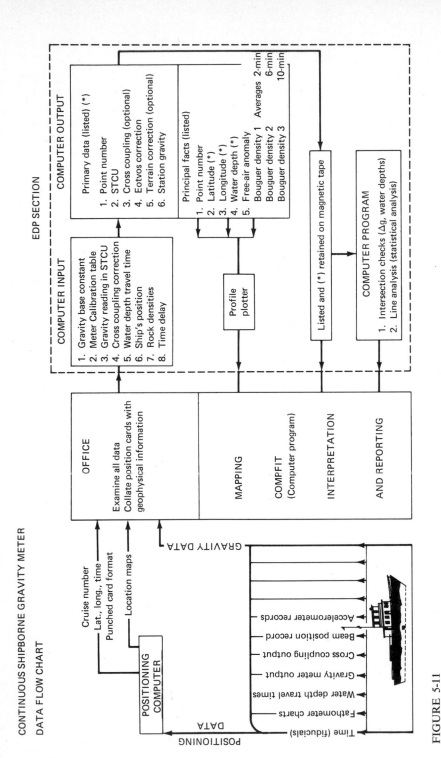

FIGURE 5-11
Data flow chart for continuous shipborne gravity meter operation. *(From LaFehr and Nettleton, 1967.)*

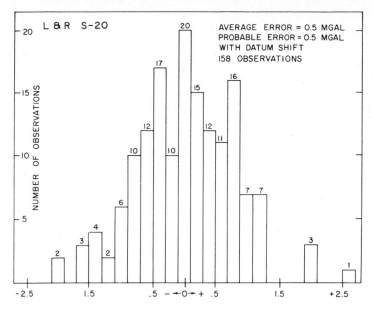

FIGURE 5-12

Histogram of errors between shipborne and water-bottom gravity-meter measurements. (*From LaFehr and Nettleton, 1967.*)

contribute to errors in the final gravity map primarily because they lead to erroneous Eötvös corrections. This problem has been treated in some detail by Coons and Smalet (1967), where it was demonstrated that lane-count errors can be identified by irregularities in the reduced gravity maps and corrections made accordingly.

Cross Correlations for Reducing Errors

It has been shown (LaCoste, 1973) that it is possible to reduce errors due to motional effects by determination of relations between ship motions which may affect the meter and the meter output itself. Monitor records can be made of motional components such as horizontal and vertical velocities and accelerations and their products. If there are correlations between any of these monitors and the meter output, it is possible to determine factors by which monitor records should be multiplied to give corrections to be subtracted from the meter output. Any recorded variable quantity which could have an effect on the meter may serve as one of the monitors. In the example given up to seven monitors were considered, but because of the wide range in the factors by which they affected the meter only the two or three with the largest factors made appreciable contributions to the corrections. In one example, variations in output not related to gravity and amounting to over 10 mgals were effectively removed. In this case, the source was from an electrical fault within the meter.

The application of the system involves a moderately complex mathematical process which must be carried out by a computer. With the increase of computer sophistication on ships, especially those where the gravity measurement is part of a multisensor system, it should be quite possible to make on-board corrections which would not only improve the quality of the results but also point to possible defects in the meter itself which might be corrected at sea.

The system would seem to be particularly applicable to measurement of gravity in the air, where there are more kinds of possible motional effects than on a surface ship. Such effects would normally be removed by corrections based on the corresponding data from auxiliary instruments. If there were deficiencies in such instruments or the correction procedures, the corresponding analysis system might provide a means for their detection and removal.

MEASUREMENT OF GRAVITY IN THE AIR

With the successful application of gravity meters to measurements at sea, some attention has been given to the measurement of gravity in the air. From the standpoint of the instrument itself the actual measurement and recording of gravity changes, for a reasonably smooth airplane flight, is no more difficult than on a ship. However, the problems of extraneous accelerations and the auxiliary data required for their calculation and removal from the measured values are much more extreme.

At airplane speeds the Eötvös corrections become very large. For a speed of, say, 200 knots in an east (or west) course the Eötvös effect is nearly 1500 mgals. To make corrections to 1 mgal the speed must be known to the same precision as for a ship, i.e., to about 0.1 knot or to a precision of about 1 part in 2000. For a true north (or south) course the Eötvös effect is zero, but at a speed of 200 knots a deviation from this course of only about 2 minutes of arc would produce an Eötvös effect of 1 mgal.

A set of nomograph charts (Glicken, 1962) has been prepared for determination of the errors of speed and course for various values of latitude and course direction. They can be used to determine the optimum flight directions depending on the precision with which course or speed can be controlled and measured.

Because of the free-air gravity effect of 0.094 mgal/ft, a change of elevation of the plane of 10 ft will produce a gravity change of nearly 1 mgal. Furthermore, any changes of elevation with time produce vertical forces because the second time derivative of the elevation is a vertical acceleration.

The calculations and removal from the gravity measurement of these extraneous accelerations and the measurement of the position and motions of the airplane impose extremely stringent requirements on the position, speed, and elevation measurements. In spite of these difficulties some successful measurements of gravity in the air have been made.

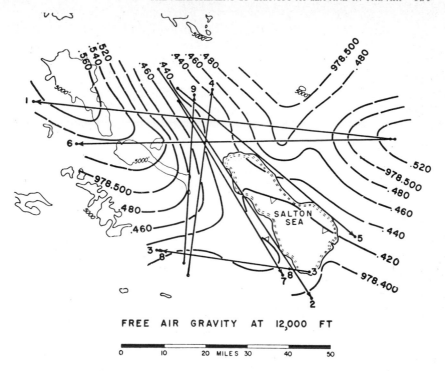

FREE AIR GRAVITY AT 12,000 FT

FIGURE 5-13
Map of gravity at 12,000 feet from airborne gravity-meter survey. (*From Nettleton, LaCoste, and Harrison, 1960.*)

The first such measurement was made in 1958 (Thompson and LaCoste, 1960) with a LaCoste and Romberg gimbal-mounted meter in a U.S. Air Force KC-135 jet tanker. The flights were over the Edwards Air Force Base in California, and the navigation was by phototheodolites on a tracking range on the ground to give both vertical and horizontal positions of the plane. From the intersection of two flight courses and also from calculation of expected gravity values from ground data it was indicated that the precision of the reduced observations was around 10 mgal.

A series of airborne traverses over the Imperial Valley of California made with a LaCoste and Romberg gimbal-mounted meter (Nettleton, LaCoste, and Harrison, 1960) produced the first (and at this time, still the only published) airborne gravity contour map (Fig. 5-13). Navigation was by photogrammetric methods, the airplane position being spotted on ground maps at 2-min intervals. Elevation changes were determined by a hypsometer.*

A careful test of the capabilities of airborne gravity measurements with a

---

* The hypsometer measures barometric pressure by determining the boiling point of a volatile liquid and can determine changes in barometric elevation to the order of 2 feet.

LaCoste and Romberg gimbal meter was made in 1961 (Nettleton, LaCoste, and Glicken, 1962) over a triangular course between Houston, Shreveport, and Baton Rouge. Flights around the triangle were made on two separate days in opposite directions. The measured gravity values over a total of some 1700 miles of flight line were compared with calculated values at the nominal flight elevation of 12,000 ft. The calculated values were based on gravity values over the general area from ground gravity-meter surveys continued upward from the ground elevation to the airplane elevation and including a correction made by an upward continuation calculation for the attenuation with height of any ground-surface gravity anomalies. Also corrections were made for vertical accelerations due to short-period changes in altitude by calculating the second time derivative of the elevation of the airplane. The primary horizontal control was by aerial photography and the techniques of photogrammetry. Vertical control was by radar altimeter supplemented by a hypsometer and by the theoretical shape of the isobaric surface. From the average for all flights the rms deviation between observed and calculated data was 6.6 mgals, and the mean deviation with regard to sign was +1.5 mgals.

Further operations of gravity meters on airplanes and even on helicopters have been carried out by the military, but the data are not generally available.

An experimental survey (unpublished) was made in 1971 with a helicopter moving at variable groundspeeds from about 17 to 60 mph. The flight pattern included a number of intersecting lines from which a map could be made and compared with a ground gravity survey. After adjustment for systematic differences between lines the average deviation between airborne and ground surveys was about 4 mgals.

One field of application of airborne gravity may be in the rapid determination of gravity over wide areas at sea not generally crossed by ships, such as the polar regions. Such an application probably would be of interest primarily for geodetic purposes (unless it has been rendered obsolete because of geodetic applications of artificial satellites).

Possible applications to petroleum prospecting might be made from a special type of low-speed aircraft (less than 100 mph). Over water a radar altimeter would permit accurate corrections for elevation and vertical acceleration. Over land such corrections would be much more difficult and probably would have to be made from barometric elevations interpolated between radar altimeter measurements over points or areas (such as lakes) of known elevation.

Terrain corrections for airborne gravity would be a rather serious problem. Such corrections would be analogous to those at sea (see page 126), where the topography is below the level of measurement, and become much larger than for a surface meter. They could be determined by a computer operation based on a digital approximation of the topography.

It seems evident that successful measurement of gravity in the air will have to use digital recording of all measured quantities together with the sophisticated techniques of filtering and cross correlation. For instance, changes in speed or course produce Eötvös effects which can be correlated and removed. Changes in elevation directly produce effects from the normal vertical gradient and also the

vertical accelerations resulting from the second derivative of the changes in elevation. Digital recording and computer processing should provide the means of
handling these complex relations quantitatively and at a rate which can keep up
with the great speed at which data are produced.*

As mentioned above, one of the problems associated with airborne gravity
measurements is the inability of the gravimeter to distinguish between the acceleration due to gravity and that due to vertical motion of the vehicle. This difficulty
could be circumvented if the measuring instrument were a gravity gradiometer,
and one such instrument has been described by Forward and Bell (1970).

---

* The previously mentioned test of an airborne gravity operation produced enough data
in about 6 h flying to require some 200 man-days for analysis by manual operations on the various
analog records.

6

ANOMALY SEPARATION AND FILTERING*†

INTRODUCTION TO GRAPHICAL AND GRID SYSTEMS

Each gravity measurement determines, at the station location, the sum of all effects from the grass roots down. A gravity map is seldom a simple picture of a single isolated disturbance but almost always a combination of relatively sharp anomalies which must be of shallow origin, of anomalies with intermediate dimensions, which may be those most probably indicative of geologically interesting sources, and of very broad anomalies of a regional nature, which may have their origin far below the section within which the geological interest lies. Therefore, gravity interpretation frequently begins with some procedure which separates the anomalies of interest from superficial disturbances on one hand and the smooth, presumably deep regional effects on the other.

Adopting momentarily a more theoretical viewpoint, we can say that a potential source will cause contributions to the potential field everywhere. Hence, the complex source distribution within the earth always gives rise to a gravitational

* This chapter was written with the collaboration of Norman Neidell.
† Portions of this chapter are quoted from Nettleton, 1968 and 1971, and are reproduced by permission of the Society of Exploration Geophysicists.

134

potential field which must be decomposed into components which can be appreciated geologically such as a deep-seated regional effect or an intrusion rising to quite shallow depths. This decomposition is at times most difficult and forms the backbone of interpretation as it relates to potential field data. Mechanisms which assist in the decomposition or, equivalently, separation of the field into geologically related components are thus an essential part of the interpretive process.

Separation or decomposition of anomalies in the real environment must be performed using data which are less than ideal. The quantitative information content of data samples which are imprecise and poorly distributed is an extremely complex matter, and only rather recently, with the advent of filter theory, has it been treated satisfactorily. In our early discussions, we shall treat the sampling question intuitively and cursorily until we have introduced sufficient background for a more detailed treatment. With these technical points in mind, we can continue the more intuitive aspects of this subject.

The anomaly-separation procedure usually consists of the removal of a smooth regional by either of two methods: (1) graphical application of intuitive methods based on the interpreter's knowledge of regional geology and of reasonable magnitudes for local anomalies or (2) numerical application of analytical methods to an array of values, usually on a regular grid. Most of the analytical approaches have been essentially linear and have attempted to emphasize or enhance certain components of the gravity field and suppress others.

The proponents of these two systems have been termed "smoothers" and "gridders" (Steenland, 1952). Traditionally, the smoother has drawn his smooth curves either on profiles or as contours on maps. These curves represent the component of the gravitational field which is to be removed. This regional is subtracted from the observed gravity map, and the resulting residual contains the components of the field which presumably are caused by mass irregularities representing geologic disturbances of interest. An empirically estimated regional gravitational field is based on the judgment of the interpreter in deciding the nature and sharpness of those components of the field which he wishes to discard. Skillfully applied, this process can produce effective isolation of anomalies and hence yield results which are simply and directly amenable to quantitive analysis.

It was once said facetiously by a geologist, that "the regional is what you take out to make what is left look like the structure." While not intended as such, this remark points a way toward useful control on the selection of a regional. Simple calculations can be made from features with the depth, form, and density contrast of geologically interesting objectives, and the resulting width and gravity amplitude can help determine the nature, particularly the rate of curvature, of the profiles or contours of the regional map which would leave such features as residuals.

Another aspect of anomaly separation is that of recognizing that a gravity anomaly may represent the overlapping effects of two or more separate sources. If the centers of mass of two (or more) sources are closer together than their depth, the anomalies merge or overlap to such an extent that the more complex origin may be unrecognizable. The separation can be improved by carrying out derivative or continuation processes, described later in this chapter, but there is always a practical limit of separation which is not much less than the depth at which multiple sources cannot be clearly recognized. Examples of this resolution problem are given by Elkins and Hammer (1938). Conversely, it is always possible to make a multiple-source interpretation having detail with smaller horizontal dimensions than the depth, but this alternative cannot be proved.

This chapter covers the various empirical and analytical systems and filter theory as applied to selecting some components of the field and rejecting others. While the discussion is largely in terms of gravity fields, most of the concepts are also applicable to magnetic fields. In fact, some of the initial developments, particularly of second-derivative calculation methods, were made for the interpretation of magnetic data.

Most of the discussions which follow treat the grid-system approaches because they require some sophistication in their technology as well as in their application. By contrast, the bland simplicity of the graphical methods is not a fair guide to their practicability, and any complete discussion of their application must include case studies and refined instances of interpretation.

The usual objective of any anomaly-selection system is to make a map on which the areas or prospects with enhanced probabilities of petroleum accumulation are outlined. The outline is commonly within a "closure," which is the expression of a geophysical quantity considered as indicative of a closed geological structure. In the earliest days of surface geological mapping, closed contours indicated a prospect for petroleum. The geophysicist, in presenting his maps to geologists must remember this background. Observed gravity or magnetic maps very often have closures which are of basement or nonstructural origin; it is a duty of the geophysicist to give fair warning to the user when these are not the probable expressions of geologic structure controlling migration and accumulation of oil.

This distinction was expressed very early in the application of geophysics by one of its pioneers, E. L. DeGolyer (DeGolyer, 1928), in a short editorial entitled The Seductive Influence of the Closed Contour. It deserves rereading now as despite the progress since that time, that seductive influence is often present yet.

HISTORY OF GRID-SYSTEM DEVELOPMENT

Historically the grid systems began in the early 1930s as methods for computing second derivatives and for downward continuation but were first published many years later (Peters, 1949). In the following years a number of other second-derivative systems were published (Nettleton, 1954). Originally, the second-derivative systems were presented as ways of calculating a certain mathematical quantity, i.e., the second vertical derivative, by a difference approximation carried out on a grid of values surrounding the point of calculation. Systems were evaluated by the closeness with which their calculated values agreed with ideal values determined for given sources (spheres, etc.). It was recognized that, theoretically, the second-derivative values for a given gravity field have sharper patterns than the original field and therefore can give better separation of features relatively close together. It was recognized also that they have an effect of removing a regional because second derivatives are closely related to the curvature of the gravity surface and that curvature is vanishingly small for a smooth regional.

The pattern of numerical values on a regular array is commonly written manually by interpolation between contours on a map. With a well-made map contoured from an adequate array of observed values, this is a quite direct and straightforward process. "Adequate" in this sense may mean observations on a regular array of points, such as on section or quartersection corners in areas covered by a regular land survey or on a preplotted regular grid for a marine survey. Much more often, contours are interpolated from an irregular pattern of observations. If the average density of stations is comparable with that of the grid points to be used, reasonably accurately interpolated values should result. More often the observations are on lines along roads or waterways and may consist of relatively closely spaced ($\frac{1}{4}$ to $\frac{1}{2}$ mile) stations around wide loops. Then, when a grid is carried within the loops, the values must be interpolated from the contours between lines of control. The resulting calculated pattern (second derivative, continuation, or filter) will reflect the choices made by the map maker, and any detail more than about one grid spacing from a line of control may be of doubtful reality.

The values at grid points can be derived by a computer using an interpolation formula between random points of control. While this process is more objective and very much faster, it is subject to the same limitations as the manual methods.

For many years the various types of grid calculations were carried out with mechanical calculators or simple adding machines. This can be done quite rapidly, a single human operator calculating 500 or more points per day with a 17-point template. The same systems are readily handled by digital computers and yield results much more rapidly than the empirical smoothing process. Also the resulting grids of numbers can be readily contoured mechanically so that the complete process, from the input of gravity data by location coordinates and gravity values on punched cards to final contoured maps can be done automatically, including, if desired, computations with a number of different parameters.

Once set up, the various grid systems (including those for which the input can be gravity stations at irregular or random locations) are fixed and automatic and therefore are not empirical and subject to human judgment in the same way that the smoother systems are. However, judgment is still very much involved through choice of parameters basic to any automated system and in the considerations leading to selection of a particular grid spacing for the survey. For the first 20 years or so of their use, second derivatives and the related upward and downward continuation were treated as mathematical operations within themselves and not as filtering functions. Millions of square miles were mapped with second-derivative contours, in units of 10^{-15} gal/cm^2 or milligals per mile squared or other units apparently without realizing that the systems used did not come close to actually determining the mathematical quantity implied.

Any grid calculation involves data extending to one or more (commonly two or three) grid intervals beyond the central point. This means that the rigorously calculated values can extend only to within the two or three grid intervals of the edge of the observed data. This blank margin can cut seriously into the mapped area if the survey is relatively narrow. In many cases the margin can be encroached upon quite reliably by empirical and intuitive extension of the contours beyond the marginal stations so that the grid values can be extrapolated far enough to extend the calculation to or near the limit of the real data. Even then half or more of the input values are based on actual observation, but the possible deterioration of marginal values should be kept in mind. A similar deterioration occurs in the application of Fourier transforms where the low-frequency components lose definition near the map edges.

In more recent years, the development of two-dimensional filter theory and its application to gravity and magnetic maps has led to the realization that the various second-derivative and continuation systems are members of a class of techniques which analyze and operate on the components of a gravity (or magnetic) field map in terms of its spatial frequencies or wavelengths. Thus the selection of anomalies to be rejected (as regionals) or retained (as residuals) is based on filter theory applied to frequencies or wavelengths, in two dimensions, on the map.

The basic operation of filter theory is convolution, which is simply a weighted moving average. Weightings are given by the filter function, and the operation can vary in the dimension of its application. For example, one-dimensional calculation of a weighted moving average corresponds to the filtering by convolution of a time series or profile. Similarly, a two-dimensional application of a weighted moving average is then equivalent to the convolutional filtering of a map of values. Although convolution is defined using calculus or continuous mathematics, its practical approximation is usually accomplished using regularly sampled (one-dimension) values or gridded values (two dimensions). Hence the grid-system calculations, which are, in fact, weighted moving averages, are readily recognized as approximations to convolution filters, and conversely convolution-filter approximations can be readily realized as grid-system calculations.

We shall first review the commonly used symmetrical grid systems in terms of the operations performed and their effects as means of separating anomalies, and then we shall treat many of the same operations in terms of filter theory. This is logically somewhat backward, but treatment in historical order makes a more familiar approach to the many who first knew the second-derivative systems. It also provides an intuitive background as an aid to understanding the abstractions of filter theory.

GRID CALCULATION SYSTEMS

This discussion considers several systems which operate on a grid of values, either as observed on a regular grid or interpolated from a contoured map. These discrete values represent the gravity or magnetic field to a degree of completeness which depends on their spacing and the amount of detail which may occur between grid points. The accuracy of the contours themselves depends also on the detail and spacing of the original observations from which they were drawn.

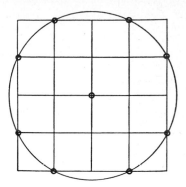

FIGURE 6-1
Rectangular grid for one-ring residual.

In all these systems a pattern consists of a center point and averages around several rings, or sets of values, at equal distance from the center. These are multiplied by coefficients derived, usually, by potential-theory analysis. A common feature of most such systems, including all grid-residual and second-derivative systems, is that they give a zero value for a region which is a plane, i.e., has no curvature, because the sum of the coefficients is zero. Therefore these systems are not handicapped by very steep regional gradients. Such steep gradients may make serious problems for a smoother, as it is difficult to recognize anomalies which are expressed as small departures from a uniform background with very close contours.

A number of somewhat different grid systems have been developed and used in producing interpretation maps. Some of the extensively used ones are described briefly in the following paragraphs.

Empirical Grid-Residual Systems

One of the simplest of the empirical systems and one that has been applied very extensively is to use the average of observed values on a circle as the regional. The residual is the difference between this average and the observed value at the center of the circle. This can be made equivalent to a grid system if the values are interpolated from contours at a regular array or grid of positions. In some cases, where stations are regularly spaced, as at section corners, the observed values can be used directly, without preliminary contouring. A system using eight points on a square grid is shown in Fig. 6-1. Another system using six points on a grid of lines at 60° is shown in Fig. 6-2.

In any center-point-and-one-ring residual method, the residual values and the nature of the residual map are directly dependent on the radius of the circle and to a more limited extent upon the number and distribution of points on the circle which are averaged. A discussion of the simple center-point-and-one-ring system with several examples which show the effect of variation of the radius of the circle on the nature of the resulting residual map has been given by Griffin (1949).

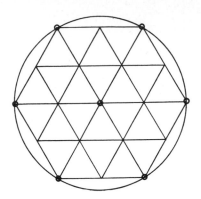

FIGURE 6-2
Hexagonal grid for one-ring residual.

Analytical Calculation of Derivatives

The grid methods were given a mathematical foundation by the application of potential theory to the calculation of derivatives, usually the second, of the potential functions (Evjen, 1936; Peters, 1949; Henderson and Zietz, 1949; Elkins, 1951; Rosenbach, 1953).* The rather involved mathematical formulations have been reduced to practical schemes of calculation, using values from a regular grid of points to determine averages around circles of different radii from a central point. These can be thought of as an elaboration of the simple one-ring method by which several rings are used instead of one, with the rings of different radii having different weights, some of which must be negative.

Figure 6-3 is an example of a grid of three rings used for second-derivative calculations. Averages at distances of S, $S\sqrt{2}$, and $S\sqrt{5}$ are determined from four, four, and eight points, respectively.

All these mathematical systems involve assumptions or empirical choices at certain stages in their reduction to practical use. These choices control the determination of the numerical coefficients by which the values read from a map are multiplied to determine the derivative values. The arbitrary nature of these final solutions has been pointed out:

From the foregoing it is evident that any number of coefficient sets may be developed. Their relative merits can be determined only by testing them. (Elkins, 1951, p. 39)†

---

* Theoretically, the second derivative is the second partial derivative with respect to the vertical, that is, $\partial^2 u/\partial z^2$ for gravity. The quantity dealt with also can be a component of the magnetic field or any other potential field to which the Laplace equation is applicable. This fundamental equation is

$$\frac{\partial^2 u}{\partial x^2} + \frac{\partial^2 u}{\partial y^2} + \frac{\partial^2 u}{\partial z^2} = 0 \qquad \text{from which} \qquad -\frac{\partial^2 u}{\partial z^2} = \frac{\partial^2 u}{\partial x^2} + \frac{\partial^2 u}{\partial y^2}$$

Thus, certain mathematical operations are carried out on the gravity values on the xy plane in order to approximate the horizontal derivatives and thus yield the desired vertical variation, that is, $\partial^2 u/\partial z^2$.

† This was true until it was realized that two-dimensional Fourier transformations make it possible to compare operators by their frequency characteristics and to reject some out of hand as being obviously inferior to others (see page 165 and Fig. 6-20).

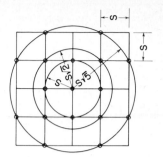

FIGURE 6-3
Rectangular grid for three-ring second-
derivative calculations.

These methods, being approximations, give results which depend to a considerable extent on the spacings used. . . . The results are more qualitative than quantitative in value but these methods have proven to be some of the better tools available for mass production analysis. (Peters, 1949, p. 304)

A number of published analytical systems for calculating second derivatives have led to different parameters and coefficients by which the averages at different distances are multiplied to give the second-derivative values. These are, in effect, variations in the weighting factors by which the gravity differences between the value at the center point and the average around a particular ring are multiplied and added to determine a second-derivative value.

All the published second-derivative systems are of the form

$$g_{zz} = \frac{C}{S^2}(W_0 g_0 + W_1 g_1 + W_2 g_2 + \cdots + W_n g_n)$$

where
$\quad g_{zz}$ = derived, i.e., second-derivative or residual, value at point 0
$\quad C$ = constant for particular system
$\quad S$ = grid spacing
$W_0, W_1, \ldots$ = weighting factors
$\quad g_0$ = gravity at point of calculation
$g_1, g_2, \ldots$ = average values around the successive rings

The weighting factors are of positive and negative sign, and the sum of the factors, $W_0 + W_1 + \cdots + W_n = 0$.

A summary of several of the published formulas, modified to be readily compared, is given in Table 6-1.

To calculate the theoretical derivative values, we must include a factor $1/S^2$, where S is the grid spacing. For instance, if r is the grid spacing on a map of scale $1:k$, then $S = kr$. A common unit for the second derivative of gravity is gals per centimeter squared; with this unit, second-derivative values are usually of the order 10^{-15}. Other units are milligals per mile squared and gammas per mile squared (for magnetics).

The original equations have been normalized by dividing through by the first weighting coefficient W_0 so that the relative weight of the g_0 term is unity. The decibel (db) value is the decibel equivalent of the multiplying factor K and will be used in the discussion of frequency analysis.

Table 6-1 EQUIVALENT FORMULAS WITH UNIT WEIGHT FOR CENTER POINT (Nettleton 1954)

Formula	Source*	db	Factor K†	W_0 $r=0$	W_1 $r=s$	W_2 $r=s\sqrt{2}$	W_3 $r=2s$	W_4 $r=s\sqrt{5}$	W_5 $r=s\sqrt{9.23}$
(1)‡		0	1	1	-1				
(1a)‡		12.0	4	1	-1				
(2)	1a	15.8	6.185	1	-1.354	$+0.354$			
(3)	1b	15.6	6.000	1	-1.333	$+0.333$			
(4)	1c	16.9	7.000	1	-1.523	$+0.571$			
(5)	2a	0.56	1.067	1	-0.125	-0.250	-0.48		
(6)	2b	-4.8	0.571	1	$+0.500$	0		-0.625	
(7)	2c	-2.8	0.710	1	$+0.364$	-0.273		-1.500	
(8)	3	1.2	1.156	1	$+0.221$	-0.385		-1.091	0.339
(9)	4	9.5	3.000	1	-0.750	-0.333		-1.175	
								$+0.083$	

* 1 = Henderson and Zietz (1949): a, eq. (10); b, eq. (13); c, eq. (15).
  2 = Elkins (1951): a, eq. (13); b, eq. (14); c, eq. (15).
  3 = Peters (1949), eq. (27).
  4 = Rosenbach (1953), eq. 16.

† The factor $K = CW_0$, where C is the multiplier in the published formula to give results in theoretical second-derivative units. In order to normalize the formulas to make the weighting of the first term unity, all the weights are divided by W_0.

‡ Center point and one ring.

It might be expected that the different formulas would give similar second-derivative values. Actually, the calculated values vary widely. Some of the formulas which give relatively more accurate values for ideal, theoretical fields tend to be unsatisfactory for actual field data because they depend on small gravity differences and emphasize irregularities. Others, which give better looking maps, do not reproduce the calculated values for ideal fields very well. In general, these differences come about primarily because a theoretically continuous gravity field is represented by a limited number of discrete points at the grid corners, leading to a necessarily incomplete or aliased* representation, which also contains noise or errors in the data. The different formulas vary in the manner in which the derivative is represented by the limited number of samples. This results from different choices for the values of the weighting factors. For instance, if the value for W_1 is relatively large and negative, it will give strong emphasis to differences between the center point and the first ring and will be very sensitive to small errors or irregularities of the map. On the other hand, if W_1 is relatively small and positive and the outer rings have negative coefficients, the derivative value depends heavily on the difference between the average over the central part of the gravity field and that of the outer circles. The resulting map will be much smoother and will tend to give a filtering effect to suppress the local variations within the area taken into account in each calculation.

The dimension of the grid spacing has a large effect on the nature of the resulting map; calculations made with a small grid spacing can accommodate smaller details, while those with larger grid spacings properly represent only the larger features. Elements of the data which cannot be properly represented by the grid spacing contribute to the effective noise level of the data by the aliasing process.

Since the second-derivative-type calculations have weighting coefficients which add up to zero, the resulting map is balanced between positive and negative areas. A simple positive anomaly on an observed map will, on the derivative map, have a central positive anomaly centered at the same place but accompanied by a flanking negative anomaly not present on the original map.

In the general mathematical expression for a second derivative the term C/S^2 converts the weighted gravity differences to second-derivative units with the dimensions of gravity divided by the square of distance. If the C/S^2 term is omitted, the weighted gravity differences add up to a quantity which is a modified residual in the dimensions of gravity, i.e., milligals.

The choice of map units, either second-derivative values or gravity differences, is dictated by the attitude and preference of the user. Making maps in second-derivative units (g/z^2) is dimensionally correct in terms of the numbers used and to the degree (which may not be close) that the formula used determines true values. For a user interested primarily in recognizing anomalies the gravity-difference

---

* "Alias" is the term used (commonly in frequency or wave-number filtering) to denote the condition where the detail of a field being analyzed is too fine to be represented by the interval at which the data are sampled, with the result that some of the fine detail reappears falsely in the sample data as large-scale features, that is, with an "alias."

units may be preferable, even though they are only empirical, because they are in dimensions more readily understood (by geologists) than the more esoteric second-vertical-derivative quantity. Since the second-derivative values are derived from the weighted differences between the center point and averages around successive rings multiplied by a constant (C/S^2), the zero contour and the maximal and minimal closures will be in the same places on second-derivative or on simple-difference maps. By appropriate choice of contour interval, the two maps can be made identical.

It may be pointed out again here that the quantity used for second-derivative maps is different in form and in units from gravity and therefore cannot be used as a basis for comparison with calculated gravity effects.

Continuation Calculations

For certain purposes, it is of interest to consider the form which a potential field would have if it were measured at a higher or a lower level. In principle the field can be continued to a different level if there is no disturbing body (mass or magnetized material) within the range of the continuation.

Upward continuation gives a smoother field and downward continuation a sharper field than that at the surface of measurement. Both processes are similar to the second-vertical-derivative calculation in that they can efficiently operate on a regular array of values. Also, in effect, they apply weighting coefficients to average values on circles of successively larger radii. Usually continuation requires data over a considerably wider area than second-derivative schemes; some use as many as 10 rings of values about the center point. In general, the radius of the field taken into consideration should increase as the magnitude of the continuation distance increases.

Unlike the second-derivative operation, continuation does not force the mean value to zero. For most continuation systems the sum of the coefficients is unity (rather than zero, as in second-derivative operations), and a surface which is plane within the area of the operation would have the same value at the level of continuation.

Downward continuation is applied for anomaly separation, in particular the resolution of overlapping effects of sources close together. In gravity or magnetics, if two separate but nearby sources are under homogeneous overlying material, the downward continuation could be made to a level closely above the sources, where the effects would be clearly separated, whereas they would be confused by overlap at the level of measurement. If continuation is carried to depths greater than the source, the continued field will begin to oscillate; this oscillation could be a criterion of depth. These results are not often obtained clearly from actual fields because, as we have indicated, the downward continuation process depends on emphasizing small effects at the measured surface. If such effects are present from shallow sources and are continued below their depth of origin, they contribute large false components to the continued field. For this reason, downward continuation has more application in magnetic than in gravity interpretation because airborne measurements made at substantial elevations above the ground and therefore remote from possible sources are relatively smooth and can be operated upon by a system of sharp coefficients without false emphasis of shallow sources. However, any sharp noise in the data, such as instrumental malfunction or unrecognized diurnal effects, would appear to be caused by shallow sources and would give false details in a downward continuation.

A rather careful application of downward continuation to both gravity and magnetic data over the same area is given by Rudman et al. (1971). They were able to delineate fairly well the boundaries and components of an intrusive body from the continued fields.

Upward continuation always gives a smoother map than the original. It has been used as a form of smoothing as a basis for anomaly separation. In airborne magnetics, upward continuation has been used to calculate the field at a higher level and the result compared with measurements actually made at that level. When control is adequate, the upward continued and measured values are almost identical.

A review of derivative and upward and downward continuations by Henderson (1960) is directed toward the application of digital computers to their calculation. The paper contains several sets of coefficients and tests of their effectiveness, including comparison of actually measured and continued magnetic fields.

A review of continuation operators by Mufti (1972) shows that it is possible to design relatively small operators, over a two-dimensional field of input values, which will give results practically equivalent to those from a much wider input field but with a greatly reduced number of numerical operations.

An application of the downward continuation process to a practical interpretation problem was made by Hammer (1947). The Moss Bluff salt dome, on the Liberty-Chambers county line in southeast Texas, has a broad, relatively flat cap rock with thickness varying from about 150 to 550 ft above salt at a depth of about 1100 ft. The dome was considered a sulfur prospect with better chances for accumulation where the cap-rock thickness is greater. The problem, then, was to map the variation in thickness from gravity data.

A detailed gravity survey was carried out with stations on a 500-ft rectangular grid. The resulting gravity map (Fig. 6-4) was continued downward to the approximate average depth of the cap-rock mass. The smooth gravity effect of the underlying salt was then subtracted to isolate gravity effects of the cap rock for the thickness calculations. A value for the effective density contrast was derived from the relation of continued gravity to cap-rock thickness at the few wells which had been drilled through the cap rock.

The resulting cap-rock-thickness map is shown in Fig. 6-5. From drilling following the gravity analysis, the average difference between predicted and actual thickness was about 75 ft. A commercial Frasch-process plant was built, following the gravity survey, to extract the sulfur found.

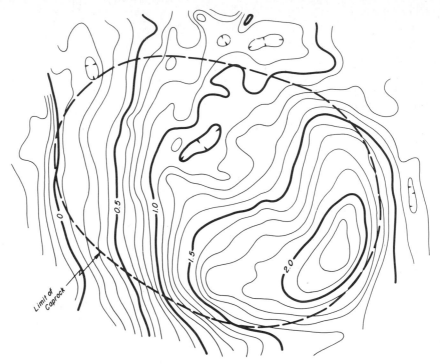

FIGURE 6-4
Observed gravity, Moss Bluff dome. (*Courtesy of S. I. Hammer.*)

COMPARISON OF GRAPHICAL AND GRID SYSTEMS

A qualitative comparison of a derivative-type grid residual and a graphical system of anomaly separation is indicated in Fig. 6-6. The line of close dots in the upper part of the figure is a hypothetical gravity profile that includes two obvious local anomalies superimposed on a much larger feature.

The nature of this particular grid-residual calculation is indicated by the short straight lines which bridge segments of the gravity profile. The ends of these lines correspond to the diameter or outer circle of a grid-residual calculation array. The arrows indicate the magnitude of the difference of the center point from the average around the circle or at the ends of the bridge. This arrow is upward if the center point is higher than the average around it and downward if its value is lower. If we take such bridges at regular intervals along the anomaly curve and plot the length of the arrows, we get the result shown by the grid-residual profile with the shaded areas marking the departures or local differences from the observed curve. It will be noted that these differences are positive over the central parts of a local anomaly and negative on its flanks.

The graphical residual process is illustrated by the "graphical regional" shown along the profile. For a relatively simple situation such as is illustrated

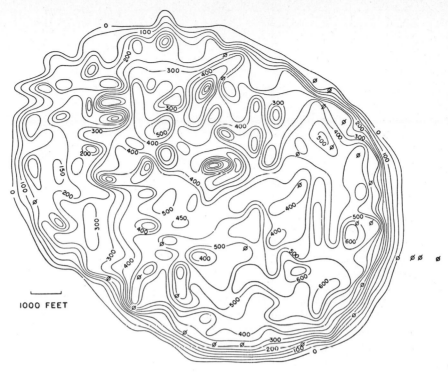

FIGURE 6-5
Cap-rock-thickness map, Moss Bluff dome. (*From Hammer, 1955.*)

here, the local departures are quite obvious, construction of a graphical residual is simple, and nearly all experienced interpreters would draw this in about the same way. The differences between this regional and the observed curve are plotted as the graphical residual at the bottom of the diagram. Note that for the two local anomalies, the positions of the peaks of the anomalies are very nearly the same as for the grid-residual curve.

The principal likenesses and differences between the anomalies as determined by the two systems are as follows:

1   Both systems show the two local anomalies and their centers at the same locations.
2   The two local anomalies are both shown as simple maxima by the graphical residual and as narrow maxima with flanking minima by the grid residual.
3   The positive amplitude is considerably greater for the graphical residual; for the grid residual, the total amplitude from the bottom of the flanking minima to the top of the maximum is approximately the same as the positive amplitude of the graphical residual. This relation is very much dependent on the choice of a radius appropriate to the size of the anomalies and the particular weightings.

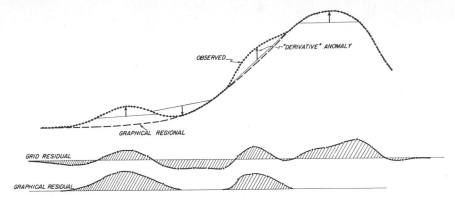

FIGURE 6-6
Hypothetical gravity profile with graphical regional; grid residual and graphical
residual below. (*From Nettleton, 1954.*)

4   The grid-residual system shows a comparatively large positive anomaly
over the apex of the regional maximum whereas, with graphical smoothing,
this feature is entirely included in the regional and no residual anomaly is
shown. This illustrates how a choice can be made in constructing a graphical
regional which will include or exclude a feature more or less completely.
Such a choice can never be made automatically by any linear grid-type calculation unless the difference in the width (or wavelength) of the anomalies
is quite large.

An example of the application of graphical and grid-residual techniques to
the same actual gravity map is shown by Figs. 6-7 to 6-10.

The gravity map (Fig. 6-7) is for an area immediately south of Houston,
Texas, which contains a small salt dome with oil production on its westerly flank.
The gravity field is dominated by a relatively strong northerly gravity increase of
about 3.5 mgals within the limits of the map.

Figure 6-8 shows this regional change expressed as smooth, east-west trending
contours at 1.0-mgal intervals. The residual remaining after subtraction of this
smooth regional is indicated by the residual contours at 0.2-mgal intervals. These
show the fairly circular residual minimum with a relief of a little over 0.6 mgal,
which is fairly well (but not exactly) centered over the dome, as indicated by the
lines of dots corresponding to the 4500- and 6000-ft contours on the salt. This
residual gives a relatively simple anomaly which would be amenable to comparison
with effects which might be calculated from an assumed form of the dome and
which could be used to determine the approximate size of the salt column.

The contours of Fig. 6-9 give values from a second-derivative calculation
carried out on gravity values interpolated on a $\frac{1}{2}$-mile grid from the contours of
Fig. 6-7. This map shows a quite sharp negative anomaly over the dome, and the
contour pattern is considerably more complicated than in Fig. 6-8. Note that the

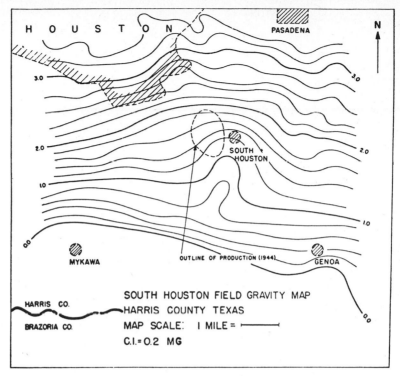

FIGURE 6-7

Observed gravity map, South Houston area, contour interval 0.2 mgal. (*From Nettleton, 1954.*)

negative anomaly is quite well centered over the dome. Note also that this negative anomaly is surrounded by a peripheral zone of positive values, indicated by the dashed line. This is the flanking reversal, which is introduced into the result by the fact that the positive and negative values add up to zero, as is indicated by the profile of Fig. 6-6.

Figure 6-10 is calculated with a simple center-point-and-one-ring average system but with the radius expanded by the factor $\sqrt{5}$ (see Fig. 6-18), so that its outer ring is at the same distance as the outer ring of the formula used for Fig. 6-9. This makes the effective outer limit of the area of calculation comparable with that of the heavily weighted outer ring of formula (7) of Table 6-1, and therefore the resulting contour pattern is very similar.

Calculations of gravity from mass anomalies could readily be made to fit the graphical residual but would not be directly applicable to the grid residual because of its positive and negative parts. As mentioned above, the effectiveness of the grid-residual system is quite independent of the general slope of the field, and grid residuals can be calculated readily where the background is very steep

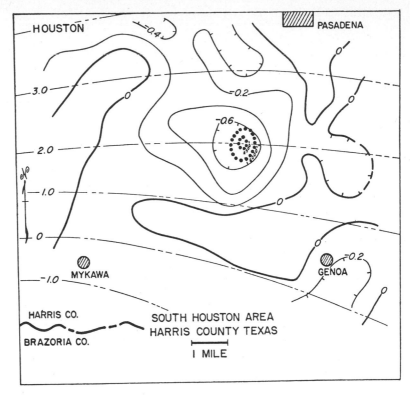

FIGURE 6-8

Regional gravity contours, at 1.0-mgal intervals, and graphical residual contours at 0.2-mgal intervals; residual contours obtained by graphical subtraction of the regional from the contours of Fig. 6-7. (*From Nettleton, 1954.*)

and where choice of a regional and subtraction to form a graphical residual might be very difficult (see example from Los Angeles Basin, Figs. 6-13 and 6-14). Other grid-residual maps which would be more amenable to quantitative analysis might result from a downward continuation operation or else be a difference map between the original data and an upward continuation operation. Also, certain trend-surface removal methods would produce maps of such character. The example of the work by Hammer (1947), discussed earlier in the chapter, using continuations adequately illustrates this point.

A major difficulty in the use of any analytical system for determination of a regional from a grid of observed values is that the system has no clear geological counterpart. For instance, it is a straightforward matter to make a series of regional surfaces by polynomial fits of successively lower order (or longer wavelength). The geology does not have a definite wavelength; one fit may seem better in one part of a map, another in a different part. A very large feature will appear to different degrees as the residuals from different orders of fit. The smoother

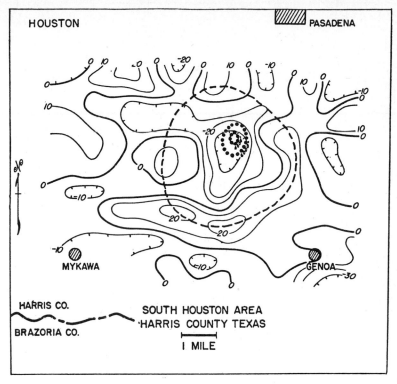

FIGURE 6-9

Second-vertical-derivative contours, calculated with $\frac{1}{2}$-mile grid spacing from map of Fig. 6-6, with weighting coefficients of formula (7) of Table 6-1. Contour interval 10×10^{-15} cgs unit. Dashed line is zone of flanking positive anomalies around central negative. (*From Nettleton, 1954.*)

could decide whether the anomaly is to be left in—all of it—or none as, for example, in Fig. 6-6. Since the earth does not behave according to any reasonably specified mathematical formula, any mathematical solution to the regional problem is an approximation. The analytical systems can be very helpful in indicating a regional, but commonsense modification is nearly always evident. Then the interpreter becomes, in part, a smoother in an attempt to make modifications to adapt the mathematical result to reasonable expectations from geological behavior.

EXAMPLE OF ANOMALY RESOLUTION

An example of gravity data which illustrates both the geological significance of a regional anomaly and the recognition of subtle but significant local anomalies is available from the Los Angeles Basin.

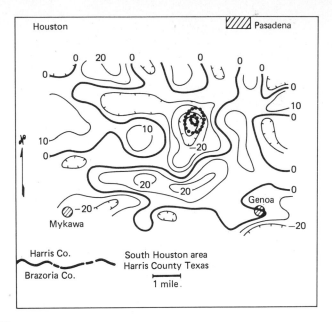

FIGURE 6-10

Second-derivative map, calculated with grid spacing $\frac{1}{2}$ mile $\times \sqrt{5}$ from map of Fig. 6-7 with weighting coefficient from formula (1a), Table 6-1; contour interval 10×10^{-15} cgs unit. (*From Nettleton, 1954.*)

Figure 6-11 is a geologic map of the basin on which the regional gravity at 5-mgal contour intervals has been superimposed. The general configuration of the gravity pattern is in accordance with expectations from the general form of the basin itself. The minimum is over the area of deepest sediments, and the gravity increases very rapidly southwest toward the basement outcrop in the Palos Verdes Hills. The sediments are all Tertiary clastics and of substantially lower density than the Franciscan schist of the basement. Quantitatively, the 60-mgal southwesterly gravity rise from the center of the minimum to the Palos Verdes Hills is consistent with that which would be expected from reasonable estimates of the total thickness of sediments and of the densities of the sedimentary and basement rocks (see calculation, page 194).

Figure 6-12 is a profile across the basin along line AA', of Fig. 6-11 and shows a geological section together with the gravity profile from Fig. 6-11. This profile shows the bottom of the gravity minimum corresponding almost exactly with the deepest sedimentation. The very strong gravity maximum over the Puente Hills structure almost certainly means that the basement rocks are involved in this structure, although not so shown on the geologic section, which extends downward only into the Tertiary sediments. On the southwest side of the basin, we see that the Dominguez structure does not produce a noticeable gravity anomaly.

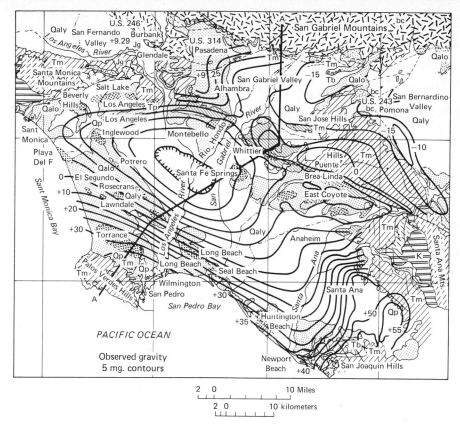

FIGURE 6-11
Surface geologic map, Los Angeles Basin. (*Modified from U.S. Geol. Surv. Prof. Pap. 190 with gravity contours at 5-mgal interval added.*)

Figure 6-13 shows a detailed gravity survey, with $\frac{1}{2}$-mile station spacing, over a part of the southwest flank of the basin which includes the areas of the Long Beach, Dominguez, parts of the Rosecrans, Wilmington, and Torrance oil fields. On this map, casual inspection would suggest that there is no gravity expression at all of these oil-field structures. However, a careful look shows that, for instance, on the northeast side of the Long Beach field the contours are a little closer together than over the field itself, and, similarly, on the northeast side of the Dominguez field the contours are crowded more closely than they are over the oil-field area. These differences are significant, as shown by the second-derivative map for this area (Fig. 6-14). This map was calculated by formula (7) of Table 6-1 with a $\frac{1}{2}$-mile grid spacing from the observed map of Fig. 6-13. This second-derivative map develops a very strong and definite positive trend which extends through the Long Beach, Dominguez, and Rosecrans fields, and also another positive trend

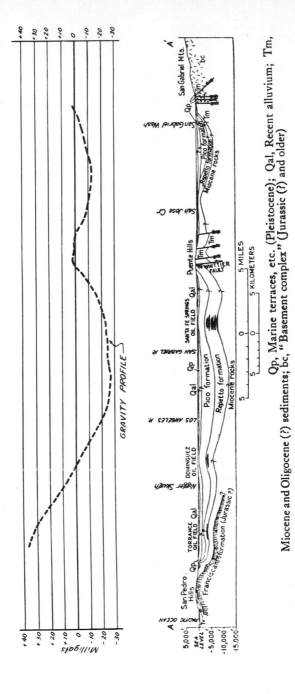

FIGURE 6-12

Cross section, line *AA'*, Fig. 6-11 (*modified from Guidebook, 16th Int. Geol. Congr., 1932*) and gravity profile from contours of Fig. 6-11.

Qp, Marine terraces, etc. (Pleistocene); Qal, Recent alluvium; Tm, Miocene and Oligocene (?) sediments; bc, "Basement complex" (Jurassic (?) and older)

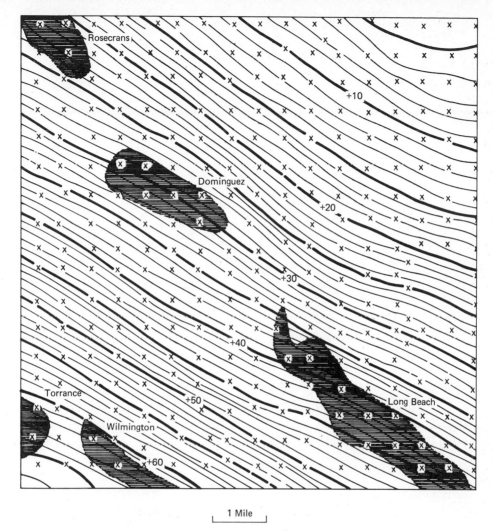

FIGURE 6-13
Gravity map, part of southwest flank, Los Angeles Basin; contour interval
0.2 mgal, stations on regular grid with $\frac{1}{2}$-mile spacing. (*From Elkins, 1951.*)

which includes the Wilmington and Torrance fields. The second-derivative calculation has brought out these features which were very obscure in the observed
gravity map because of the very strong regional gradient. Of course, they must be
present on that map to be developed by the derivative calculation, and their
presence is indicated by the previously mentioned variations in contour spacing.
It should be pointed out also that the negative flanks of the positive trends are parts
of the total anomaly in the same way that previously mentioned negative aspects
of second-derivative maps have been pointed out.

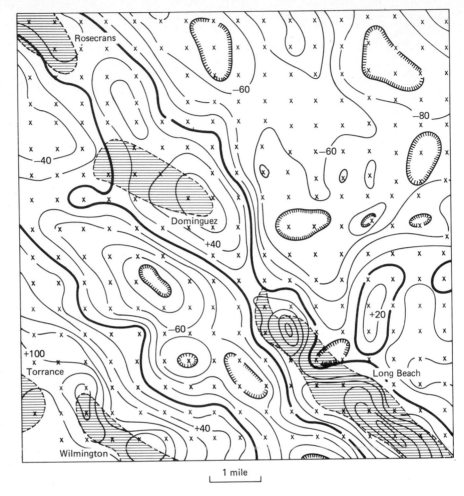

FIGURE 6-14
Second-vertical-derivative map, Los Angeles Basin; calculated from gravity of Fig. 6-13 by formula (7) of Table 6-1; contour interval 20×10^{-15} cgs unit. (*From Elkins, 1951.*)

While a test has not been carried out, it is almost certain that residuals from a least-squares fit or from anomalies from a frequency filter, with optimum choice of parameters, would clearly resolve the same features. These processes are discussed in the following section.

The separation of anomalies from overlapping effects of other anomalies and from trends ultimately rests with the basic accuracy of the data, the precision with which they are recorded, the density with which observations are made, and the distribution of the data. No amount of computer processing can resolve information which has not been preserved in the sampled data.

FILTER THEORY IN ANOMALY SEPARATION

The foregoing considerations of anomaly recognition have been presented in the historical order in which these concepts were applied. In more recent years two basic developments have radically changed the methods of calculation and the basic concepts of the nature of the processes used. These are:

1 The development of digital computers and their almost complete domination of the numerical-reduction processes; such operations applied on a regular grid of input values are very natural for a computer and give speed improvements of several orders of magnitude.

2 The realization that most of the linear grid processes used are the equivalent of two-dimensional filters and that the numerical processes, Fourier transforms, and concepts of space and frequency domains of filter theory are directly applicable to the theory and analysis of all the old grid systems.

The following discussion of filter theory and applications includes a review of the several grid systems in terms of the equivalent filter concepts as well as some perspective of analytic methods in general. All wave-number filters as described here attempt in some measure to perform anomaly separation in the same manner as the smoothers. In many cases a separation of this type is more definite than the problem requires. Mere recognition of the location of an anomaly may constitute all the information that is needed, particularly if the anomaly is a subtle one. Derivative filters, for example, were introduced to assist in defining anomalies and hence accomplish a qualitative separation of anomalies as opposed to a quantitative one.

The open literature does not represent fully the efforts made in the analysis of derivative filters, but the works of Swartz (1954), Mesko (1966), and Dampney (1966) are of particular significance. Zurflueh (1967) gives some interesting examples of applications to measured magnetic fields. Fuller (1967) has made frequency analyses of several of the grid systems considered previously, which will be discussed later in this section (page 163). As with other wave-number filters, both the data-domain viewpoint and the spatial-frequency viewpoint may be adopted. The applications of interest here evolved from the spatial point of view, and so in line with the historical approach this will also serve as the starting point of our discussion.

While the basic method of the smoothers is straightforward and unvarying, one can recognize three main categories of analytic approaches used by the gridders. First, there are the wave-number filters which include derivatives, upward and downward continuation, spatial bandpass operators, and directional enhancement or suppression mechanisms. Next there are the trend-surface* analysis and removal methods. These utilize a least-squares sum or some other measure to fit

---

* As used here, a trend surface denotes any surface, i.e., geological, gravity, magnetic, etc., which indicates the general direction and magnitude of rates of change. It corresponds with the regional of the smoother or to a low-order term in a surface-fitting or frequency-domain analysis.

functions like polynomials or Fourier series to the observations. Both the wave-number filters and the trend-surface methods can be approached from either a statistical or a deterministic viewpoint. Finally, there are the inverse modeling techniques, which attempt to identify directly the elements which contribute to the observed field.

All three approaches in some degree offer analytic vehicles for automatically emulating the objectives accomplished empirically by the smoother. In the discussions which follow, the basic organization is by category; however, two continuing themes are also pursued. The similarities and differences between the analytic methods and the smoothers' approach are explored in some detail. Also, it will become evident that some of the diversity appearing in the methods employed by the gridders stems from different interpretations of the meaning of anomaly separation.

GENERAL WAVE-NUMBER FILTERS

The petroleum industry has examined the use of wave-number filtering techniques for gravity and magnetic data since about 1955. Unfortunately, the bulk of such studies remains unpublished. In the open literature the works of Dean (1958), Byerly (1965), Darby and Davies (1967), Fuller (1967), Zurflueh (1967), and Kanasewich and Agarawal (1970) provide some sampling of the progress which has been made.

Wave-number filters change the distribution of the wave numbers within the data. Each wave number corresponds to a spatial-frequency component and is, in fact, its reciprocal. The concept underlying such filtering is that there is a spatial-frequency-domain representation of the potential field data which although different in form from those data is informationally its equivalent. Further, for certain of the characteristics through which anomalies might be separated, the frequency-domain viewpoint offers some distinct operational advantages. While it is not necessary to utilize the frequency domain directly to achieve these benefits, the discussions which take this approach are the more instructive.

Several basic reasons make wave-number (or reciprocal of wavelength) filtering useful:

1 It is readily applied, by computer, to large quantities of potential-field data.
2 Potential fields are well behaved in the spatial frequency domain.
3 Anomaly patterns tend to separate clearly in the frequency domain (Spector and Grant, 1970). This point may be considered controversial; an experienced interpreter can see the regional just as clearly in the space domain.
4 Regional separation based on this approach can be automated more readily.

The Fourier Transform

The reversible Fourier transform allows the easy transitions back and forth between the spatial data domain and the spatial frequency or wave-number domain. In fact, because the Fourier harmonics are orthogonal, the Fourier-transform operation is precisely equivalent to determining the best least-squares harmonic trend. This property will be considered again when trend-surface methods are discussed. At this point, we are concerned with gaining an understanding of the spatial frequency domain and those of its features which bear on the separation of anomalies.

To avoid subtle questions, all the data to be treated will be assumed to be "properly" sampled at a regular grid interval. Specifically, the highest spatial-frequency component of significant amplitude will be lower than the spatial frequency corresponding to the reciprocal of twice the grid interval, which is the definition of the Nyquist frequency where aliasing begins.

If we consider a profile of sampled gravity values, the discrete approximation to the Fourier transform is simply a sum of harmonic sinusoids which, at the sampled points, are equal to the sampled values. Each harmonic has two constants associated with it, an amplitude and a phase angle. The amplitude gives a measure of its absolute contribution to the sum, while the phase angle fixes the position of the coordinate origin of the variable for the particular harmonic with respect to the coordinate origin used in the transform calculation. A plot of the amplitudes against the harmonic number or equivalently against the linear or angular frequency is called an "amplitude spectrum," while a similar plot of the phase angles is called a "phase spectrum." The power spectrum is like the amplitude spectrum except that the squares of the amplitude values are plotted.

Figure 6-15 portrays the essentials of the Fourier transform of a hypothetical data profile. The grid spacing is S, and the amplitude and phase spectrum for $\pm M/2$ harmonic components are shown, where M is the number of data samples. Recall that each harmonic component has the continuous analytic form $A_n \sin (W_n X + \Phi_n)$, where A_n and Φ_n are the amplitude and phase functions to which we refer. The W_n are the discrete angular frequencies against which we plot, while X, the spatial distance variable, would be expressed in units of the grid spacing S. Note that W_n, the angular frequencies, are simply related to the linear frequencies by a scale factor of 2π. The subscript n is used to count the harmonics and the harmonic number. In the case of profiles, if our objective is the separation of anomalies, two types of information are available: (1) the contributions of each of the harmonics are given, and (2) their relative alignment is indicated. Since each harmonic corresponds to a spatial frequency, and hence a spatial wavelength or physical size parameter, some reference is thus provided for the identification of anomalies of differing extent or size.

Two-dimensional Fourier transforms of gridded map data give analogous information, but in this case a direction of alignment accompanies the harmonic information. A regularly sampled data map is represented as a double sum of

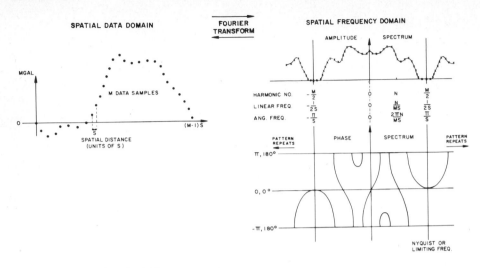

FIGURE 6-15
Fourier transform of data profile.

sinusoids which are harmonics related to the dimensions of the map along two orthogonal coordinate axes.

In more elementary form, each term of the double summation corresponds to a "corrugated" surface. A typical corrugated surface is shown in Fig. 6-16, where the parameters characteristic of the surface are also noted. Each such surface has an azimuth of orientation with respect to the two coordinate axes of the analysis, an amplitude, a projection of spatial wavelength and corresponding frequency along each coordinate axis, and a phase angle which specifies the alignment of the corrugated surface with respect to the coordinate origin. Figure 6-16 also shows how all this information is presented in the two-variable spatial frequency domain.

In no published work up to this time have modifications of phase played any role in the separation of anomalies. On the other hand, the amplitude spectrum and the directional properties have been utilized by a number of authors and in a variety of ways.

Separation of Anomalies in the Frequency Domain

The process of making use of frequency-domain characteristics for separating anomalies is illustrated by the idealized gravity map shown in Fig. 6-17a, in which the low-frequency trend is associated with an east-southeasterly orientation. The higher-frequency anomalies show no directional preference. In the spatial frequency domain the data components corresponding to the trend and to the anomalies are in the main clearly separable.

Figure 6-17b shows the hypothetical amplitude spectrum corresponding to

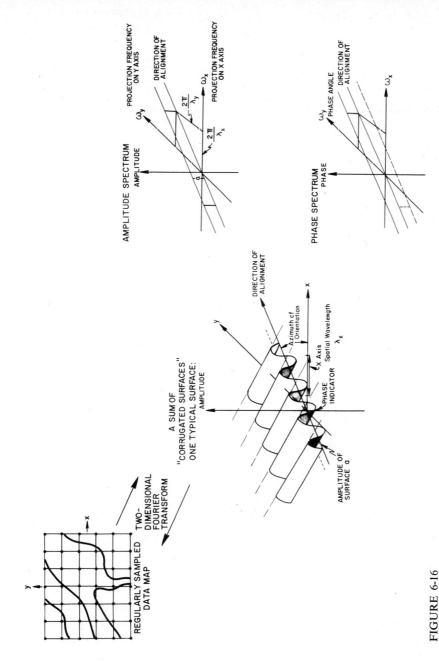

FIGURE 6-16
Representation of two-dimensional Fourier transform of regularly sampled potential-field data.

161

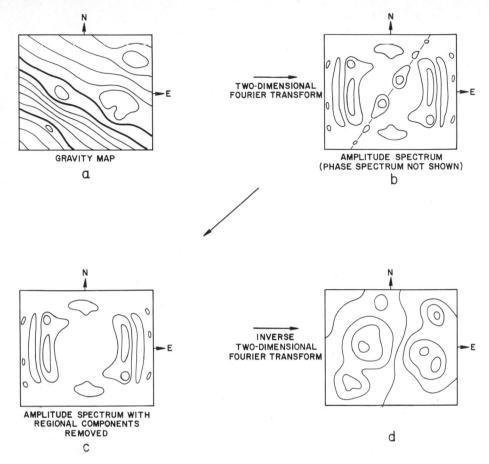

FIGURE 6-17
Idealized separation of regional and anomalies using frequency-domain characteristics. Long wavelengths, i.e., small wave numbers, are plotted at the center of the diagram; hence the data involving the regional appear near the center of the amplitude-spectrum plot; recomposing the spatial-data map from the wavelength (or wave number) requires phase information like that indicated in Fig. 6-16.

the gravity map. Lower-spatial-frequency contributions associated with long wavelengths (recall the reciprocal relationship) are developed toward the center of the contour plot. Similarly, the higher-frequency contributions are found as one progresses to the edges of the same plot. Note that the trend in the low-spatial-frequency content develops at right angles to the contour levels in the original data map, just as an interpreter would enter it.

The indicated procedure is then straightforward. Use the two-dimensional Fourier transform to calculate the spatial frequency domain. Remove those

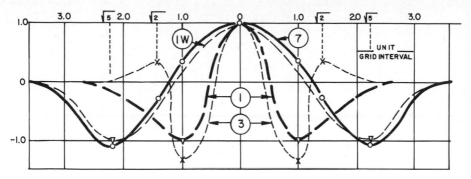

FIGURE 6-18

Profiles of second-derivative formulas with normalized weights, from Table 6-1. The normalized weights are plotted against the radii of the circles to which they apply.

elements in the spatial frequency domain corresponding to the regional effect (Fig. 6-17c). Apply an inverse Fourier transform, and the resulting map (Fig. 6-17d) should now show only the anomalies. Analytically, the procedure described is wave-number filtering and is equivalent to applying multiplicative weights in the spatial frequency domain, where for this example the weights are unity for information to be preserved and zero for information to be deleted. We should add that, alternatively, a fully equivalent operation of convolution filtering can be used to achieve the same objective without ever leaving the spatial data domain, but it is important to understand both viewpoints. Note also that the phase information, although unmodified and not depicted, has been used in order to accomplish the transitions back and forth to and from the spatial frequency domain.

In fact, the separation of anomalies described cannot be exactly accomplished owing to the distributed spatial-frequency spectra which are a physical characteristic of all anomalies. That is, just as the potential field of a source never completely vanishes, the spatial-frequency spectrum can never completely vanish. There are always contributions to all frequencies. The particular distribution of spatial frequencies will vary with distance to the source and the source characteristics, but it will never consist of a single-frequency component or even a small number of components unless we are prepared to consider sources which are infinite in extent. Hence the practical science of filtering in this instance takes on many aspects of an art.

Grid Operators as Frequency Filters

The general ideas of grid operators as frequency filters can be expressed by plotting the relative weights of the coefficients as a function of their radial distance from the center point of calculation. Figure 6-18 shows the relative weights for formulas (1), (3), and (7) of Table 6-1, with the curves numbered accordingly. It will be noted

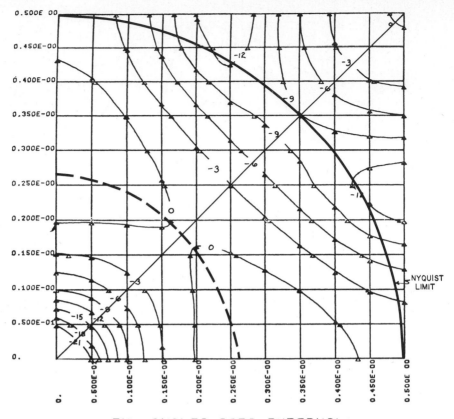

FIGURE 6-19

Sample of Fuller frequency-response chart showing contours of equal attenuation. This chart is for formula (7) of Table 6-1. Profile 7 of Fig. 6-20 is plotted
from the contours on the diagonal line across this chart. Outer circle is at the
Nyquist limit of 0.5 cycle per data interval. Dashed line is zone of maximum
anomaly emphasis. (*Modified from Fuller, 1967.*)

that the curve for formula (3), with alternating negative and positive weighting
factors, has a relatively sharp form corresponding to a shorter wavelength and
this formula therefore tends to fit anomalies with a diameter of about 2 grid units
and consequently emphasizes relatively small features. On the other hand, formula
(7), which has positive coefficients in its central part and negative in its outer part,
gives a much broader waveform and tends to emphasize features with a diameter
of about 4 grid units.

The frequency characteristic of a filter also can be expressed in the two-
dimensional xy frequency field as contours of attenuation expressed in decibels
(Fuller, 1967). Figure 6-19 is one of many such charts in Fuller's paper. This
example corresponds to the weighting factors and grid distances of formula (7) of

Table 6-1. The horizontal and vertical scales are in cycles per data interval. The lowest numbers, toward the lower left corner of the diagram, correspond to long wavelengths; the limiting numbers (0.50) on the horizontal and vertical axes of the diagram and on the arc connecting them correspond to a wavelength of only two grid data intervals. This is the minimum which can be reliably expressed by the grid because of the Nyquist limit, at which aliasing begins for components with certain orientations.

On the spatial-frequency chart (Fig. 6-19) for formula (7) the greatest emphasis corresponds to the maximal zone bounded by the zero contour and indicated by the dashed line; the axis of this maximum is at coordinates of approximately 0.25 cycle per data interval, which is the reciprocal of the 4 grid units mentioned above. Hence, as we have noted, one can appreciate the effects of a filter in both the spatial frequency domain and the data domain itself.

Curves 1 (Fig. 6-18) (for the center-point-and-one-ring system) and 3 (for the Henderson and Zeitz formula with the first coefficient strongly negative) are similar and would produce similar maps. Curve $1W$ is the same as curve 1 but with the horizontal scale widened by $\sqrt{5}$. This gives its circle the same radius as the outer circle of curve 7, leading to a very similar curve; calculations produce similar maps. This similarity of maps is demonstrated by Fig. 6-9, calculated with formula (7), and Fig. 6-10, calculated with the center-point-and-one-ring system with radius expanded by $\sqrt{5}$.

Fuller includes diagrams similar to Fig. 6-19 for several of the grid operators in Table 6-1. By plotting profiles of the frequency characteristics from the spatial-frequency plots it is possible to get a feeling in the frequency domain for the nature of the various operators. The frequency-response curves (Fig. 6-20) are taken from the frequency-domain contours on profiles diagonally from the lower left to the upper right corner, as illustrated by Fig. 6-19. The profile on that particular diagonal line is shown by curve 7 of Fig. 6-20. Profiles from Fuller diagrams for formulas (1) and (3) are shown also. To compare these frequency diagrams in a manner similar to that used for comparing the second-derivative formulas of Table 6-1 they can be normalized by making the first coefficient equal to unity. This means that the numbers from the contours of the frequency diagrams must be raised or lowered by the decibel equivalent of the factor K of Table 6-1, which is determined from db $= 20 \log K$ (Fuller 1967; p. 662). Thus profile 7 of Fig. 6-20 can be normalized by increasing it by 2.8 db (the equivalent of the factor 0.710) to give the curve shown as $7n$. The curve for formula (1) (Fuller's fig. 15) which corresponds to a normalized second-derivative operator, is plotted as curve 1 of Fig. 6-20. To compare this with the normalized formula (7) the horizontal scale is divided by $\sqrt{5}$, accounting for the smaller numbers of cycles per data interval, to give the curve $1W$. Thus these two normalized curves, $7n$ and $1W$, are very close together, which is another way of expressing the similarity of the profiles for formulas $1W$ and (7), as shown by Fig. 6-18 and by the similarity of the maps, Figs. 6-9 and 6-10. Similarly the frequency-response profile for formula (3) is plotted as curve 3, Fig. 6-20. When it is normalized by dividing by the factor 6.00

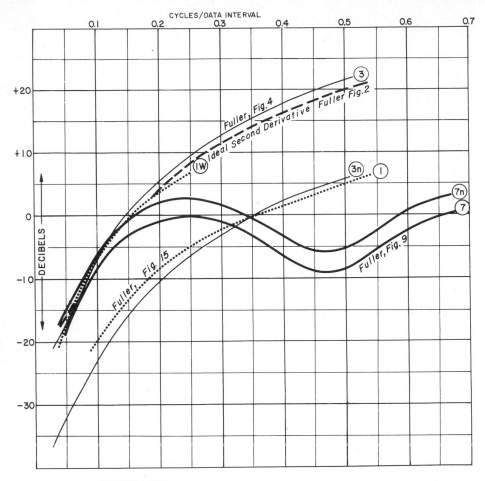

FIGURE 6-20
Frequency-response profiles from attenuation contours of Fuller charts (1967), on diagonal from lower left to upper right corners. Ordinates are attenuation in decibels. Abscissas are frequencies in cycles per data interval. Curve numbers, in circles, are the numbers from the normalized equations in Table 6-1; numbers with *n* are normalized by raising or lowering the decibel equivalent of the amplitude factor *K*, that is, fourth column of the table.

(equivalent to lowering the curve by 15.6 db), it is quite close to formula (1), which again corresponds with the similarity of the waveform expression of these two formulas as shown by Fig. 6-18.

Directional Sensitivity

Nearly all the Fuller frequency-response charts have some directional sensitivity. If they had none, the contours would be circles, as shown by Fuller's fig. 2, which is for an ideal second-derivative operator. The profile for this operator is included in Fig. 6-20. The directional variation can be relatively large. For instance, in Fig. 6-19, along the x and y axes, at 0.5 cycles per data interval, the attenuation is about -1 db. On the diagonal at the same distance it is about -9 db. This difference of 8 db corresponds to a factor of 2.5. This distortion disappears near the origin (less than about 0.15 cycle per data interval in Fig. 6-19), where the contours are nearly circular.

The distortion can affect the form of relatively small features which are in the outer part of the attenuation field, i.e., with more than about 0.25 cycle per data interval, where the attenuation along the diagonal is greater than along the x and y axes. The danger in this sort of distortion is not so much in its effect on a single anomaly as in its possibility of emphasizing or even creating nonexistent trends in the high-gain directions.

An example very likely influenced by such trend distortion is given in the study by Affleck (1963). He made statistical rosettes of trend directions from some 50,000 anomalies from more than 1 million square miles of aeromagnetic surveys. These were partly from observed maps but more from second-derivative maps, computed from Peters' equation (27), which corresponds to equation (8) of Table 6-1. Its frequency-response chart is given by Fuller's figure 6, and its general pattern is quite similar to that of Fig. 6-19. At about 0.4 cycle per data interval the attenuation is about 6 db greater along the diagonal than on the x and y axes. Nearly all Affleck's rosettes, in particular the one from all surveys, show strong spikes in north-south and east-west directions. Since the grid patterns used are nearly always oriented with the cardinal directions, a systematic emphasis of these directions by the calculating system could contribute to this statistical result, particularly since the effect is more dominant in the outer (high-frequency) part of the frequency-response diagram. Hence the implied result should be interpreted in the light of such effects.

Of course the consideration of profiles is not sufficient to understand wave-number filters which emphasize or suppress data components with specified directional properties. Here the full wave-number or spatial frequency domain is needed, and plots like Fig. 6-19 are useful for making comparisons.

The above illustrations have indicated generally the use of wave-number filters and their role in anomaly separation. By means of the Fourier transformation and diagrams of the wave-number or spatial frequency domain, the role of filters and their effects have been considered both in the spatial data domain and in the frequency domain.

GENERAL RELATIONS OF GRID OPERATORS TO WAVE-NUMBER FILTERS

In the spatial frequency domain the various grid operators approximating the second derivative can be approximated in much the same way wave-number filters were treated in Fig. 6-20. Meskó (1966) made such a comparative study, and his

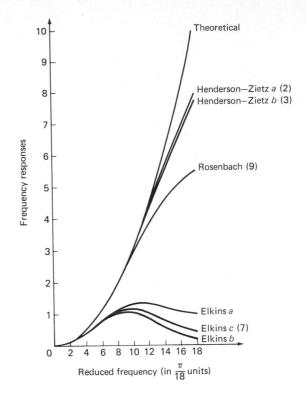

FIGURE 6-21
Frequency (also amplitude) response of some second-derivative formulas. The curves show the average behavior. The numbers on several of the curves correspond to the formula numbers of Table 6-1. (*From Meskó, 1966.*)

results are illustrated in Fig. 6-21. Note that the Elkins operators, which by experience have proved to be the most effective of those operators examined in the diagram, do not provide the theoretical derivative enhancement at the high spatial frequencies. Since common data errors and inaccuracies affect the higher frequencies most severely, this feature helps prevent the unwarranted exaggeration of noise. This same effect is shown in Fig. 6-20 by the flattening of curve 7 [Elkins, 1951, equation (15)] compared with the very high gain of the ideal second derivative at frequencies greater than about 0.25 cycle per data interval.

Meskó also recognized that the uniformly gridded operators must have certain directionality properties owing to their cartesian form of regular sampling. Figure 6-22 shows the difference of the Elkins formula spatial-frequency response as a function of direction with respect to the fundamental grid. This is another way of expressing the directional sensitivity shown by the Fuller diagrams, as discussed earlier.

The theoretical derivative amplification of the zero frequency component is zero, making the means value of derivative-filtered maps equal to zero. In the

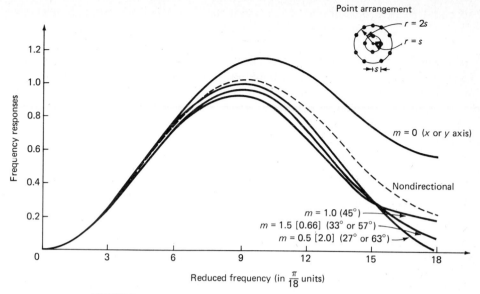

FIGURE 6-22
Curves intersected from the frequency-response surface of Elkins' (1951) formula
[formula (7) of Table 6-1] by planes at the angles shown for each. (*From Meskó, 1966.*)

space domain, as noted, positive and negative areas must balance. It is this feature
of the derivative filters which both limits their use in the quantitative separation of
anomalies and necessitates caution in ascertaining the reality of indicated anomalies. Fajklewicz (1965) studied the anomalies which resulted from the derivative-
filter method and warned that from two to four fictitious anomalies were produced
for each real anomaly.

Foundations of the more recent viewpoints toward continuation are given in
the works of Henderson (1966), Nagi (1967), and Roy (1967). Basically, the amplitude effect of the continuation operator in the spatial frequency domain is $e^{h\omega}$,
where ω is a spatial-frequency component in angular units and h is the continuation
distance which will in general represent some integral number of spatial sampling
intervals s. For downward continuation h will have a positive value, while for
upward continuation h will be negative.

We shall not examine in detail the assumptions and conditions attendant
upon the simple expression given for the continuation operator in the spatial frequency domain. Instead, it is worthwhile to contrast the continuation operation
with the derivative operation which has already been discussed. Downward continuation and the ideal derivative operator share the characteristic of very great
amplification of higher spatial-frequency components. (See the curve for the ideal
second-derivative operator in Fig. 6-20.) The amplification supplied by the ideal

second-derivative operator is proportional to ω^2, while for continuation an exponential gain dependent on $h\omega$ is indicated.

Clearly, then, both the derivative operator and downward continuation are most sensitive to data errors and require that the data-sampling grid be chosen with some care. Also as a practical matter, both require some curtailment of the high-frequency amplification to avoid amplifying noise components excessively.

Unlike the derivative operation, however, the continuation operation does not force the mean value of the data to zero, as the sum of the coefficients is unity, and so will not create false anomalies by this mechanism (note the amplification for ω^2 and $e^{h\omega}$ when $\omega = 0$). As a result, the data produced by a continuation operation are still in units of milligals and suited to quantitative methods of anomaly separation or source identification. The salt-dome cap-rock evaluation from downward-continuation calculations (page 145) is an example of this point.

ANALYTICAL METHODS FOR DETERMINATION OF THE REGIONAL

A quite different approach to anomaly separation is the application of surface fitting. Such mathematical methods can be used under the assumption that a mathematical trend surface describes the regional gravity. In this method, a least-squares fit or some equivalent operation is carried out to determine a potential field surface, which is fitted to the observed gravity map. The closeness of fit depends on the degree or order of the calculation.

The general idea of the surface-fitting technique is illustrated schematically by Fig. 6-23. The illustration is for a single profile line only, whereas the operation is carried out in two dimensions to fit a surface rather than the single line. Observed gravity is represented by curve G. A first-degree fit would be a straight line, line 1. A second-degree fit could have one reversal and two line crossings, as indicated by line 2. A third-degree curve could have two reversals and three line crossings, as shown by line 3, and a fourth-degree curve could have three reversals and four line crossings, as shown by line 4. In each of these cases the calculated curve is fitted so that the sum of the squares of the differences between it and the observed curve is a minimum. The differences between the calculated and observed values are the residuals. As higher orders are calculated, the fit between the calculated and observed curves becomes closer and the residuals become smaller. A series of maps can be prepared for different degrees of fit. At higher degrees, the smaller and sharper residuals emphasize smaller details of the original map. It is possible also, if desirable, to map the difference between a higher- and a lower-order fit. By using different combinations, a great variety of maps can be made, depending on the use desired and the nature of the features chosen for emphasis.

The rather involved mathematical procedure of surface fitting has become a routine operation only with the application of high-speed digital computers. With the computer it is possible to carry out the entire operation, including the contouring of the maps, quite rapidly. The initial input can be the horizontal coordinates and gravity values at the individual stations on a set of punched IBM cards.

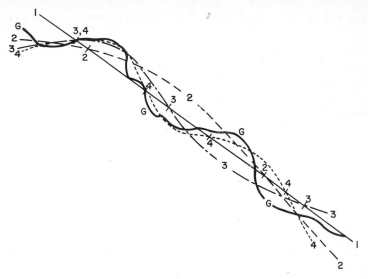

FIGURE 6-23
Schematic curves illustrating least-squares surface-fit technique. Curve *G* represents an observed gravity profile. Curves 1, 2, 3, and 4 represent fits of successively higher orders. The residual for a given order is the difference of the observed from the corresponding surface fit.

Before reviewing some examples which illustrate more practical elements of the trend-surface filtering methods, it should be helpful to examine in greater generality such methods themselves. If we knew of a simple mathematical function which was similar in form to the smooth regional contribution of a set of gravity values, we could remove its effect quite simply. By assuming that trend functions can be approximated well by polynomials and polynomial surfaces of low order, we are appealing to rather general theorems of mathematics which indicate that well-behaved functions can be represented by polynomials. Since the polynomial is recognized as an approximation, some measure of the goodness of approximation or fit is needed, and this is most often taken to be the residual sum of squares.

Skeels (1967) pointed out one very obvious self-defeating feature which is inherent in the procedure described. If the data comprise both trend and anomalies, all fitting methods for trend identification will be in some error due to the presence of the anomalies. His suggestion of deleting from the computations those portions of the data easily recognized as being influenced by anomalies is of course a compromise between the extreme poles of view taken by the gridders and smoothers.

Despite this difficulty and others which we shall note, the various methods of surface fitting have proved of value. Where the least-squares criterion of goodness of fit has been employed, the trend line or surface is usually represented by either orthogonal functions or nonorthogonal functions. The distinction is an interesting one and is best explained by illustration. The usual polynomial fits, as described for instance by Coons, Mack, and Strange (1964), use ordinary nonorthogonal terms. Here, the coefficient belonging to any low-order term is successively modified as higher-order fits are made. By contrast, the coefficients of the low-order terms remain unchanged as one progresses to higher-order fits if the terms are orthogonal. Grant and West (1957) and Van Voorhis and Davis (1964) described surface-fitting approaches to removing trends from gravity data which follow these ideas.

Among the common least-squares methods using orthogonal components is the Fourier fit or Fourier analysis. If for a particular set of data the trend function appears to be better represented by a sum of corrugated surfaces, as shown in Fig. 6-16, than by algebraic polynomials, the Fourier fit should be used. The Fourier residual maps are of course analogous to the polynomial ones. Following the influential work of Bullard, Hill, and Mason (1962), Fourier fits have been widely used for trend removal in magnetic surveys. This is particularly true for marine surveys, where the influence of sea-floor spreading and reversals of the earth's field have imprinted a cyclic trend. In recent times, however, the Fourier approaches are finding increasing application for use with gravity-survey data (see Odegard and Berg, 1965).

Use of other criteria for measuring the goodness of fit apart from least squares has not received much attention, though some is clearly warranted. In fact, it is easy to show, using the convolution theorem and Plancherel's theorem of the equivalence of least-squares fits in the spatial data domain and Fourier-transform domain, that for deeper source distributions the least-squares criterion simply responds only to source parameters as they affect the observation mean values (see the appendix to this chapter, page 186). Other criteria deserving of consideration are the sum of residual absolute values and the minimization of the maximum residual.

Our discussions in the main have skirted around the issue of data sampling; i.e., regularly sampled or gridded surveys have been assumed and most explicitly for the considerations of wave-number filters. Theoretical studies of the role of nonuniform sampling are far less definitive than those which relate to regular sampling. Hence, as a practical matter, the gridder faced with a nonuniform data survey appeals to his intuition much as the smoother does to determine an interpolation method for producing a uniform grid for subsequent analysis. Commonly, this amounts to simple writing of values at a regular grid from a contour map constructed from the original data on a random pattern of points or of lines, as discussed above (page 139).

There is much debate about, and also much misunderstanding of, the role which interpolation methods should play in forming optimum grids from irregularly sampled surveys. Some portion of these difficulties relates to early "analytic" approaches discussed in a previous section, where the interpolation problem was not separated from the techniques for analysis. In such methods one or more rings were centered at a sample value to be reduced, and the reduction scheme employed averages from observations in the various bands defined by the rings. Isolation of the analysis method from the interpolation mechanism represented genuine progress; however, the entire subject has been treated inadequately in the literature. As a general statement, those interpolation methods conforming most closely to the physical equations governing the data will perform best.

EXAMPLE OF TREND-SURFACE ANALYSIS BY LEAST SQUARES

Turning now to an illustration of the surface-fitting method, we shall look at the work of Coons, Woollard, and Hershey (1967), who investigated the Mid-Continent Gravity High. This is one of the strongest gravity anomalies in the United States, with very sharp relief and with amplitudes of well over 100 mgals. It is not a good example from the standpoint of petroleum exploration but serves quite well to illustrate the application of surface fitting.

Figures 6-24 and 6-25 illustrate the process as carried out over a portion of the survey in north-central Iowa. In Fig. 6-24 the set of four maps shows the original Bouguer anomaly map and the surface fits calculated for the seventh, tenth, and thirteenth orders. It will be noted that as the order becomes higher, the complexity of the pattern increases. The axis of the central maximum, as drawn on the Bouguer anomaly map, is repeated, for comparison, on each of the surface-fit maps. With the higher orders, the details of the maps conform more closely with this axis.

The four maps of Fig. 6-25 show the Bouguer anomaly map repeated and the residuals for the three different orders. These residuals are the differences between the surfaces shown in Fig. 6-24 and the Bouguer anomaly map. It will be noted that as the order of fit increases, the complexity of the residual increases but the total relief of anomalies decreases. The seventh-order residual is composed largely of the central maximum and its flanking minima with relatively little detail within these features. In the thirteenth-order residual, there is a great deal of internal detail, and many local features are brought out which are not apparent in the seventh-order map. The anomalies marked *A* and *B* on the thirteenth-order map are local minima which develop from quite inconspicuous saddles on the Bouguer map.

Profiles from diagonal lines across the maps of Figs. 6-24 and 6-25 are shown by Fig. 6-26. Each set shows the Bouguer gravity (the same for all three sets), the fit profile (from Fig. 6-24), and the residual profile (from Fig. 6-25). There is a tendency for the system to introduce residual periodic features, as the wavelength of the order attempts but does not completely succeed in fitting the observed gravity. On the thirteenth-order residual, the local features 1 and 2 are of this nature, where the fit does not completely match the flanks of the large central anomaly.

Since the surface-fit technique is a least-squares process, the fitting surface, which is the regional, is a surface which will have both positive and negative differences from the observed; the residual maps are therefore balanced between positive and negative areas. This is similar in character to the calculation of second-

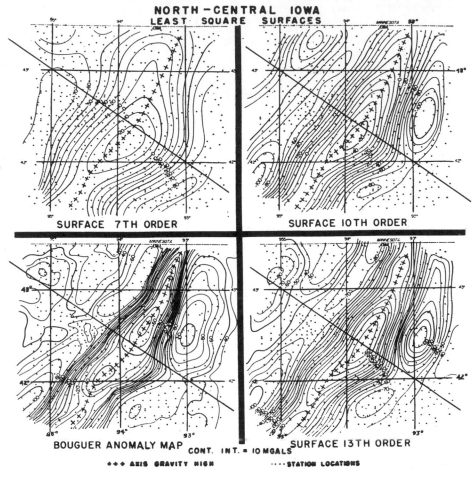

FIGURE 6-24
Example of least-squares surface fit; lower left map is Bouguer gravity. Other maps show gravity calculated with successively higher orders of fit. (*From Coons, Woollard, and Hershey, 1967.*)

derivative or grid residual maps with central maxima and flanking minima as shown by Fig. 6-6. For this reason, the residual anomalies from the surface-fitting techniques cannot be used directly for quantitative geological interpretation by calculating gravity effects to compare with the observed anomaly.

As shown by this particular example, the system is quite effective in picking out local anomalies. It is not completely objective because the results depend to a considerable degree on the operator's judgment and discretion in deciding which size or type of anomalies to emphasize and the corresponding "order" of the operation. Since the process is so highly automated, it is largely computer time

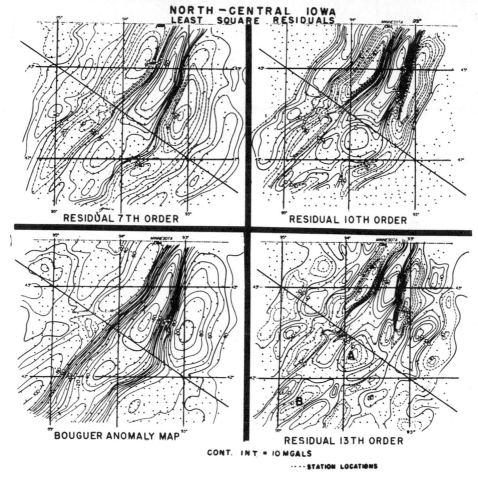

FIGURE 6-25
Example of least-squares residual; lower left is observed gravity. Others are residuals calculated by subtracting the successively higher orders of fit of Fig. 6-24 from observed gravity. (*From Coons, Woollard, and Hershey, 1967.*)

and the efficiency of the program which determine how many and which maps to produce and use. In this way the system is quite comparable with the second-derivative calculation in that, for those applications also, the general nature of the resulting maps depends very greatly on a human judgment of the grid spacing and the weighting of coefficients used. Second-derivative maps calculated with a suitable choice of grid spacing and coefficients would give results very similar to those shown by the least-squares residuals of an order corresponding to a similar size of unit area and would have almost exactly the same centers of maximal and minimal closures.

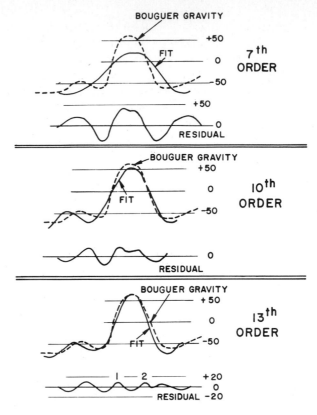

FIGURE 6-26
Profiles from Figs. 6-24 and 6-25, showing increasing closeness of fit and smaller residuals with higher degrees of least-squares surface-fitting calculations.

By an appropriate choice of the order of the residual it is possible, by the least-square fit technique, to isolate or emphasize anomalies of almost any magnitude desired within the limits of the definition by the spacing and numerical accuracy of the original data. If location and approximate amplitude of anomalies is the principal objective of the operation, the system can be very effective. If quantitative evaluation of particular anomalies by calculation from geological interpretations or assumptions is to be carried out, it is necessary to go back to the original Bouguer gravity map and perform some sort of graphical regional-residual operation to determine more precisely the magnitude and form of the isolated anomaly which is to be accounted for. The maps from this surface-fitting technique may be very helpful in choosing the particular anomalies to analyze and also in providing some control on the choice of a regional as this may be approximately indicated by a low-order surface fit.

The least-squares fit method has been widely referenced in the literature. In the early development of analytic techniques for defining the regional, polynomials were sometimes used because they could be readily applied using hand calculations (but are now done largely by computer). However, as the example shows, they do not often represent regional geology in a satisfactory way. Analytical methods that use some information related to regional geology are more satisfactory.

In our consideration of continuation filters, some comparison was made with the derivative filters. It was noted that the results of processing with continuation filters could be used for quantitative interpretation methods. As a parallel, it is likely that for the surface-fitting method, more quantitatively meaningful results might be obtained if other criteria for the goodness of fit of the trend surface were substituted for the least-squares sum.

STATISTICAL APPROACH TO WAVE-NUMBER ANALYSIS

The problem of separating anomalies from trend or background using filtering techniques is complicated by the continuous nature of both the anomalies and the trend in the frequency domain. This point was emphasized in our general discussion of wave-number filters. Hence while the application of a wave-number filter can be an automated process, the burden of designing a geologically meaningful filter remains. This problem, though couched in mathematical terms, is precisely the same as the one facing the smoother who is using graphical methods.

In general, the applications of wavelength filters to potential-field data are predicated on having sufficient data for the statistical properties of the data to be reliably sampled. This is particularly relevant in the application of spectral methods. For example, the "phaseless" technique of Spector and Grant (1970), which defines the regional and residual in the frequency domain, requires about 10 residual anomalies to be included in the map for the amplitude spectrum to become reliable.

In advanced filter theory a class of techniques seeks to separate signal from noise by utilizing statistical properties which are known *a priori* and which characterize signal and noise respectively. By analogy, the statistical approaches to wave-number filters for anomaly separation equate the anomalies to signal and the trend function to noise of predominantly low frequency. The intent of drawing such unusual parallels is to make the transfer of techniques from filter theory to anomaly separation possible. By such methods one seeks an objective means for separating anomalies from trends wherein the physical realities of the data are entered through statistical descriptors.

The work of Strakhov (1964) and its extensions by Naidu (1966, 1967) are precisely of this nature. Also, the work of Spector and Grant (1970) incorporates elements of this thinking. Figure 6-27 illustrates the objectives of the statistical filtering approaches using a simple spatial-frequency component. In essence, a statistical basis is sought for accomplishing the indicated separation as a counterpart to the intuitive separation accomplished by the smoother.

Strakhov's particular vehicle is the Wiener filter, which is nothing more than an extension to the least-squares method for continuous linear models. In fact,

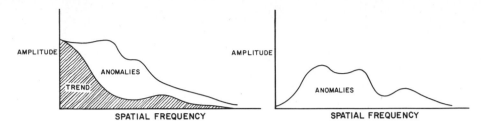

FIGURE 6-27
Idealized complete separation of trend from anomalies in the wave-number domain. Amplitude spectral distinctions accomplished on a statistical basis.

the discrete approximations needed for implementing the theory turn out to be a least-squares solution. The statistical fundamentals of this approach are subject to some question, and the reader is encouraged to satisfy his own intuitions in regard to pursuing such methods.

The basic philosophy of a statistical viewpoint toward trend-surface analysis is summarized also by Fig. 6-27, except that the vehicle for obtaining the separation is the use of a trend surface rather than a wave-number filter. Most commonly the statistical properties are summarized by means of the autocovariance function or its spatial Fourier counterpart, the power spectrum. Much of the work in this area has been concerned with magnetic data (see Horton, Hempkins, and Hoffman, 1964; Gudmundsson, 1966, 1967); however, it has applicability to gravity data as well.*

If we have a trend surface and a superimposed collection of anomalies, then if information about their respective statistics is available, it can be utilized for assisting the separation of the two components. Specific examples of this procedure using surface fitting are not readily available in the literature; however, studies which examine the statistical information have appeared.

The spatial autocovariance or normalized autocorrelation function is nothing more than a simple, multiplicative comparison of a profile or map against itself using also displaced positions. A Fourier transform of this statistical measure gives a spatial power spectrum, and we have already encountered this quantity in discussing wave-number filters (page 162).

Figures 6-28 and 6-29 show an aeromagnetic map and the two-dimensional autocovariance function. The preponderance of anomalies aligning with an almost north-south direction is especially clear from the statistical information. Statistics of trend surfaces could be quantitatively compared with the map statistics to isolate the anomaly contributions. There is, of course, much development work of this type which may be profitably performed.

* Airborne magnetic data which are from observations at elevations of 1000 to several thousand feet above the ground and which are continuous are relatively much less disturbed by local influences than gravity data. For that reason it is usually practical to use operators with more nearly ideal characteristics which would tend to overemphasize the high-frequency components of gravity maps or respond to high-frequency noise.

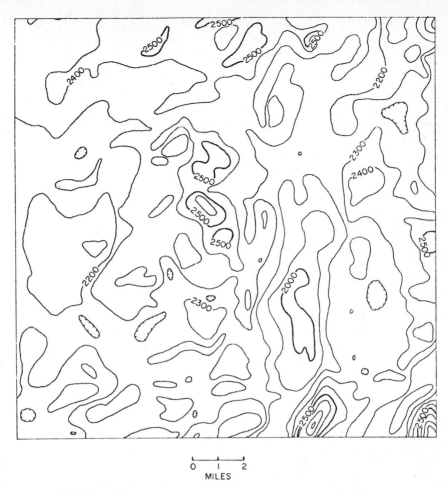

FIGURE 6-28
Aeromagnetic-data contour map. (*After Horton, Hempkins, and Hoffman, 1964.*)

Nonlinear and Inverse Modeling Approaches

Experience with digital computing has evolved approaches toward anomaly separation far more ambitious both in terms of objectives and computing requirements than either the trend-surface fitting techniques or linear wave-number filters. Two mainstreams of development may be identified, one an outgrowth of filter theory and the other drawing its origins from surface fitting.

The work of Naudy and Dreyer (1968) illustrates the progress in applications of nonlinear filtering. A nonlinear filter is best described by its effects. For example, a linear filter applied to the corrugated surface in Fig. 6-16 can change only its orientation, position with respect to the coordinate origin, and amplitude. A

FIGURE 6-29
Autocovariance (normalized autocorrelation function) of aeromagnetic data. (*After Horton, Hempkins, and Hoffman, 1964.*)

nonlinear filter could further change its fundamental frequency and in fact even alter the sinusoidal cross section so that it consists of many spatial frequencies in superposition. Further, a linear filter applied to anomaly and trend together will give the same result as summing the results of applying that filter to the anomaly and trend taken separately. A nonlinear filtering operation does not necessarily have this commutative property and can give quite different outputs for the two circumstances described.

Very often the uses for nonlinear filters take advantage of certain of their special properties. In the anomaly-separation problem they can emulate in a quantitative manner the technique employed qualitatively by the smoothers. As an example, if a map of gravity data is reduced and crude anomalies formed by some simplified trend removal, additional enhancement of the anomalies can be provided using a nonlinear approach. A prerequisite for this technique is that the contribution of the anomaly on the average at any data point exceed the contribution of the unremoved trend.

Suppose that on the average at each data point the contribution of the anomalies is N times greater than the contribution of unremoved trend, where N is greater than unity. If the amplitude at each data point is subjected to the simple nonlinear operation of squaring, the anomalous contributions receive an amplification of N^2 in contrast to unit amplification for the unremoved trend. There is of course a cross-product term involving both anomalies and trend; however, for values of $N = 5$ or greater, which are rather easily attainable on the average, the selective amplification is quite large. Of course the alteration of the data then makes them require special consideration if subsequent quantitative computations are to be performed, to take into account the effects of the distortion.

Modeling and Iterative Systems

The second approach to the separation problem using nonlinear methods has much in common with the fitting of surfaces. If instead of modeling the trend with some general functional representation, the entire anomaly-causing mass distribution could be modeled, the trend portion and more localized density-contrast elements of the survey could be identified from their very origins. This objective is indeed ambitious, and so, for practical purposes, the applicable model is somewhat simplified and might be taken as a mass distribution on a plane at depth, or a contrasting body of polygonal cross section, or else a body which can be represented in good approximation by the sum of a distribution of rectangular blocks standing above some reference plane at depth.

Bott (1967), Tanner (1967), Cordell and Henderson (1968), Dampney (1969), and Kunaratnam (1972) explore these approaches quite extensively. The fit of values calculated from a model to observed (or residual) data is involved. As before, questions about the criterion for goodness of fit become applicable. Limitations on the interpretive and quantitative applications are no longer of such great moment because the meaning of the models as the very source of the survey data makes their use very straightforward.

Both magnetic and gravity data are accommodated by these approaches, and much of this work has been concerned with essentially two-dimensional or cross-sectional studies. While the authors frequently refer to the data they use as anomalies, it should be recognized that source components will be computed for such unremoved trend as is present in the data and will appear in the resulting model.

Tanner calls his approach "automated interpretation" and iterates on a linearization of the subsurface model. The resolution of his model depends on the density of observations. He can compute more blocks and consequently describe the subsurface better if there are more data points. His method is two-dimensional but can be extended to the third dimension. Figure 6-30 is a test of Tanner's method using a theoretical model. Such tests are often misleading owing to the unusual, noise-free conditions and great precision of the theoretical input data.

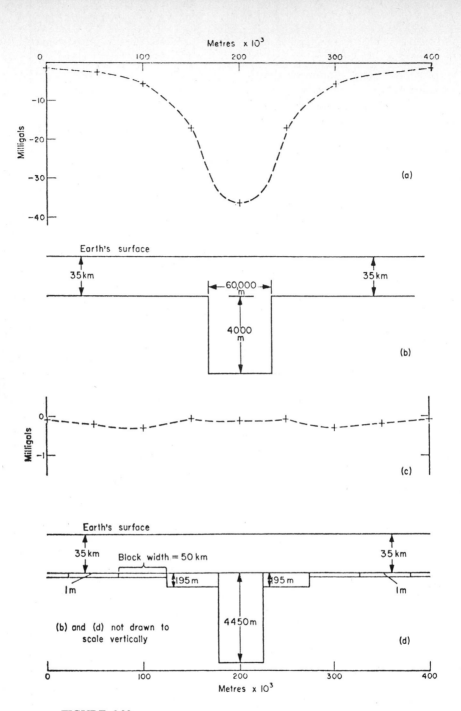

FIGURE 6-30
Accuracy test for model recovery using Tanner's nonlinear method of automatic interpretation. The structure assumed is shown in (b) and the anomaly it produces in (a). The model recovered is shown in (d) and the remaining residuals in (c). (*From Tanner, 1967.*)

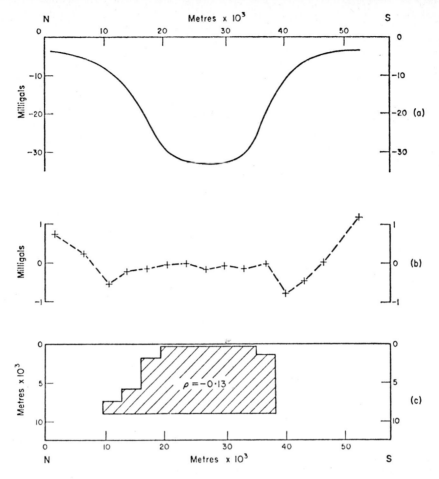

FIGURE 6-31
Interpretation of Weardale granite (northeastern England) by Tanner method. The model has an assumed extension of 10,000 m in either direction normal to the profile. Anomaly is given by (a), residuals after application of model in (b), and computed model in (c). (*From Tanner, 1967.*)

Nevertheless, such a test serves to indicate in some quantitative measure the degree to which the source may be modeled.

In Figure 6-31 the Weardale granite is modeled from a gravity profile. Note that the errors in fitting the model are largest at the ends of the profile. This is physically reasonable because such data points are most likely to be affected by source density contrasts not included in the model. Cordell and Henderson (1968) carried a similar approach to a three-dimensional model, and one of their results is shown in Fig. 6-32. Note that a reference plane for the model building was selected at a depth 8 km below the plane of reduced data.

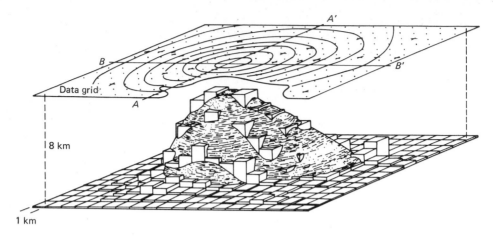

FIGURE 6-32
Three-dimensional nonlinear modeling. Gassman model is recovered after 20
iterations based on 44 data control values. (*After Cordell and Henderson, 1968.*)

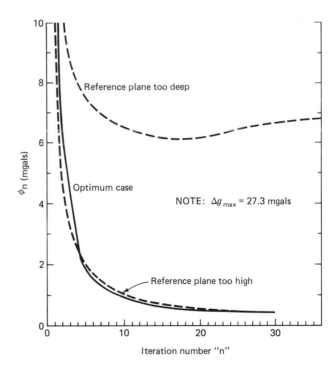

FIGURE 6-33
Plot of fit criterion against iteration for spherical model. (*From Cordell and
Henderson, 1968.*)

These authors did consider the model-fitting criterion and concluded that the largest absolute error was a better practical standard than an rms value (related to sum of squares). Figure 6-33 shows the convergence of the iterative process for a variety of situations. If the reference plane is selected at an excessive depth, the convergence appears to be very poor.

An example of an application of a three-dimensional iterative calculation to the interpretation of a salt-dome gravity anomaly is given on page 266.

Dampney (1969) in determining an equivalent source distribution on a plane at depth was less demanding in the detail of information required. In this case the nonlinear technique was employed to produce a more accurate reduction scheme so that techniques such as continuation could give more reliable results.

It must be kept in mind that all these modeling methods require density information as part of the input, along with some depth parameter. The resulting model can vary widely, depending on these factors, and still give a calculated curve or map in close agreement with the data to be fitted. (See page 447 and Fig. 16-7.)

It should be clear from results like those of Fig. 6-32 and the salt-dome example that nonlinear and inverse modeling approaches offer a potentially fruitful mechanism for quantitative separation of anomalies, both from trends and from each other.

APPENDIX TO CHAPTER 6

THE LEAST-SQUARES NORM AND DEEP POTENTIAL SOURCES*

A sum of squares is often selected as the criterion for establishing the performance of an analysis without consideration of any alternate performance measures and even without regard to its suitability for the particular problem. This discussion will demonstrate that for mass distributions on a plane at depth, least-squares modeling is a poor technique which tends to give accurate results only for the distribution mean value.

Let us take a planar mass distribution which is uniform in one variable. If such a distribution is $m(x,\alpha_i)$, where the constants α_i express the details of mass variation in the coordinate x, the vertical component of the anomalous gravity field along x at an elevation h above the plane will be

$$g(x,\alpha_i) = \int_{-\infty}^{\infty} \frac{Kh}{\gamma^2 + h^2} m(x - \gamma, \alpha_i) \, d\gamma \tag{1}$$

where K is an appropriate constant and, as stated in our profile, parallels the direction of variation of the mass distribution. The integral (1) is developed by integrating the effects of horizontal line elements along the x axis to synthesize the planar distribution. Recall that for a single line element of mass related to dm the vertical component of gravitational force is

$$\frac{Kh \, dm}{\gamma^2 + h^2}$$

Fortuitously (1) is also of the form of a convolution; hence in the Fourier frequency domain it has the equivalent expression as a product of transforms, or

$$G(\omega,\alpha_i) = \pi K e^{-h|\omega|} M(\omega,\alpha_i) \tag{2}$$

where ω = angular spatial-frequency variable
  $G(\omega,\alpha_i)$ = Fourier transform of $g(x,\alpha_i)$
  $M(\omega,\alpha_i)$ = Fourier transform of $m(x,\alpha_i)$

For large values of the source-plane depth h, the function $e^{-h|\omega|}$ tends rapidly toward the character of a Dirac delta function for a sampled system, i.e., a simple spike of near infinitesimal width. We shall thus write $e^{-h|\omega|}$ as $\delta(\omega)$ and make use of its approximate properties accordingly in the continuous case.

If gravity observations $g_0(x)$ have been made along the profile to be used for modeling, the least-squares fitting criterion is simply the minimization of

$$s = \int_{-\infty}^{\infty} [g(x,\alpha_i) - g_0(x)]^2 \, dx \tag{3}$$

* This appendix was written by Norman Neidell.

Plancherel's theorem gives us an equivalent expression written in terms of frequency-domain quantities, or

$$s = \int_{-\infty}^{\infty} [G(\omega,\alpha_i) - G_0(\omega)][G(\omega,\alpha_i) - G_0(\omega)]^+ \, d\omega \qquad (4)$$

where $+$ denotes the complex conjugate. Now substituting (2) into (4) and using the delta-function-like properties of $e^{-h|\omega|}$ gives

$$
\begin{aligned}
s = \int_{-\infty}^{\infty} &\{\pi^2 K^2 \delta(\omega)\delta(\omega)M(\omega,\alpha_i)M^+(\omega,\alpha_i) \\
&- \pi K \delta(\omega)[G_0(\omega)M^+(\omega,\alpha_i) + G_0^+(\omega)M(\omega,\alpha_i)] \\
&+ G_0(\omega)G_0^+(\omega)\} \, d\omega \\
\approx \ &\pi^2 K^2 M(0,\alpha_i)M^+(0,\alpha_i) - \pi K[G_0(0)M^+(0,\alpha_i) + G_0^+(0)M(0,\alpha_i)] \\
&+ \int_{-\infty}^{\infty} G_0(\omega)G^+(\omega) \, d\omega \\
\approx \ &\pi^2 K^2 M^2(0,\alpha_i) - 2\pi K G_0(0)M(0,\alpha_i) + \int_{-\infty}^{\infty} G_0(\omega)G^+(\omega) \, d\omega \qquad (5)
\end{aligned}
$$

Note that the real nature of the zero-frequency component has been used.

The normal equations to recover least-squares values for h and α_i are

$$\frac{ds}{dh} = 0 \qquad (6)$$

$$\frac{ds}{d\alpha_i} = 0 \qquad (7)$$

It is clear that (6) has no meaning because h does not appear (at least to the order of our approximation) in the least-squares sum. Hence virtually no information is available about the source-plane depth. Now, from equation (7) we have

$$M(0,\alpha_i) = \frac{G_0(0)}{\pi K} \qquad (8)$$

Equation (8) tells us only that we have knowledge of the parameters α_i, which specify the mass distribution only insofar as they affect the mean-value observations.

The lesson of this exercise should be clear. Other criteria for fitting the data should receive consideration. The example used by Cordell and Henderson to compute the results given in Fig. 6-33 further verify our results. Recall the failure of their method if a reference plane at too great a depth was selected (Fig. 6-33); also, they concluded that other error measures led to results superior to those obtained from the least-squares method.

7

CALCULATION OF GRAVITY EFFECTS

GRAVITY EFFECTS OF GEOMETRIC FORMS

The computation of the gravity effects of some simple geometrical models can be very useful in quantitative interpretation. The models may not be geologically realistic, but usually approximate equivalence is sufficient to determine whether the form and magnitude of the calculated gravity effects are close enough to those observed to make the geological postulate reasonable. If little or no factual information other than the gravity data is available, the comparison of the effects of such simple models with those observed may give as close a fit to the observed data as is justified by other uncertainties of the problem. Further interpretation then becomes a matter of substituting a reasonable equivalent geological configuration for the too regular ideal geometric form.

It is convenient to reduce the geometric part of the formulas for simple geometric bodies to dimensionless expressions in terms of the ratio x/z, where x is usually the horizontal distance from the center of the body to the point at which the effect is to be calculated and z is a depth parameter of the body. A curve is then plotted for this geometric term and is applicable to any body of the form under consideration. The other factors are assembled in a numerical coefficient which is

188

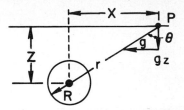

FIGURE 7-1
Notation for gravity effects of sphere and
horizontal cylinder.

determined by the dimensions and density contrast of the particular body. The
calculated effect at a given point is the product of a value read from the general
curve multiplied by the coefficient for the particular body. For convenience, the
numerical coefficients and conversion factors are combined so that the calculated
effects are in milligals for dimensions in 1000-ft (kilofeet, kft) units and kilometers.
Several of these form curves and some other tools are included in an appendix at
the end of this chapter, as Figs. 7A-1 to 7A-11.

The Sphere

The sphere, the simplest geometric form for which gravity effects can be calculated,
will serve to illustrate the methods which are applied to the various forms considered. Therefore the mathematical steps are shown in somewhat greater detail
than for the other forms. Masses in nature are never very nearly spherical in form,
but in any case where the horizontal dimensions of the body are less than the depth
to the center of mass, the approximation as a sphere will give surprisingly accurate
results.

The gravity effect of a homogeneous sphere is exactly the same as it would be
if its entire mass were concentrated at a point at its center. Therefore for the sphere
(Fig. 7-1) of mass m at the point P at a distance r from its center, $g = Gm/r^2$,
where G is the gravitational constant. This, however, is directed toward the center
of the sphere.

We are always interested in the vertical component g_z, which will be

$$g_z = g \cos \theta = g \frac{z}{r} = \frac{Gmz}{r^3}$$

If R is the radius of the sphere and σ its density (or density contrast), its mass
anomaly is

$$m = \tfrac{4}{3}\pi R^3 \sigma$$

and
$$g_z = \frac{G4\pi R^3 \sigma z}{3r^3} = \frac{4\pi G\sigma R^3}{3} \frac{z}{(z^2 + x^2)^{3/2}}$$

To reduce the geometric term to a dimensionless form in terms of x/z we
divide through by z^3 and

$$g_z = \frac{4\pi G\sigma R^3}{3} \frac{z}{z^3(1 + x^2/z^2)^{3/2}} = \frac{4\pi G\sigma(R^3/z^2)}{3} \frac{1}{(1 + x^2/z^2)^{3/2}}$$

This gives g in gals when r, R, x, and z are in centimeters and σ is in grams per cubic centimeter. If we change the units to express g in milligals, express r, R, x, and z in kilo feet, and include the values of the numerical constants (π and $G =$ 6.6732, page 12), the equation becomes

$$g_z = 8.520\sigma \frac{R^3}{z^2} f_1 \frac{x}{z}$$

$$= K_1 f_1 \frac{x}{z}$$

where $K_1 = 8.520\sigma R^3/z^2$ is the particular constant for the sphere being calculated and gives the gravity value over its center and $f_1(x/z)$, the geometric or form curve for all spheres, is given by curve 1 of Fig. 7A-1. If units are kilometers,

$$K_1' = 27.94\sigma \frac{R^3}{z^2}$$

As an example, suppose we wish to calculate the gravity profile for a sphere with a radius R of 3000 ft, depth to center of 5000 ft, and density contrast of 0.25 gm/cm³. Then the value of K_1 (remember that R and z are in kilofeet) is

$$K_1 = \frac{8.52 \times 0.25 \times 27}{25} = 2.30$$

Over the center of the sphere, where $x = 0$ and $f_1(x/z) = 1$, its gravity effect is 2.30 mgals. Gravity effects at other distances can then be written down by inspection of curve 1 of Fig. 7A-1, as illustrated for a few points by Table 7-1.

As another example, the gravity effect at the point over the center of a sphere with a depth to the center of 1 mile, a radius of 1000 ft, and a density contrast $\sigma = 0.25$ is 0.0765 mgal. The total excess mass that this body represents is 33 million tons. The comparatively small effect that this enormous mass produces at the surface shows why gravity measurements must be made to the order of 0.1 mgal or better to be sufficiently accurate for the demands of geophysical exploration.

Table 7-1

Distance x, ft	$\dfrac{x}{z}$	$f\left(\dfrac{x}{z}\right)$	Gravity, mgals
0	0	1	2.30
2500	0.50	0.71	1.63
5000	1	0.35	0.81
10,000	2	0.089	0.21

We can estimate depths from the form of the anomaly curve by calculating the half-amplitude width $x_{1/2}$. This is the value of x/z for which the value of g is half its maximum. Thus, for the sphere, we set

$$\frac{1}{(1 + x^2/z^2)^{3/2}} = \frac{1}{2}$$

and solve for x/z (or by inspection of curve 1 of Fig. 7A-1) and find

$$x_{1/2} = 0.766z \quad \text{and} \quad z = 1.305x_{1/2}$$

It is usually convenient to measure the double half-width, i.e., the horizontal distance between the two sides of the curve where the value is half the maximum ($2x_{1/2}$), from which $z = 0.652(2x_{1/2})$.

We can now use this value of z to calculate the radius of the spherical mass approximation. From the relation $g_0 = 8.52\sigma r^3/z^2$ (for dimensions in kilofeet) we have

$$R = \sqrt[3]{\frac{z^2 g_0}{8.52\sigma}}$$

Then the depth to the top of the sphere is $T = z - R$. A chart for the relations between R, z, and g_0 is given in Fig. 7A-2. For calculating R it is necessary to assume (or guess) a value for σ. This is not very critical because the cube root greatly reduces the range. For example, a change in density by a factor of 2 produces a change of about 25 percent in the calculated radius.

The Horizontal Cylinder

Another useful simple geometric form is the horizontal line element or horizontal cylinder, which is roughly equivalent to structures that are several times longer in one horizontal dimension than the other. The gravity effect of a homogeneous cylinder is the same as if the mass were concentrated on a line along its axis.

The vertical component of gravity from a horizontal line element of infinite length can be shown by integration of the effect for a point mass to be

$$g_z = \frac{2Gmz}{r^2}$$

where m is now the mass per unit length. If we replace the horizontal line element by a cylinder of density σ, the mass per unit length is

$$m = \pi R^2 \sigma$$

and

$$g_z = 2\pi G\sigma R^2 \frac{z}{r^2} = 2\pi G\sigma R^2 \frac{z}{x^2 + z^2}$$

Expressing the above equation in terms of the ratio x/z, we have

$$g_z = \frac{2\pi G\sigma R^2}{z} \frac{1}{(x/z)^2 + 1}$$

Substituting numerical values for the various constants gives

$$g_z = \left(12.78\,\frac{\sigma R^2}{z}\right)\frac{1}{(x/z)^2 + 1} = K_2 f_2\,\frac{x}{z}$$

where

$$K_2 = 12.78\,\frac{\sigma R^2}{z} \qquad \text{for distances in kilofeet}$$

$$K_2' = 41.93\sigma\,\frac{R^2}{z} \qquad \text{for distances in kilometers}$$

and

$$f_2\,\frac{x}{z} = \frac{1}{x^2/z^2 + 1}$$

which is plotted as curve 2 of Fig. 7A-1.

This curve can be used for calculation in the same way as that for the sphere. Suppose, for instance, that we take a cylinder with the same depth (5000 ft), radius (3000 ft), and density contrast (0.25) as the sphere used above. Then

$$K_2 = \frac{12.78 \times 0.25 \times 9}{5} = 5.75$$

Gravity effects at a few distances are given in Table 7-2.

We can estimate depths in the same way as discussed above for the sphere, for which we have

$$x_{1/2} = z \qquad z = \tfrac{1}{2}(2x_{1/2}) \qquad R = \sqrt{\frac{zg_0}{12.78\sigma}}$$

A chart for relations between R, z, and g_0 for horizontal cylinders, with $\sigma = 0.25$, is given in Fig. 7A-3.

Here, also, the square-root term reduces the sensitivity of the value of R to uncertainty in σ, but not as much as for the sphere. A change in σ by a factor of 2 makes a change in R by a factor of 1.4.

Table 7-2

Distance x, ft	$\dfrac{x}{z}$	$f_2\left(\dfrac{x}{z}\right)$	g, mgals
0	0	1	5.75
2500	0.50	0.80	4.60
5000	1	0.50	2.87
10,000	2	0.20	1.15
15,000	3	0.10	0.57

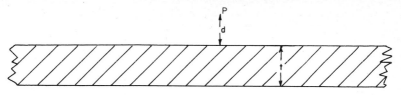

FIGURE 7-2
Notation for gravity effect of infinite horizontal sheet.

The Horizontal Sheet

For a sheet of infinite horizontal extent, depth d, thickness t, and density contrast σ (Fig. 7-2), the gravity at a point p above the sheet is

$$g = 2\pi G\sigma t \begin{cases} = 12.78\sigma t & \text{for } g \text{ in mgals, } t \text{ in kft} \\ = 41.93\sigma t & \text{for } t \text{ in km} \end{cases}$$

The gravity is independent of the depth d to the sheet as long as that depth is small compared with the horizontal extent of the sheet. This is the same as the quantity calculated for the simple Bouguer correction, page 21.

Horizontal variations in σt, with horizontal extent large compared to the depth d, result in changes in surface gravity that are almost directly proportional to these variations. Thus, the variations in the gravity effect corresponding to changes in thickness of a horizontal sheet give the order of magnitude for gravity effects applicable to many geological situations where the horizontal extent of the variations is large compared to their depth.

The factor $g = 12.78\sigma t$ (for t in kilofeet) indicates the thickness necessary to account for a given gravity effect. This means that for unit density contrast it takes a thickness of $1000/12.78 = 78$ ft to produce a gravity effect of 1 mgal. This expression applies to a sheet whose horizontal extent is infinite compared with depth or thickness, and a finite sheet has to be thicker to produce the same effect. From the fault curves of Fig. 7A-4 it is evident that it takes a distance of 6 times the depth (from $-3z$ to $+3z$) to reach about 80 percent (from 0.1 to 0.9) of the theoretical total magnitude for an infinite width, and for this width it would require $78/0.80 = 97$ ft of unity density contrast to have a gravity effect of 1 mgal. Thus, as a practical and easily remembered basis for estimating orders of magnitude to be expected from broad but limited density anomalies we may use a figure of 100 ft of thickness of material with unit density contrast to have a gravity effect of 1 mgal (rather than the theoretical value of 78 ft for a sheet of infinite horizontal extent). Then, for a given gravity effect and density contrast, the approximate required thickness is $t = (100/\sigma)g$, where t is in feet, σ is the density contrast, and g the maximum gravity anomaly in milligals. As an example of application of such an estimate we may take the total gravity effect of the Los Angeles Basin. From the gravity profile across the basin (Fig. 6-12), with some allowance for a regional gravity decrease toward the right (north) end of the profile the gravity relief over

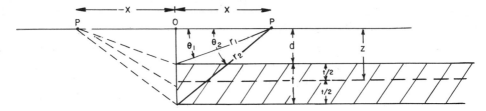

FIGURE 7-3
Notation for gravity effect of a vertical fault.

the center of the basin is approximately 60 mgals. The density contrast is not known definitely, but the Pliocene and Miocene clastic deposits would be expected to be of relatively low density with an average of perhaps 2.2 to 2.4, increasing with depth. The underlying Franciscan metamorphic basement should have a density of around 2.6 to 2.7. These values suggest that the density contrast between the sediments and the basement as a whole should be in the neighborhood of 0.3 to 0.4. If we use a figure of 0.35, we have a thickness of $100/0.35 = 300$ ft/mgal, so that the thickness of the sediments in the basin for the 60-mgal anomaly would be $60 \times 300 = 18,000$ ft. This is in quite reasonable agreement with the basement depth which might be inferred from the section (Fig. 6-12).

The Vertical Fault

A vertical fault (Fig. 7-3) is equivalent to the horizontal sheet which is cut off at a vertical plane below point O (Fig. 7-3). The expression for the gravity at point P, which depends on x being positive to the right of point O and the angles being measured counterclockwise (Nettleton, 1940, p. 112), is

$$g = 2G\sigma \left[x \ln \frac{r_2}{r_1} + \pi t - (t + d)\theta_2 + d\theta_1 \right]$$

In this form, the expression cannot be readily normalized* (expressed in ratio of distance to depth) as for the sphere and horizontal cylinder. This can be done, however, for two special cases.

CASE 1: THIN SHEET For this case, the material is considered as being condensed onto a thin sheet, or lamina, of mass σt per unit area and at a depth z, corresponding to the center of the original slab (Fig. 7-3). Consider the lamina first as a thin sheet of thickness t, so that $d = z - t/2$, $t + d = z + t/2$. Then,

$$g = 2G\sigma \left[x \ln \frac{r_2}{r_1} + \pi t - \left(z + \frac{t}{2} \right)\theta_2 + \left(z - \frac{t}{2} \right)\theta_1 \right]$$

$$= 2G\sigma \left[x \ln \frac{r_2}{r_1} + \pi t - z(\theta_2 - \theta_1) - \frac{t}{2(\theta_2 + \theta_1)} \right]$$

---

* Hammer (1974) gives a normalized formula, but it is much more complex than that given here for the special cases considered.

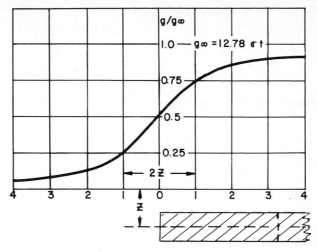

FIGURE 7-4
Fault gravity effect and depth parameter.

As the thickness of the slab decreases to zero,

$$\ln \frac{r_2}{r_1} = 0 \qquad r_1 = r_2 = r \qquad \theta_1 = \theta_2 = \theta \qquad \theta_2 - \theta_1 = 0$$

and

$$g = 2G\sigma t(\pi - \theta) = 2G\sigma t \left(\pi - \tan^{-1} \frac{z}{x} \right)$$

$$= 2G\sigma t \left[\pi - \left(\frac{\pi}{2} - \tan^{-1} \frac{x}{z} \right) \right]$$

$$= 2\pi G\sigma t \left(\frac{1}{2} + \frac{1}{\pi} \tan^{-1} \frac{x}{z} \right)$$

$$= K_3 f_3 \left(\frac{x}{z} \right)$$

where

$$K_3 = 12.78\sigma t \qquad \text{for } t \text{ in kft}$$

or

$$K_3' = 41.93\sigma t \qquad \text{for } t \text{ in km}$$

and

$$f_3 \left(\frac{x}{z} \right) = \frac{1}{2} + \frac{1}{\pi} \tan^{-1} \frac{x}{z}$$

which is plotted as the density sheet or lower curve of Fig. 7A-4.

For a depth criterion for a fault we note that, for the lamina, at $x/z = 1$, the amplitude is 0.75 and would be 0.25 at $x/z = -1$ on the lower half of the curve. Thus, if we can divide the total fault curve into quarter-amplitude sectors (Fig. 7-4), the distance between these points is twice the depth to the center of the

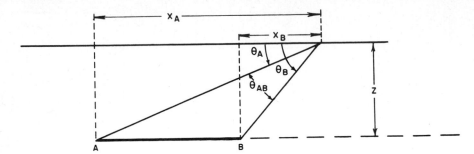

FIGURE 7-5
Notation for gravity effect of sheet of finite width.

density contrast. To illustrate how the result for the semi-infinite sheet can be used to derive the result for a sheet of finite extent in the x direction, note that the finite sheet can be obtained by subtracting two semi-infinite sheets whose edges are at A and B, respectively (Fig. 7-5). Then

$$g_{AB} = g_A - g_B = 2G\sigma t(\pi - \theta_A) - 2G\sigma t(\pi - \theta_B)$$

$$= 2G\sigma t(\theta_B - \theta_A)$$

$$= 2\pi G\sigma t \, \frac{\theta_B - \theta_A}{\pi}$$

$$= 2\pi G\sigma t \left(\frac{1}{\pi} \tan^{-1} \frac{x_A}{z} - \frac{1}{\pi} \tan^{-1} \frac{x_B}{z} \right)$$

From the next to last line above we observe that the anomaly due to the finite sheet is proportional to the angle (in radians) subtended by AB, that is,

$$g_{AB} = 2G\sigma t \theta_{AB} = K t \theta_{AB}$$

Table 7-3 lists values for K, to give g_{AB} in milligals with different units for t and θ_{AB}.

Table 7-3

t, in	θ_{AB}, in	K
Kilofeet	Radians	4.07
	Degrees	0.0710
Kilometers	Radians	14.95
	Degrees	0.261

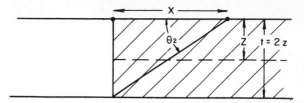

FIGURE 7-6
Notation for thick fault to surface.

CASE 2: THICK FAULT TO SURFACE   The other fault model which can be normalized is that for a thick fault with its bottom at depth $2z$ and its top at the surface (Fig. 7-6). For this case, the quantities of Fig. 7-3 become

$$d = 0 \qquad d + t = t = 2z \qquad \theta_1 = 0 \qquad \theta_2 = \tan^{-1}\frac{t}{x} = \frac{\pi}{2} - \tan^{-1}\frac{x}{t}$$

$$r_1 = x \qquad r_2 = \sqrt{x^2 + t^2}$$

Then

$$g_2 = 2G\sigma\left[x\ln\frac{\sqrt{x^2 + t^2}}{x} + \pi t - t\left(\frac{\pi}{2} - \tan^{-1}\frac{x}{t}\right)\right]$$

$$= 2\pi G\sigma t\left(\frac{1}{\pi}\frac{x}{t}\ln\frac{\sqrt{1 + x^2/t^2}}{x/t} + \frac{1}{2} + \frac{1}{\pi}\tan^{-1}\frac{x}{t}\right)$$

$$= K_4 f_4\left(\frac{x}{t}\right)$$

where

$$K_4 = 12.78\sigma t \qquad \text{for } t \text{ in kft}$$

or

$$K_4' = 41.93\sigma t \qquad \text{for } t \text{ in km}$$

and

$$f_4\left(\frac{x}{t}\right) = f_4\left(\frac{x}{2z}\right) = \frac{1}{\pi}\frac{x}{t}\ln\frac{\sqrt{1 + x^2/t^2}}{x/t} + \frac{1}{2} + \frac{1}{\pi}\tan^{-1}\frac{x}{t}$$

We can plot $f_4(x/t)$ with the abscissas doubled (since $t = 2z$) to make the curve on the same horizontal scale as that for the thin lamina. This gives the upper ("thick fault to surface") curve of Fig. 7A-4, which is of the same form as the expression for the thin-lamina approximation except for the logarithmic term. This term goes to zero at $x = 0$ and at $x = \infty$. We see that the maximum difference between the lamina and the thick fault to the surface is at a distance of $x/z = 0.5$, where it amounts to only about 4 percent of the total effect. Thus, in spite of the extremes of the approximation, the curves for the thin and thick faults are not greatly different; these extremes embrace the entire range of possible faultlike occurrences with vertical contact. The errors of the fault approximation are included in those examined by Hammer (1974).

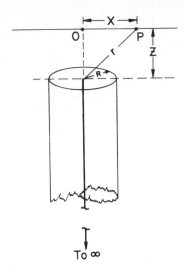

FIGURE 7-7
Notation for gravity effect of vertical
cylinder.

To ∞

The Vertical Line Element

A vertical line element can be used to approximate a vertical cylinder for which the horizontal dimension is considerably less than the depth to the top. Appropriate geologic examples are a narrow salt dome or a volcanic plug. The formula for the case in which the bottom of the line element is at infinite depth is simply (Fig. 7-7) $g = Gm/r$, where m is the mass per unit length, which, for the cylinder of radius R, is $\pi R^2 \sigma$. Since $r = \sqrt{x^2 + z^2}$, we have

$$g = \pi R^2 G\sigma(x^2 + z^2)^{-1/2} = \frac{\pi R^2 G\sigma}{z} \frac{1}{\sqrt{1 + x^2/z^2}}$$

and the normalized equation becomes

$$g = K_5 f_5 \left(\frac{x}{z}\right)$$

where $f_5(x/z) = (1 + x^2/z^2)^{-1/2}$, which is shown as curve $A5$ and is for a cylinder of infinite depth to the bottom. Putting in the numerical values for G and π and for conversion of units gives

$$K_5 = \frac{6.39 R^2 \sigma}{z} \qquad \text{for } R \text{ and } z \text{ in kft}$$

$$K_5' = \frac{20.96 R^2 \sigma}{z} \qquad \text{for } R \text{ and } z \text{ in km}$$

For a cylindrical column of finite length (top at depth z_1, bottom at depth z_2) the effect calculated for depth z_2 is subtracted from that calculated for z_1.

This approximation has been examined in detail by Hammer (1974), who shows that it deteriorates rapidly as the ratio of the radius to depth to top of the cylinder is greater than about $\frac{1}{2}$.

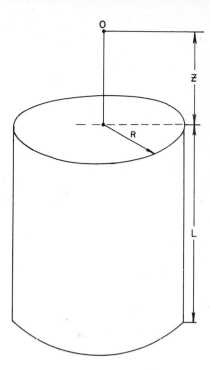

FIGURE 7-8
Notation for gravity effect of a vertical
cylinder for a point on its axis.

The Vertical Cylinder for a Point on Its Axis

This form is useful for a quick approximation of the maximurn gravity effect over
the center of a roughly cylindrical body such as a salt dome or igneous intrusion.
The attraction at a point O on the axis of a vertical cylinder, with the notation of
Fig. 7-8, is

$$g = 2\pi G\sigma[L + \sqrt{R^2 + z^2} - \sqrt{R^2 + (z + L)^2}]$$
$$= K_6[L + \sqrt{R^2 + z^2} - \sqrt{R^2 + (z + L)^2}]$$

where $K_6 = 12.78\sigma$ for dimensions in kilofeet and $K_6' = 41.93\sigma$ for dimensions in
kilometers. In this form the expression cannot be normalized, but it can for the
special case where the point of calculation is on the upper surface of the cylinder
(Fig. 7-9); then $z = 0$ and

$$g = K(L + R - \sqrt{R^2 + L^2}) = K\left(1 + \frac{R}{L} - \sqrt{\frac{R^2}{L^2} + 1}\right)$$

This form is useful for estimating the deficiency in the gravity effect of a relatively
broad density sheet, such as a limited basin compared with that for infinite width.
Figure 7A-6 gives a curve for the ratio of the gravity effect of a disk of limited radius
to that for the one of infinite radius. Let us apply this to the center of the Los
Angeles Basin. The approximate radius of the deeper part is 10 miles and the
sedimentary thickness about 3 miles, to give the R/L ratio of 3.3, for which the
gravity effect is read from the curve as about 83 percent of what it would be if

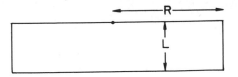

FIGURE 7-9
Notation for gravity effect at a point on
the surface of a circular disk.

infinitely broad. Thus, we are led to the same approximately 80 percent figure which we used above in estimating the thickness of this same basin on the basis of a horizontal sheet of limited extent.

For the special case of the cylinder reduced to a thin disk, $g = 2\pi G\sigma t$ $(1 - \cos \alpha)$ (see discussion of solid angles, page 203), which is the basis for the calculation of the solid angle chart (Fig. 7A-9) and the three-dimensional dot chart (page 204 and Fig. 7A-11).

The Vertical Sheet

The gravity expression of a two-dimensional vertical sheet is useful as an approximation of the effect of a vertical dike. The approximation is made that the material of the dike is considered as being condensed onto a thin lamina, but this approximation is satisfactory if the thickness is substantially less than the other dimensions and the distance to the plane of calculation.

The gravity effect, with the notation of Fig. 7-10, is (Nettleton, 1942)

$$g = 2G\sigma t \ln \frac{r_2}{r_1} = 2G\sigma t \ln \frac{\sqrt{x^2 + z_2{}^2}}{\sqrt{x^2 + z_1}} = K_7 f_7(x,z)_2$$

where $\qquad K_7 = 2 \times 6.67 \times 10^{-8} \times 2.3026\sigma t^* = 30.7 \times 10^{-8}\sigma t$

This gives g in gals for t in centimeters. For other units,

$$g = \begin{cases} 30.7\sigma t f_7(x,z) & \text{for } g \text{ in mgals, } t \text{ in km} \\ 9.36\sigma t f_7(x,z) & \text{for } g \text{ in mgals, } t \text{ in kft} \end{cases}$$

The quantity $f_7(x,z)$ is not as easily normalized as some of the previous expressions. However, it can be normalized for certain special relations, as follows. Let $z_2 = nz$. Then

$$g = K_7 \log \frac{\sqrt{x^2 + n^2 z^2}}{\sqrt{x^2 + z^2}} = \frac{K_7}{2} \log \frac{x^2 + n^2 z^2}{x^2 + z^2}$$

$$= \frac{K_7}{2} \log \frac{n^2 z^2 (1 + x^2/n^2 z^2)}{z^2 (1 + x^2/z^2)} = \frac{K_7}{2} \left(2 \log n + \log \frac{1 + x^2/n^2 z^2}{1 + x^2 z^2} \right)$$

* The factor 2.3026 changes the natural logarithm to the common logarithm with base 10.

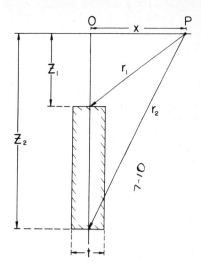

FIGURE 7-10
Gravity effect of a vertical body (dike)
approximated as a vertical sheet.

At $x = 0$, $g = g_0 = K_7 \log n$ and $K_7 = g_0/(\log n)$. Then

$$g = \frac{g_0}{2 \log n} \left(2 \log n + \log \frac{1 + x^2/n^2 z^2}{1 + x^2/z^2} \right)$$

$$= g_0 \left(1 + \frac{1}{2 \log n} \log \frac{1 + x^2/n^2 z^2}{1 + x^2/z^2} \right)$$

and

$$\frac{g}{g_0} = 1 + \frac{1}{2 \log n} \log \frac{1 + x^2/n^2 z^2}{1 + x^2/z^2}$$

$$= 1 - \frac{0.5}{\log n} \log \frac{1 + x^2/z^2}{1 + x^2/n^2 z^2}$$

which gives normalized values for a given value for n. Curves for g/g_0 for vertical sheets with ratios of depths to bottom and top of $\sqrt{10}$, 10, and $\sqrt{1000}$ are given by Fig. 7A-7. The value of g_0, that is, the gravity over the center of the dike is

$$g_0 = \begin{cases} 9.37 \sigma t \log n & \text{for } t \text{ in kft} \\ 30.7 \sigma t \log n & \text{for } t \text{ in km} \end{cases}$$

The errors in the approximate formula have been examined by Hammer (1974). Curves are given for the errors for various ratios of thickness (width) to depth.

Gravity Gradient for Sloping Contrast

A rather common interpretation problem is that of estimating quantitatively the slope of a density contrast which could account for an extensive gravity gradient. For small slope angles (error only 3 percent for a slope of 1 in 5) the change in

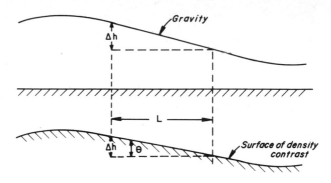

FIGURE 7-11
Diagram and nomenclature for gravity effect of a sloping density contrast.

gravity along the horizontal surface (Fig. 7-11) can be approximated by considering the effect of a slab with thickness equal to the change in elevation of the contact:

$$g_1 - g_2 = \Delta g = 2\pi G \sigma \, \Delta h$$
$$= 12.78\sigma L \tan \theta \qquad \text{for } L \text{ in kft}$$

and
$$\frac{\Delta g}{L} = \begin{cases} 12.78\sigma \tan \theta & \text{in mgals/kft} \\ 41.93\sigma \tan \theta & \text{in mgals/km} \end{cases}$$

If the structural slope is expressed as S ft/mile, then

$$\Delta h = LS \qquad \text{ft}$$

and
$$\Delta g = 0.01278\sigma LS$$

so that the gravity gradient in milligals per mile is

$$\frac{\Delta g}{L} = 0.01278\sigma S$$

For example, a slope of 100 ft/mile into a basin with a density contrast of 0.2 between sediments and basement would produce a gradient effect of

$$\frac{\Delta g}{L} = 0.01278 \times 0.2 \times 100 = 0.256 \text{ mgal/mile}$$

If the slope and gravity gradient (in milligals per mile) are known, the required density contrast at the sloping surface is

$$\sigma = \frac{\Delta g}{L} \frac{1}{0.01278S} = \frac{\Delta g}{L} \frac{78.3}{S}$$

We can apply this to our familiar Los Angeles Basin example (Fig. 6-12). The gravity change over the steep gradient in the left part of the figure is about 60 mgals in a distance of about 10 miles, or 6 mgal/mile. The basement slope over the linear

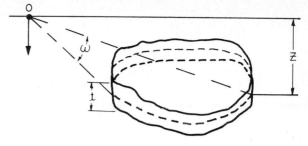

FIGURE 7-12
Diagram and nomenclature for calculation of gravity effect of a finite horizontal tabular body by the equivalent solid angle.

sector (at the word "Jurassic") projected to the surface is about 10,000 ft in 6.5 miles or 1500 ft/mile. From these data the estimated density contrast at the basement surface is

$$\sigma = \frac{6 \times 78.3}{1500} = 0.31$$

which is not too different from the value (0.35) estimated (page 194) from another combination of data for the basin.

Calculation by Solid Angles

The lamina, or density-sheet, approximation by solid angles provides a convenient method of calculating gravity effects of three-dimensional bodies. If a horizontal slab of material with any outline (Fig. 7-12) is replaced with a central lamina having the same outline, the vertical gravity effect at point O is directly proportional to the solid angle subtended at that point by the boundary of the lamina. Since the infinite sheet which subtends a solid angle of 2π has an effect $2\pi G\sigma t$, the lamina will have an effect $\omega/2\pi$ as large, namely

$$g = \omega G\sigma t$$

For g in milligals,

$$g = \begin{cases} 2.03\omega\sigma t & \text{for } t \text{ in kft} \\ 6.65\omega\sigma t & \text{for } t \text{ in km} \end{cases}$$

The approximation is quite satisfactory for thickness up to about half the depth.*

---

* The accuracy of the approximation of a thick circular slab by a density sheet or lamina depends on the ratios R/z and t/z. The errors are maximum for a point on the axis of the circle and, for $t/z = 0.5$, they are

R/z	0.0	0.5	1.0	2.0
Error, %	−6	−4	−2	−0.5

For points off the axis these errors decrease rapidly.

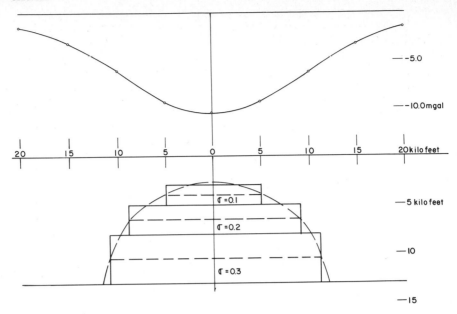

FIGURE 7-13
Division of domal body into slices for calculation of gravity effects by solid angles and calculated curve.

Calculation of solid angles for even simple geometrical bodies is mathematically difficult, but they have been calculated for circles (Masket et al., 1962). A chart giving solid angles for circles is given by Fig. 7A-8. This chart is convenient for calculating gravity effects for any body such as a salt dome or volcanic plug which can be approximated by horizontal circular disks. The thicknesses of the disks can be increased with depth to nearly $t/z = 0.5$ (see table in previous footnote) and still have errors of less than 5 percent.

A numerical example of the calculation by solid angles of circles for a simple model is shown by Fig. 7-13. The assumed model is a broad domal structure (salt dome) with a symmetrical circular form. It is approximated by three disks with the dimensions and other parameters given by Table 7-4.

Table 7-4

Layer	z	t	R	z/R	ω_0	σ	g_0, mgals
1	4000	2000	5000	0.500	3.48	0.1	1.40
2	6500	3000	9000	0.722	2.60	0.2	3.17
3	10,500	5000	11,000	0.954	1.95	0.3	5.94
							‾‾‾‾
							10.51

In Fig. 7A-8 points along the axis of the disks correspond with points along the bottom of the chart ($x/z = 0$ for which values are given in detail by Fig. 7A-9). These are the ω_0 values of Table 7-4, each corresponding to the z/R value for the corresponding disk. Vertical lines on the chart correspond to horizontal profiles, with the solid angles given by interpolating between the curved lines, or contours, of constant ω. Interpolating between contours along a vertical line above the z/R value for a given disk at points corresponding to the x/z value gives the series of ω values for that disk as listed in Table 7-5. Multiplying the g_0 values by ω/ω_0 gives the gravity values at the corresponding horizontal distances. The values at each x are added horizontally to give the total gravity of the last column in Table 7-5, which is plotted as the calculated gravity profile (Fig. 7-13).

The process may seem complicated but is really quite simple. The calculations of Fig. 7-13 were made by the writer with a slide rule in about 40 min. A computer, once set up, could do the equivalent calculation in a few seconds, but for a quick check and to help in guiding a choice of model parameters and when a computer with appropriate program is not immediately available, a quick approximation such as this is often very useful.

METHODS FOR MORE DETAILED CALCULATIONS

The simple formula for spheres, cylinders, faults, and other geometric models can serve very well to give the approximate magnitudes of calculated gravity effects for comparison with the observed values and for quantitative checks of geological possibilities to account for observed gravity anomalies. There are, however, many interpretation problems in which more detailed calculations are useful. Such problems arise, for instance, when it is desired to determine the gravity effect from a geological situation which is partially defined by known stratigraphy and drilling and where the gravity expression is complex but well defined or when general effects over large areas are to be interpreted.

Table 7-5

x	Layer 1			Layer 2			Layer 3			
	x/z	ω	g	x/z	ω	g	x/z	ω	g	Total g
0	0	3.48	1.40	0	2.60	3.17	0	1.95	5.94	10.51
5000	1.25	3.00	1.21	0.770	2.20	2.88	0.475	1.75	5.33	9.42
10,000	2.50	1.00	0.40	1.54	1.25	1.52	0.95	1.35	4.11	6.03
15,000	3.75	0.28	0.11	2.31	0.50	0.61	1.43	0.80	2.44	3.16
20,000	5.00	0.10	0.04	3.08	0.21	0.26	1.90	0.40	1.22	1.52
25,000	6.25	0.06	0.02	3.85	0.14	0.13	2.38	0.23	0.70	0.85
40,000	10.00	0.014	0.005	6.15	0.03	0.036	3.81	0.06	0.19	0.23

There are a number of ways in which more detailed calculations can be carried out. In general they may be divided into two-dimensional and three-dimensional problems. Two-dimensional calculations are applicable to situations where the contours are roughly parallel. Then a profile which can be correlated with a sub-surface cross section is representative of conditions extending for relatively large distances perpendicular to the line of section. Three-dimensional methods are required when the anomalies are irregular or do not have a strong elongation in one direction. Greatly simplified methods can be used where anomalies are roughly equidimensional and can be approximated by circular symmetry.

Two-dimensional Calculation Systems

These methods can be applied to a problem where conditions perpendicular to the line or profile are essentially constant for distances on each side of the profile line of about 2 or 3 times the depth of the section calculated. While details differ, in several useful systems the mass to be calculated is considered as being made up of horizontal line elements (or cylinders) of infinite length in the direction perpendicular to the section.

In one form, a chart is made up in which units of area are proportioned so that they represent elements of mass having equal effects at the origin of the chart. Thus these units increase in size with their distance from the origin. If these elements of area are made small enough, each such area can be represented by a single dot at its center. Then, to determine the gravity effect it is necessary only to draw an outline of the body being considered and count the number of dots within the area with the center of the chart at each point where the gravity effect is desired. The dot count at each point is multiplied by a factor which depends on the design of the chart, the scale of the cross section, and the density contrast. By repeating the operation for a number of points along the profile, the variation of gravity can be determined to whatever degree of detail is required to give the calculated gravity effect. Such a chart is illustrated by Fig. 7A-10.

Hubbert (1948a) describes a graphical two-dimensional line-integral method which adds elements around the periphery of the body. A variation which has elements of Hubbert's chart and of that described above, together with detailed theory of design and adaptation to magnetic calculations, is given by Millett (1967). A mechanical planimeterlike device has been used (Nettleton, 1940, p. 118) which gives the gravity effect when the stylus traces the body outline.

A number of other chart designs have been published in the geophysical literature, some of which are more convenient for certain conditions or shapes of bodies for which gravity effects are desired. All require some sort of measurement or count within the area or around the periphery of the cross section of the body being computed and summation of effects for each point of calculation. While the operations are basically simple and can be carried out readily by un-trained personnel, they are quite time-consuming and tedious. These methods are applicable when a simple section is to be computed at a moderate number of points. They can be considered as intermediate between the very simple calculations made with horizontal cylinder or fault formulas discussed previously (pages 191 to 194)

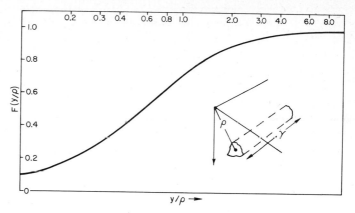

FIGURE 7-14
End correction $F(y/\rho)$ for reducing calculations made with two-dimensional methods to their value for finite length of body. Note that if length of body (on each side of plane of calculation) is twice the distance to the point of calculation, the effect is about 90 percent of that for infinite length.

and those made with electronic computers. Where the section to be computed has several layers with different densities and must be computed at a large number of points, and where sections must be successively modified on the basis of calculations to obtain a reasonably close fit with the anomaly being interpreted, it is almost necessary to use the computer to make the calculations feasible (see page 209).

Three-dimensional Calculations

When the body for which the gravity effect to be calculated is not relatively uniform in one direction and is too complex to be approximated by simple geometric bodies, three-dimensional calculations must be carried out. In general, such calculations are an order of magnitude more difficult or time-consuming than two-dimensional approximations.

One relatively simple approach is to make end corrections for a two-dimensional or dot-chart type of calculation. This recognizes the excess in the effect calculated for the ideal two-dimensional condition when the length perpendicular to line of section is limited, as illustrated in Fig. 7-14. The end correction is the ratio of the gravity effect for a two-dimensional body of a limited length (y in the figure) to the effect which would be calculated for infinite length and is given by the relation

$$\frac{g}{g_\infty} = \frac{1}{\sqrt{1 + (\rho/y)^2}} \equiv F\left(\frac{y}{\rho}\right)$$

where y is the half-length of the element perpendicular to the plane of the section and ρ is the distance from the center of the element to the point of calculation. This function is shown by the curve of Fig. 7-14.

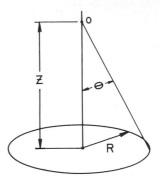

FIGURE 7-15
Notation for solid angle of circle for
point on axis.

Solid Angles for Irregular Forms

It has been shown (page 203) that the vertical gravity effect of a horizontal thin
lamina is proportional to its solid angle, and a chart (Fig. 7A-8) gives solid angles
for circles. For three-dimensional problems, where the body cannot be approximated by circular layers, the solid angles for laminae of irregular outline can be
determined by a polar dot chart.

The design of such a chart is quite simple. The solid angle subtended at a
point O on the axis of a circle (Fig. 7-15) is

$$\omega = 2\pi(1 - \cos\theta) = 2\pi\left[1 - \cos\left(\tan^{-1}\frac{R}{Z}\right)\right]$$

From this relation we can design a horizontal chart or graticule which divides
up the area out from the center of the circles into increasingly larger units so that
each has the same effect at the point of calculation (Fig. 7-16). Then the areas are
replaced by dots and the gravity effect of a lamina with any outline is determined
by counting the number of dots within the outline with the center of the chart at
the point where the effect is desired.

A chart of this kind is correct only for the vertical distance or unit depth for
which it is designed (the distance z of Fig. 7-15). This means that when an outline
or contour is drawn as the boundary of the lamina to be computed, the scale of the
outline must be chosen or changed so that at the scale used the depth to the plane
of the lamina is equal to the unit depth. Usually the chart is laid out and drawn
for a relatively large unit depth, and other charts are made by photographic reduction to convenient smaller unit-depth values. A chart of this design with a total of
1277 dots to $R = \infty$ is shown by Fig. 7A-11. Then, since for thickness in kilofeet,
$g = 12.77\omega/2\pi$, the gravity effect is given by $g = 0.01\sigma nt$, where σ is the density
contrast, n is the dot count, and t is the thickness in kilofeet.

The difference in gravity effect between a circular lamina and one with an
irregular outline depends greatly on the depth. If the horizontal dimensions are
considerably greater than the unit depth of the chart and the ratio of maximum to
minimum horizontal dimensions is 2:1 or more, the effects would be grossly

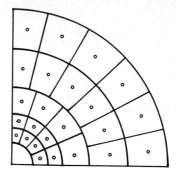

FIGURE 7-16
Layout of polar solid-angle chart.

different from those of a circle. As the horizontal dimensions become less than the depth, the differences decrease rapidly.

CHARTS FOR CALCULATION OF GRAVITY EFFECTS

The form curves and other diagrams for gravity calculations are assembled in the appendix (page 217) at the end of this chapter. They are reproduced as full-page illustrations so that they may be used as quantitative tools to make quick approximations of gravity effects. Such approximations may be sufficient for answers to many interpretation questions or as controls for the limiting parameters of computer calculations.

GRAVITY CALCULATIONS BY DIGITAL COMPUTER

After about 1960, methods of using digital computers for gravity calculations began to appear in the geophysical literature and are now used routinely in certain academic institutions and in oil and service company offices. Similar but somewhat more complex systems are applied to magnetic calculations.

The basic systems may be classified as (1) two-dimensional, which calculate a gravity profile from a cross section; (2) three-dimensional, which calculate gravity effects from a three-dimensional body over a plane surface; and (3) iterative, which determine a body form, limited by certain initial parameters, which will account for a given gravity field.

Talwani, Sutton, and Worzel (1959) describe a two-dimensional system operating on a geological cross section. The body (or bodies) for which effects are calculated is defined by a closed outline, or polygon, which has as many straight-line segments as are necessary to make a reasonably close equivalent of the actual outline used. The inputs are the horizontal and vertical coordinates of the apices of the polygon and of the point for which the calculation is desired and, of course, the density contrasts. The system is readily adapted to a complex cross section with rocks of different densities. For example, the writer found it quite effective and practical in a study of gravity and structure over a complex overthrust section in the Canadian Rockies.

Talwani and Ewing (1960) describe a three-dimensional system in which the body is approximated by horizontal laminae (similar to the approximations described above for computation by solid angles). Each horizontal section or slice of the body is represented by a contour at its center. The contour is approximated by a closed polygon. The computer determines, for each point of calculation the gravity effect of each layer or polygon and adds them together.

Danes (1960) describes a system in which the body is represented by a system of regular, square columns or prisms. The gravity effect of each column is calculated for a finite depth to the top and infinite depth to the bottom. A finite bottom limit is made by subtracting the effect from that depth to infinity. The geometry of the body is digitized by writing, on a regular grid, the depth to the top of each of the square prisms which approximate the top surface of the body. If the bottom surface is irregular, its depth must also be written for each prism. Commonly, as in salt-dome calculations, the bottom is a plane surface at constant depth, and effects from that depth to infinity are subtracted from each prism. Then, for each grid point, the sums of the gravity effects of all the prisms are determined, which gives a pattern of calculated values which can be contoured and compared with a theoretical or actual gravity map.

An extensive review of computer methods for calculation of gravity effects is given by Talwani (1973). This includes calculation of geometric forms, including more exact determination of several of those for which approximate methods are given earlier in this chapter.

Cordell and Henderson (1968) describe an iterative system which, within limits, is an approach to a direct interpretation method, which calculates structure, rather than the usual indirect method of calculating gravity effects from a given structure, comparing calculated with observed gravity, modifying the form, repeating the calculations, etc. Because of the inherent ambiguity of gravity effects, as noted below (page 241), it is necessary to impose certain limits on the system, and the result is very much dependent on these controls (see page 185 for a general discussion in terms of filter theory).

The system uses a regular grid of square columns or prisms, similar to those of Danes. The observed or input gravity field is represented by a regular grid of values, each value being over the center of a unit column. The other controls are a common depth or plane at the bottom of all the columns (or a central plane for a symmetrical body) and a density contrast. The calculation starts from an initial distribution of depths, which may be on a uniform plane or can be calculated as simple Bouguer slabs from the input gravity values. The sums of the effects of all columns are calculated at each grid point and compared with the input gravity at that point. From the differences between calculated and observed effects the process automatically makes a modification of the height of each column. This new model is calculated and compared with the input at each point, and column heights are again modified, etc. Thus, the procedure repeats, or iterates. The calculation continues until it is stopped at a given number of iterations or, if desired, until a certain average difference between calculated and observed values is achieved.

The original Cordell and Henderson system has been the starting point for the development of a number of commercial systems, which have included a provision for variable density, have modified the algorithms to achieve faster operation, and have become a very useful and practical tool of gravity interpretation.

The iterative calculation of structure must be used with care. Its widest application has been in salt-dome interpretation. It must be preceded by some sort of anomaly isolation or removal of regional effects which would lead to a broad distortion of the pattern. It is very sensitive to relatively sharp anomalies which can lead to sharp spikes in the height of columns where the bottom depth of the column is too great to give such sharp anomalies. The choice of bottom depth (depth to "mother" salt in salt-dome problems) and, of course, the density contrasts, are critical, and some cut-and-try initial procedures may be necessary to give reasonable results. Where contacts on the body (salt) are available, as from drilling, it is possible in some systems to use them as controls. If the control parameters are grossly in error, the calculation may oscillate or give unreasonably sharp local features, which can be used as a criterion to modify the controls.

On the whole, the iterative systems are very useful but need to be used with care and common sense to avoid interpretations which give almost exact agreement with the input gravity but which are geologically unreasonable. An example of application of this system to a salt-dome anomaly is given below (page 266).

A review of inverse methods as applied to the interpretation of magnetic and gravity methods is given by Bott (1973). Such methods can be used to determine distribution of density or magnetization and have possibilities and limitations similar to those of the iterative methods described above. A particular application to which they are well suited (in this case, a magnetic rather than a gravity problem) is that of determining the widths and distribution of bands of normal and reversed magnetization corresponding to bands of magnetic intensity observed over and on both sides of mid-ocean ridges.

The usefulness and in fact absolute necessity of the computer for certain data analysis and filter applications have been pointed out in Chap. 6. The application of computers to calculations of gravity effects and to modeling bodies to account for given gravity effects can be timesaving by one or more orders of magnitude. A word of caution is in order. A very close fit of a calculated effect with observed data is not a criterion for the correctness of the model (see Fig. 16-7). The parameters of depth and density contrast can be varied widely and still give a close fit (see Fig. 8-1).

For many relatively simple quantitative interpretation problems, it is often possible, by sphere, cylinder, or fault models, to fit the observed data within the limits of their precision and spacing. Any further refinement of the interpretation, including known or estimated density contrasts, must come from geological assumptions or probabilities. The corresponding mass, width, and depth values are limited by the constraints imposed by the simple analysis. Refinement by application of a computer process would produce a model having a gravity effect which would fit the observations as closely as desired; added detail which depends on differences smaller than those defined by the original observations is fallacious. Thus a computer-modeled solution may give an impression of reliability which is unwarranted.

Grant (1972) concluded his Gravity and Magnetic Review with the following remarks on computer applications:

This much at least can be said. Judging by the material that has been published within the past six years, there can be no doubt about the overwhelming swing that has taken place toward the processing of potential field maps by computers. In some areas, the commitment has been total. However, there is probably almost as much danger in an over-commitment to this view as there is in reacting against it; and we may all live to see a backward oscillation or two before the optimum point of balance is found. The computer offers the capability of performing complex arithmetical tasks with tremendous speed, but it lacks the power to make judgments when several alternatives have equal mathematical validity. The task of the immediate future will be to learn how best to use this enormous calculating capacity without sacrificing the skills in discrimination and judgment of which, at the present time, only a human being is capable. The answer, if there is one, is most likely to be found in the field of computer graphics.

ESTIMATION OF TOTAL MASS

There is one quantity in gravity interpretation that is not subject to the ambiguities mentioned throughout the discussion of possible sources. This unambiguous quantity is the total mass anomaly. It is particularly useful in estimating the total mass of ore bodies in the application of very detailed gravity surveys to mineral prospecting.

It can be shown (Hammer, 1945; LaFehr, 1965) that, by Gauss' theorem

$$M = \frac{1}{4\pi G} \int_S \int g_n \, dS$$

where M = total mass producing the anomaly
  G = gravitational constant
  g_n = gravity component normal to surface S which encloses the anomalous mass

Since gravity measurements are made on the earth's surface, we can consider the enclosing surface as a horizontal plane above and a hemisphere around and below the anomalous mass. Then the gravitational flux through the plane of measurement is half the total, and we can write

$$M = \frac{1}{2\pi G} \int \int g_n \, dS$$

This means that if we determine a volume for which the horizontal dimensions are lengths and the vertical dimension is gravity, the total mass causing an anomaly is proportional to this volume. The volume can be estimated by digitizing the anomaly map, i.e., dividing it into squares, estimating the average anomaly value in each square, and adding up the unit areas times gravity values to get the total anomaly volume.

If numerical values are substituted for constants in the equation for the anomalous mass, we obtain

$$M = 2.39 \times 10^6 \iint g_n \, dS$$

in the cgs system, where M is in grams, g in gals, and S in square centimeters. To permit the use of more practical units note that

$$M = K \sum gS^2$$

where

$$K = \frac{1}{2\pi G} \frac{R_g R_s^2}{R_m} = 2.39 \times 10^6 \frac{R_g R_s^2}{R_m}$$

where R_g, R_s, and R_m are ratios of the mass, gravity, and map-scale units used to represent grams, gals, and centimeters, respectively.

Values of K for several combinations of commonly used units are given in the Table 7-6. These values of K, of course, give the *anomalous* mass. For the total mass of the anomalous body they must be multiplied by the density factor $\sigma_a/(\sigma_a - \sigma_c)$, where σ_a is the density of the anomalous material and σ_c is the density of the normal, or "country rock," material.

Unfortunately, for quantitative calculation, seemingly small errors in the integration quantities can lead to very large errors in the total mass calculation (Hammer, 1945, p. 57).

The "tailing-off error" results when the integration is not carried to the outer limits of the anomaly (Fig. 7-17). The "datum error" (Fig. 7-18) results from an error in the zero-anomaly level, usually a regional of some kind used for the isolation of the anomaly. These errors can be reduced by carrying the integration from the center out to an estimated fraction of the total anomaly amplitude and applying a correction factor for the undetermined fringes. Curves for the mass deficiencies resulting from incomplete anomalies can be calculated for assumed mass forms but vary widely between different forms.

Table 7-6

M	g	S^2	K
Tons, 2000 lb	mgals	miles^2	6.81×10^7
	mgals	kft^2	2.44×10^6
	0.01 mgal	m^2	0.2634
Tons, metric	mgals	hectares	2.39×10^5
	mgals	km^2	2.39×10^7

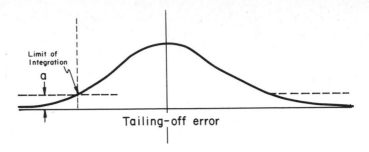

FIGURE 7-17
Tailing-off error. (*From Hammer, 1945.*)

In Fig. 7-19, let g_0 be the total anomaly amplitude and g_i the anomaly magnitude at the limit of integration. Then it can be shown (Hammer, 1945, p. 57) that for a point mass (sphere)

$$M_c = M(1 - r^{1/3})$$

where M_c = calculated mass
     M = actual mass
     $r = g_i/g_0$ = ratio of anomaly amplitude at limit of integration to maximum amplitude

For a line element (horizontal cylinder) the relation is

$$M_c = M \frac{2}{\pi} \cos^{-1} r^{1/2}$$

Curves for these two relations are given in Fig. 7-20. For these two cases the relations are independent of depth.

The relations for bodies of other shapes are quite different, especially for shallow bodies. LaFehr (1965) gives a number of examples.

An example of application of a total mass estimate can be made on the gravity curve of Fig. 7-13, for which gravity was calculated by solid angles of

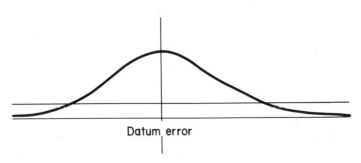

FIGURE 7-18
Datum error. (*From Hammer, 1945.*)

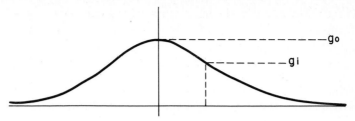

FIGURE 7-19
Notation for curves of Fig. 7-20.

circular disks. The maximum gravity g_0 is 10.51 mgals. The integration was carried out to distances corresponding to gravity values of 8, 6, 4, and 2 mgals with the results given in Table 7-7. Units are kilofeet and tons.

For the value from the widest integration, the calculated anomalous mass is $11,800 \times 2.34 \times 10^6 = 2.76 \times 10^{10}$ tons. The total anomalous mass of the model (Fig. 7-13) from which the curve was calculated is derived in Table 7-8. The total anomalous mass unit times 62.5 (number of pounds in 1 ft^3 of unit density) and divided by 2000 gives 2.31×10^{10} tons.

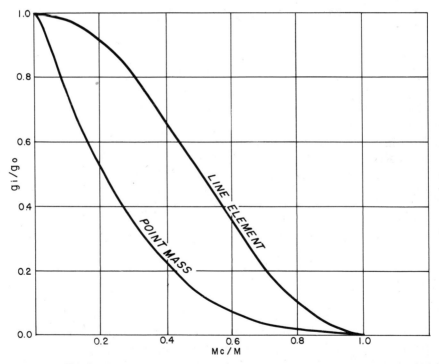

FIGURE 7-20
Effect of width of integration on apparent total mass. (*From LaFehr, 1965, with axes reversed.*)

The fact that the mass value (2.76×10^{10} tons) calculated from the integration is about 20 percent too large is explained by the use of the correction factor curves for a point mass (sphere) as the approximation for a wider body. The optimum correction curve for this example would lie between that for the point mass and that for a width-to-depth ratio of unity (LaFehr, 1965, fig. 2). However, the integration does provide a basis for testing total anomalous mass for a given model to give an approximate check of dimensions and density contrasts of a proposed geologic solution.

Table 7-7

Gravity contour	$\dfrac{g}{g_0}$	Radius, kft	Volume kft$^2 \times$ mgals	Correction factor $\dfrac{M}{M_c}$ (Fig. 7-20)	Multiplier $\dfrac{M_c}{M}$	Corrected total, kft$^2 \times$ mgals
8.0	0.762	7.2	1529	0.090	11.1	17,000
6.0	0.571	10.0	2586	0.178	5.61	14,500
4.0	0.381	12.2	3396	0.275	3.62	12,300
2.0	0.190	18.0	5046	0.438	2.34	11,800

Table 7-8

Disk	Radius, kft	Thickness, kft	Vol, 10^9 ft^3	Density contrast σ	Mass units, 10^9 ft$^3 \times \sigma$
1	5.0	2.0	157	0.1	15.7
2	9.0	3.0	761	0.2	152.2
3	11.0	5.0	1900	0.3	570
Total					738

APPENDIX TO CHAPTER 7

CHARTS FOR CALCULATION OF GRAVITY EFFECTS

This appendix contains the working charts or calculation tools (Figs. 7A-1 to 7A-11) corresponding to the several approximation methods given in the preceding text. They are reproduced full-page size and with the coordinate grid lines included so that they can be used quantitatively.

There is a tendency for these approximate methods to be considered obsolete because of the extensive use of computers for calculating gravity (and magnetic) effects and for direct derivation of model forms by iterative processes. However, computers are not always available or programmed for a particular calculation. Furthermore, the approximate methods can be used to estimate limiting parameters for computer applications. Also, in a great many cases, use of these methods will give results quickly which match observations within the reliability or spacing of the original data.

FIGURE 7A-1

Gravity effects of spheres and horizontal cylinders. These normalized curves give the variation in gravity effect laterally from a point on the surface over the center of spherical and horizontal cylindrical sources. For magnitudes in milligals multiply values from the curves by the following factors:

Form	Curve	Distance units	Factor
Sphere	1	kft	$8.52\sigma R^3/Z^2$
		km	$27.94\sigma R^3/Z^2$
Cylinder	2	kft	$12.78\sigma R^2/Z$
		km	$41.93\sigma R^2/Z$

For examples of application, see pages 190 and 192.

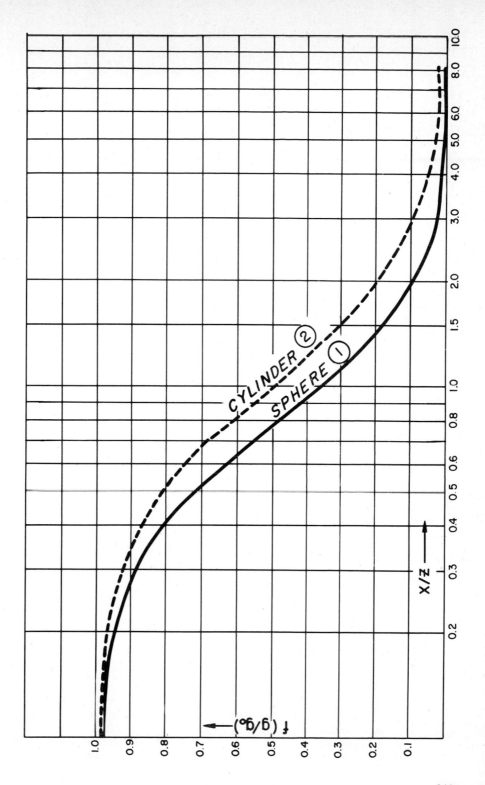

FIGURE 7A-2

Chart to estimate size of spherical sources. This chart is used to estimate the radius R of a spherical source of a roughly circular anomaly. The data used are the gravity magnitude g_0 at the center of the anomaly and its half-width $X_{1/2}$. The chart is calculated for a density contrast of 0.25 g/cm³. For another density contrast σ, the radius R_1 would be $R_1 = \sqrt[3]{0.25/\sigma}$.

For example, an anomaly has a half-width of 8000 ft and gravity relief at its center g_0 of 2 mgals. At the 8000-ft horizontal distance go up to intersect the straight diagonal line ($Z = 1.305X_{1/2}$) to give the depth to center of 10,500 ft. At this depth move to the left to the intersection with the 2-mgal line on the bottom scale, which is between the curves for $R = 4000$ and $R = 5000$, to give an interpolated value of about $R = 4600$ ft. The depth to the top of the body would be $Z - R = 10,500 - 4600$ or about 6000 ft.

A spherical approximation is fairly close for oblate bodies with ratios of dimensions of less than about 2:1. The half-width dimension should be measured in the direction at which it is a minimum.

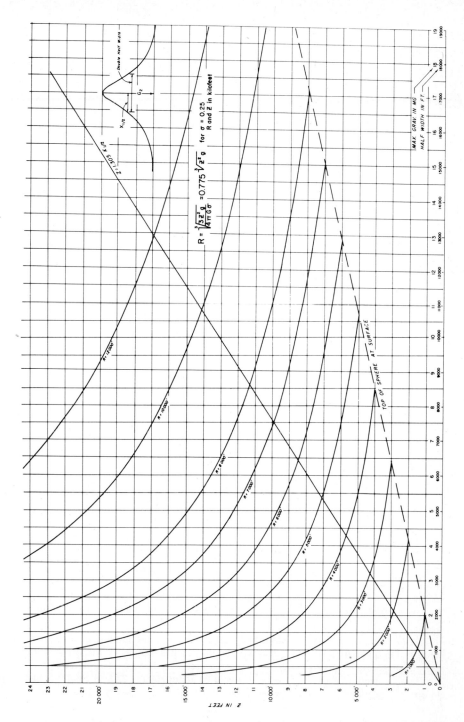

$$R = \sqrt[3]{\frac{3z^2 g}{4\pi G\sigma}} = 0.775 \sqrt[3]{z^2 g} \quad \text{for } \sigma = 0.25$$

R and $\bar{z}$ in kilofeet

FIGURE 7A-3

Chart to estimate size of cylindrical sources. This chart is used to estimate the radius R of a cylindrical source producing a roughly two-dimensional anomaly. The data used are the gravity-anomaly magnitude g_0 on the anomaly axis and half-width $x_{1/2}$ (see insert on Fig. 7A-2). The chart is calculated for a density contrast of 0.25 g/cm^3. For another density contrast σ, the radius R_1 would be $R_1 = \sqrt{0.25/\sigma}$.

For example, an anomaly has a half-width of 8000 ft, and gravity relief g_0 at its center is 2 mgals. For a horizontal cylinder $Z = x_{1/2}$, so the depth is 8000 ft. At this depth on the vertical scale, move to the right to 2 mgals on the bottom scale to give an interpolated radius of about 2200 ft. The depth to the top of the body would be about $8000 - 2200 = 5800$ ft (which is quite close to the depth estimate for a spherical body with the same gravity anomaly and half-width values).

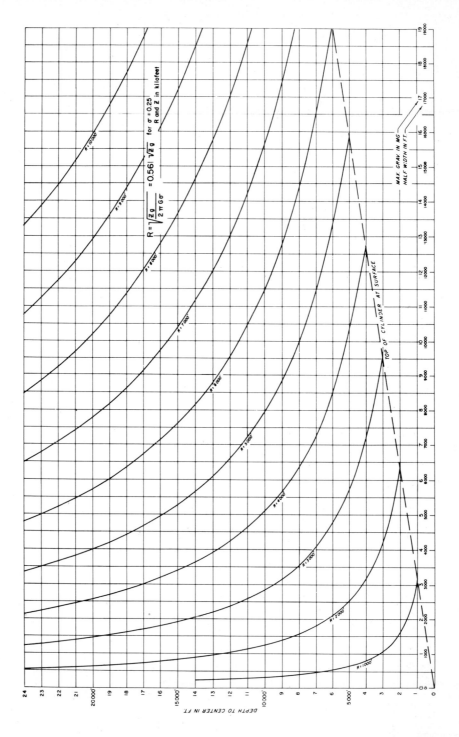

$$R = \sqrt{\frac{Z g}{2 \pi G \sigma}} = 0.56i \sqrt{\frac{Z g}{\sigma}} \quad \text{for } \sigma = 0.25$$
R and Z in kilofeet

R=10,000

R=9,000

R=8,000

R=7,000

R=6,000

R=5,000

R=4,000

R=3,000

R=2,000

R=1,000

TOP OF CYLINDER AT SURFACE

MAX. GRAV. IN MG
HALF WIDTH IN FT.

DEPTH TO CENTER IN FT.

24 23 22 21 20,000' 19 18 17 15,000' 14 13 12 11 10,000' 9 8 7 6 5,000' 4 3 2 1 0

0 1000 2000 3000 4000 5000 6000 7000 8000 9000 10,000 11,000 12,000 13,000 14,000 15,000 16,000 17,000 18,000 19,000
0 1 2 3 4 5 6 7 8 9 10 11 12 13 14 15 16 17 18 19

FIGURE 7A-4

Gravity effects of thin and thick fault blocks. The curves are for two extreme conditions indicated by the insert diagram. For one extreme the mass of the fault block is assumed to be condensed onto a thin sheet at its center, at depth Z, to give the lower curve. For the other extreme the fault extends from the surface to depth $2Z$, to give the upper curve.

The curves are for the effect from the fault edge over the uplifted (positive-density-contrast) side, and begin with a value of one-half the total effect. The effect on the downthrown side is antisymmetrical, asymptotic to zero, and represented by the g/g_∞ values in parentheses.

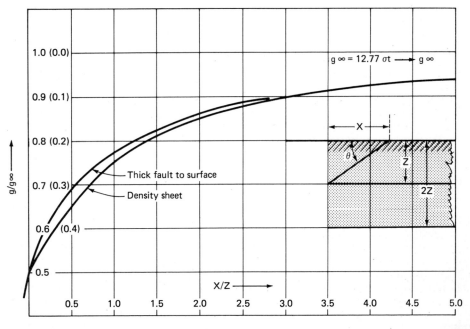

FIGURE 7A-5

Gravity effect of a vertical line element. This can be used as the approximation of the gravity effect of a narrow cylindrical body, such as a salt dome or volcanic plug, with its mass concentrated on a vertical line on its axis. The approximation deteriorates rapidly as the ratio of radius to depth to top is greater than about 0.5. (For error curves, see Hammer, 1974.)

The curve is for the bottom of the mass at infinite depth. For a finite depth, the effect for an element with its top at the lower level is subtracted.

The magnitude of the gravity effect, in milligals, over the center is given by

$$g_0 = \begin{cases} \dfrac{6.39 R^2 \sigma}{Z} & \text{for } R \text{ and } Z \text{ in kft} \\[2em] \dfrac{20.96 R^2 \sigma}{Z} & \text{for } R \text{ and } Z \text{ in km} \end{cases}$$

where R = radius of cylinder
Z = depth to top
σ = density contrast

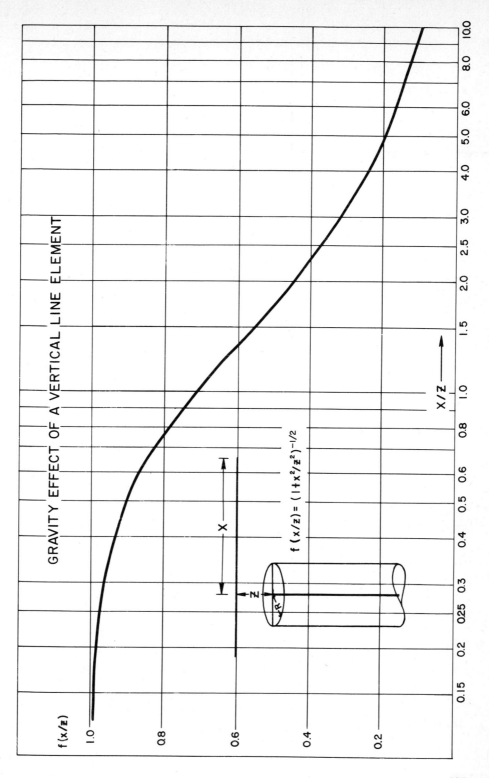

GRAVITY EFFECT OF A VERTICAL LINE ELEMENT

$f(x/z)$

$f(x/z) = (1 + x^2/z^2)^{-1/2}$

X/Z

FIGURE 7A-6

Gravity effect of a thin disk at the center of its surface. The curve is for the variation of the gravity effect of a short cylinder or disk with variation of its radius. It can be used as an approximation for the anomaly of a roughly equidimensional geological basin. For application to the Los Angeles Basin, see page 194.

The quantity given, g/g_∞, is the ratio of the gravity effect at the center of the basin to that of a body of infinite radius, for which

$$g = \begin{cases} 12.78\sigma L & \text{for } L \text{ in kft} \\ 41.93\sigma L & \text{for } L \text{ in km} \end{cases}$$

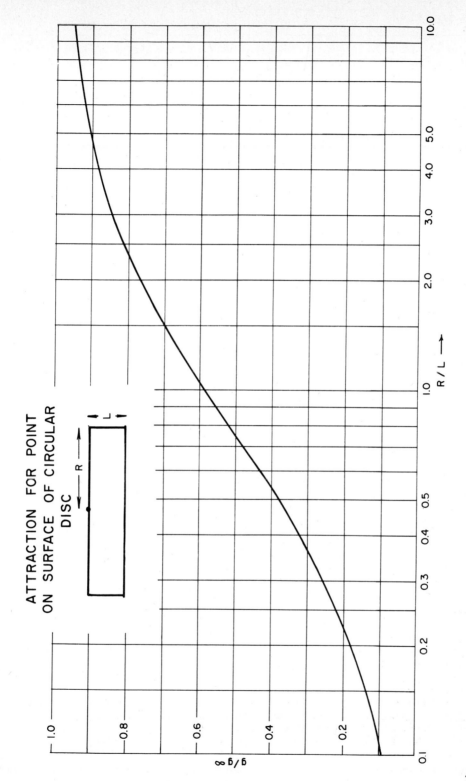

ATTRACTION FOR POINT ON SURFACE OF CIRCULAR DISC

g / g_∞

$R / L \longrightarrow$

229

FIGURE 7A-7

Gravity effect of a vertical thin sheet. The gravity effect for a thin sheet can be used as an approximation for a dike or any vertical two-dimensional body for which the thickness is small compared with the other dimensions.

Gravity over the axis is given by

$$g_0 = \begin{cases} 9.37\sigma t \log n & \text{for } t \text{ in kft} \\ 30.7\sigma t \log n & \text{for } t \text{ in km} \end{cases}$$

where σ = density contrast
  t = thickness
  n = ratio of depth to bottom to depth to top (see diagram on curve sheet)

The three curves are for values of n of $\sqrt{10}$, 10, and $\sqrt{1000}$. The horizontal coordinate is values of x/z, where z is the depth to the top of the dike. The vertical coordinate is values of g/g_0 and shows how gravity decreases away from the axis.

For intermediate values of n approximate interpolation between the curves can be made, using appropriate values of $\log n$.

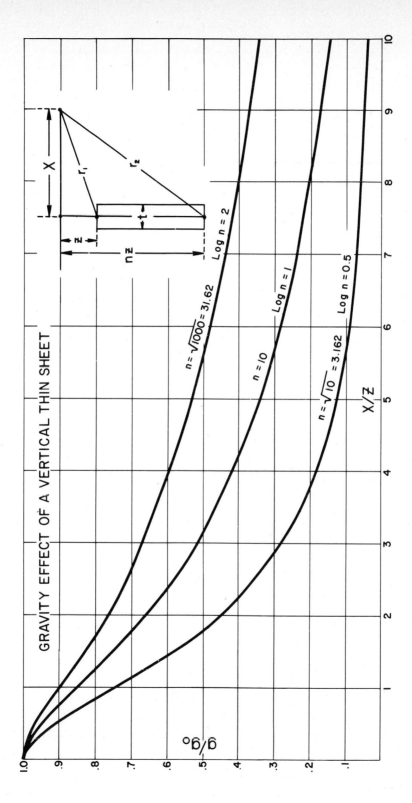

GRAVITY EFFECT OF A VERTICAL THIN SHEET

231

FIGURE 7A-8

Solid angle of a circle for a point above its center. This chart is convenient for determining the central or g_0 value in calculations of models by solid angles or for a quick estimate of the maximum gravity over the center of a model composed of one or more circular disks. Values from this chart correspond to the bottom margin ($x/z = 0$) of Fig. 7A-9.

Numerical values are given by

$$g_0 = \begin{cases} 2.03\omega\sigma t & \text{for } t \text{ in kft} \\ 6.65\omega\sigma t & \text{for } t \text{ in km} \end{cases}$$

where g_0 = gravity, mgals
ω = solid angle read from chart
σ = density contrast
t = thickness of disk represented by circular lamina at its center

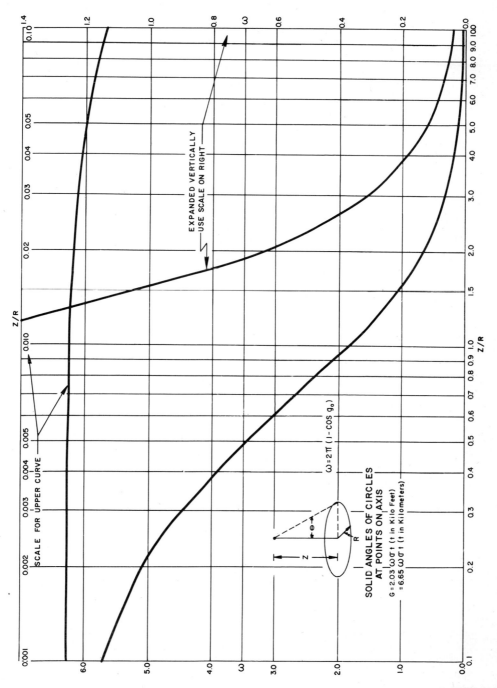

SOLID ANGLES OF CIRCLES
AT POINTS ON AXIS

$$\omega = 2\pi (1 - \cos g_o)$$

$G = 2.03 \, \omega \sigma t$ (t in Kilo Feet)
$= 6.65 \, \omega \sigma t$ (t in Kilometers)

Z/R

EXPANDED VERTICALLY
USE SCALE ON RIGHT

SCALE FOR UPPER CURVE

FIGURE 7A-9

Chart for solid angles of circles. Gravity effects for bodies with roughly equal horizontal dimensions can be approximated by one or more circular disks. The approximation is correct to within less than about 5 percent for thickness of one-half the depth or less.

The lines on the chart are contours of equal values of solid angle. A vertical line on the chart corresponds to a horizontal profile outward from the surface projection of the center of the circle. Values of x/z where this vertical line crosses a contour multiplied by Z for a particular layer give the horizontal distance to the contour value of the solid angle. For an example of application, see page 203.

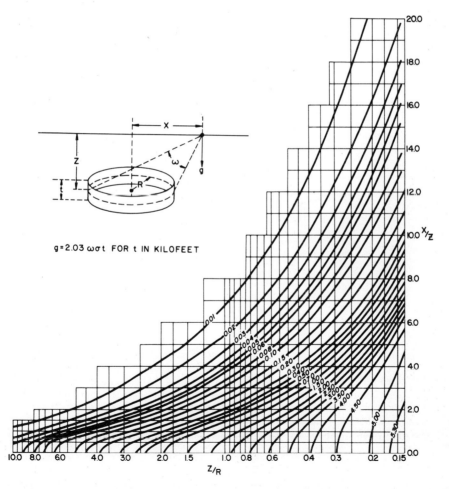

$g = 2.03 \, \omega \sigma t$ FOR t IN KILOFEET

FIGURE 7A-10

Two-dimensional dot chart for a vertical cross section. This chart can be used to calculate the gravity effect from the cross section of an elongated (two-dimensional) body. The section must be drawn with equal vertical and horizontal scales. An outline of the section is drawn at scale $1:k$, where k is the scale ratio. The section, on transparent film, is laid over the chart with the chart center at the point of calculation and the number of dots within the outline counted. The chart is then moved to another point and the count repeated.

The gravity value, in milligals, is $g = 10^{-5}nk\sigma$, where n is the dot count, k the scale ratio of the outline, and σ the density contrast. The solid dots have a value of unity. The open circles have a value of 0.1.

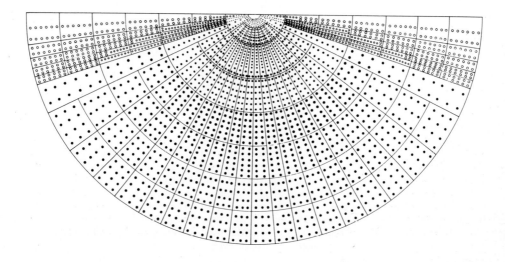

FIGURE 7A-11

Polar dot chart for a three-dimensional lamina. This chart can be used for the calculation of the gravity effect of a lamina or sheet with any outline. The outline must be drawn at a scale such that its depth below the plane of calculation corresponds to the unit depth shown on the chart.

The chart is designed with the total number of dots to infinity, that is, solid angle of 2π, equal to 1277, so that the gravity effect in milligals at the center of the chart is $g = 0.01n\sigma t$, where n is the number of dots within the outline, σ is the density contrast, and t is the thickness of the lamina in kilofeet. The numbers on the circles are the total number of dots within that circle to save counting for those circles which are completely within the body outline.

The approximation is good to about 5 percent or less if the thickness is less than about half the depth.

This chart is useful for estimating gravity effects over the center and beyond the edges for irregularly shaped bodies or as a general control of magnitudes in setting up more complex or computer calculation processes.

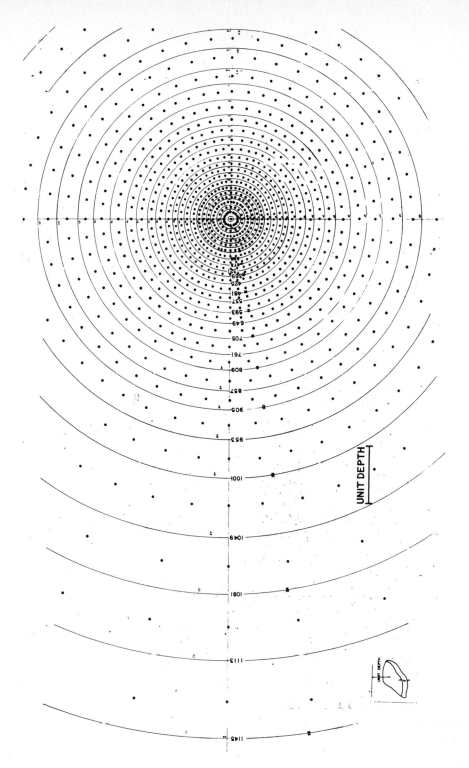

UNIT DEPTH

8

INTERPRETATION OF GRAVITY DATA

INTRODUCTION

The usual purpose of a gravity interpretation is to add to the geological information below the area covered. The degree and the certainty with which this can be done are highly variable and depend on the intensity of coverage, the precision of the measurements, and the amount of other geophysical and geological information available.

If an interpretation is to be quantitative to any degree at all, there must be density information from measurements or, more commonly, from inferences based on the general nature of the rocks, which can be learned from existing geological studies or maps of the area, if any. Before undertaking the analysis of a survey in a new area, an interpreter should acquaint himself with whatever geological information is available.

For a reconnaissance survey, often with lines of stations several miles apart, only rather general geological inferences are possible. Usually the anomalies are broad-scale features such as closed maxima or minima or alignments or trends of maximal and minimal axes. Depending on their relief and the distances between axes, these may be interpreted as being indicative of structural features or trends which may be attributed to deformation in sediments or to density contrasts in the basement or both. Tests for possible depth and structural relief, based on reasonable estimates or guesses for probable density contrasts, can be made from appropriate simple models such as the sphere, cylinder, or fault (Chap. 7). The approximate extent and, to some degree, the probable depth range of a geological basin can be estimated, although basement depths nearly always are more reliably estimated from magnetic than from gravity data.

With surveys having greater detail, smaller anomalies often become apparent, and their evaluation may become the primary objective of an interpretation. Such anomalies nearly always become much more obvious on a residual or filtered map of some kind, e.g., that of the Los Angeles Basin (page 153). An interpretation may be concerned with the analysis of a single anomaly surveyed in detail. Such is the case for many of the salt domes of the Gulf Coast. As will be pointed out later, to be quantitative such interpretations require data other than gravity even if the amount and detail of the gravity survey are as great as can be usefully applied.

AMBIGUITY IN INTERPRETATION

Any interpretation of gravity data is subject to two limitations that must always be kept in mind:

1 The inherent ambiguity in the possible sources of a given gravity anomaly
2 The complete dependence of gravity anomalies on the existence and magnitude of horizontal variations in the densities of the rocks, together with the form, magnitude, and depth of the boundaries of the density anomaly

These limitations are treated in some detail in the following paragraphs.

The various systems of calculating gravity effects from a given mass distribution, as outlined in Chap. 7, give definite, single-valued results when the outlines and density contrasts are defined. The reverse is not true; for a given distribution of gravity there is no single and unambiguous distribution of mass which will have a calculated effect corresponding with that observed. Over a relatively deep mass anomaly, even though that anomaly is very local, the gravity effects are spread out over horizontal dimensions which are several times the depth. On the other hand, a broad and shallow source will produce a gravity anomaly which is not much wider than the source itself. For a given width of anomaly there is a maximum possible depth for the source. This depth is that of the point source (or sphere) which will produce an anomaly of that width. Between this maximum depth and the surface, there is a cone of possible sources, as illustrated schematically by Fig. 8-1. The same anomaly could be accounted for completely by a thin lenticular body (3 in the figure) at a very shallow depth or by an intermediate narrower and thicker body (2). The spherical mass (1) is the deepest possible solution, as any deeper mass would produce a broader anomaly, but there is an infinite variety of possible solutions at shallower depths. Also, combination solutions are possible, for which parts of the total mass anomaly are in different bodies at different depths.

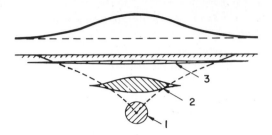

FIGURE 8-1
Cone of sources. The sphere (1) is the deepest body which can approximately account for the gravity anomaly shown. Shallower and broader bodies, such as 2 and 3, also could account for the anomaly. All would have the same total mass anomaly.

The only property that the several possible solutions have in common is that their total-mass anomaly (the product of the volume times the density contrast of the anomalous material) must be the same for each complete solution. This corresponds to the total-mass calculation, page 212.

A rather extensive treatment of the ambiguity problem has been given by Skeels (1947). Figure 8-2 shows how a very wide range of configurations of a single density contrast, the basement surface, can give the same gravity anomaly, depending on the assumed depth to that surface. The same anomaly also can be accounted for by a heavy mass within the basement material, i.e., an "intrabasement" contrast combined with a very small anomaly on the basement surface. Figure 8-3 is another example of how the sharpness of a feature which will account for a given gravity anomaly depends on its depth. Here, a vertical step or fault and a shallower but much smoother density surface produce the same calculated gravity effect.*

Many other examples could be given, but those shown should suffice to point out clearly that no exact solution for the source of a given gravity anomaly can be obtained without recourse to information other than gravity. This may range all the way from simple geologic reasonableness and common sense to other geophysical data, e.g., seismic or magnetic surveys, or one or more drilling contacts with the principal source of density contrast responsible for the anomaly.†

DENSITIES AND DENSITY CONTRASTS

The second uncertainty that is always involved in any quantitative evaluation of gravity results is that gravity anomalies depend on density contrasts which are seldom accurately known. Furthermore, they depend on horizontal variations in

---

* This diagram is not quite accurate because the total relief of the anomaly from the smooth flexure should be the same as that of the vertical fault for the same density contrast. The total anomaly for the 3000-ft displacement with 0.3 density contrast should be 11.25 mgals (see page 193), rather than about 8 mgals within the diagram. The remainder of this total relief is in the asymptotic portion of the fault curve beyond the limits of the figure.

† Theoretically, if the source of an anomaly can be ascribed to a single surface at which the density contrast is defined, a single control point can define the entire surface. The relevant geological example is a simple salt dome. If the gravity anomaly is completely defined and the density-depth relation is known, a single drilling contact on the salt would permit a calculation of the complete salt-sediment boundary.

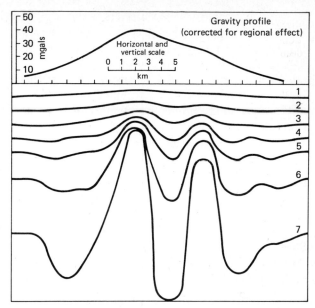

FIGURE 8-2
Various solutions of a basement profile to satisfy the gravity profile shown.
(*From Skeels, 1947.*)

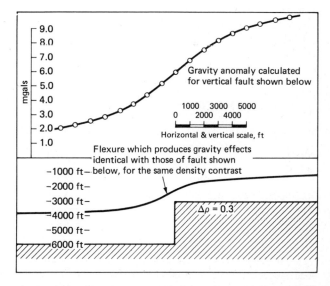

FIGURE 8-3
Gravity profile for a vertical step fault and, as an alternative solution, a flexure
which would produce an identical gravity effect. (*From Skeels, 1947.*)

density. We could have rock layers with strong density contrasts between them, but if these layers are perfectly horizontal, they will cause no gravity anomalies. On the other hand, if rock layers with differing densities are deformed, as in Fig. 8-4, an anomaly will be produced.

Referring to Fig. 8-4, let us assume that the layers 1 to 4 have successively higher density values d_1, d_2, d_3, and d_4. The flat-lying strata at the margins of the figure will produce no gravity anomaly there because there is no horizontal variation. Over the central part of the figure, these layers are disturbed by a structural uplift which produces the density contrasts indicated by the various hachured areas. In the upper part of the figure, layer 2 is uplifted into the normal zone for layer 1 and produces a density contrast $d_2 - d_1$. Other density contrasts would be produced as indicated on the diagram. In this instance, where each layer is of successively greater density, all the density contrasts are positive, and the sum of all of their effects produces the positive anomaly indicated by the gravity profile.

From this example it is evident that any displacement of normally flat-lying layers with different densities will produce some sort of density contrast which will cause a gravity disturbance. The magnitude and form of the gravity effect depend on the details of the density contrasts involved, i.e., their magnitudes, vertical relief, depth, and horizontal extent.

Horizontal density contrasts also can result from facies changes, as from sand to shale or limestone to dolomite. Resulting gravity effects usually are quite diffuse because such changes occur mostly over horizontal distances which are several times the thickness of the geological unit involved. Thus, the primary property of the rocks which relates gravity anomalies to geology is the densities of the various components of the geologic column. The resulting density differences produce the gravity anomalies. The purpose of the survey is the expectation that the analysis of these anomalies will lead to a better understanding of the nature, structure, or attitude of the rocks which produce them.

ROCK DENSITIES

The densities of rocks are controlled by three factors, the grain density of the particles or matrix forming the rock mass, the porosity, and the fluid in the pore space.

For the common rock-forming minerals, the grain densities are not widely variable. For pure quartz (SiO_2) the density is 2.65, and for calcite ($CaCO_3$) the density is about 2.72. The clay minerals are more variable but are generally in the range 2.5 to 2.8. Thus for the more common sedimentary rocks, sandstones, limestones, and shales, the densities of the basic materials of which they are formed do not vary widely, and their densities are determined to a very large extent by their porosities. For most rocks below the near surface water table we can assume that the pore space is filled with water. The density of the water probably is somewhat greater than unity because of salt and other minerals in solution. Also lighter fluids, oil or gas, may be present. However, these factors affect only the pore-space portion of the rock volume, and usually, unless accurate data on mineralogy

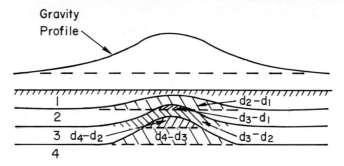

FIGURE 8-4
Density layers, density contrasts, and gravity anomaly.

and fluids are available, density estimates from porosity are made assuming that the pore space is filled with water with density 1.0.

Figure 8-5 is a simple chart which gives rock density as a function of porosity. Each line represents the density of the rock with the grain density indicated and with the pore space dry and saturated with water of density 1.0. The numerical values of Fig. 8-5 are derived very simply from the relations

$$d_r = \begin{cases} d_g(1 - P) & \text{for dry rocks} \\ d_g(1 - P) + P & \text{for rocks saturated with water of unit density} \end{cases}$$

where d_r = bulk density of rock
d_g = grain density of rock-forming minerals
P = fractional porosity

This discussion and the chart are valid for the common sedimentary rocks. Some important exceptions which may be encountered in gravity surveys are due to the chemical nature of the rock material. The most common and important are those encountered in salt domes. Rock salt (halite) has a very low grain density, about 2.165 for pure sodium chloride, and practically no porosity; it forms the general mass of the dome. Quite commonly the salt is overlain by a cap rock, usually (in the Gulf Coast) composed of a lower layer of anhydrite ($CaSO_4$), with a density of about 2.9, and an upper layer of limestone.

Anhydrite and dolomite, with grain densities of 2.8 to 3.0 and often with very low porosity, are common components of chemical (evaporite) sedimentary deposits. Also, salt and other low-density evaporites, e.g., sylvite, may be major components of diapiric structures and be the source of large negative gravity anomalies. The lowest-density material affecting gravity surveys of which this writer is aware is diatomite (in the southwestern part of the San Joaquin Valley, California), which can have a density as low as 1.0 or slightly less. Some minerals, e.g., magnetite, have densities up to 5.0 but are not a factor in petroleum gravity surveys.

Generally, the igneous and metamorphic rocks, including most basement

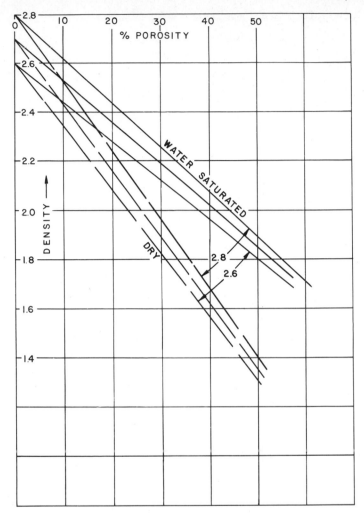

FIGURE 8-5
Relation of density to porosity for dry (*lower*) and for water-saturated (*upper*)
rocks. Curves are shown for grain densities of 2.6, 2.7, and 2.8.

rocks, have very low porosities and may contain minerals of higher density than
those of the sediments, so that there is commonly a density increase from the sediments to the underlying basement. The light-colored igneous rocks, e.g., granites
with large quartz content, tend to be of lower density than the darker igneous rocks.
While most igneous materials are of higher density than the sediments, there are
some overlaps. For instance, some granites with densities about 2.6 to 2.7 may be
lighter than a nonporous dolomitic limestone with density of 2.8. Thus, it is possible (but rare) for a granite knob buried by a heavy limestone to cause a negative
gravity anomaly.

SOURCES OF DENSITY INFORMATION

Since densities and density contrasts are the fundamental factors controlling gravity anomalies, it is important to have as much information about densities in the geologic section involved as feasible. Often, especially in newly explored areas, where the lithology of the section may be quite unknown, it is impossible to obtain factual density data, and possible variations in density have to be inferred from whatever general information is available or from the nature of the gravity anomalies themselves.

Table 8-1 gives densities of common rocks and is helpful in indicating probable density contrasts when only the general nature of the rocks is known from descriptions of outcrops or of cores from drilling which may be given in geological reports. Unfortunately, such reports rarely give densities of the different formations, and only educated guesses can be made from the geological description and the densities of the table. Therefore, any definite information on probable densities is always useful in making gravity interpretations.

Several sources of density data may be available in different areas under different circumstances and with widely differing reliability. Any possibility of their use should be pursued in connection with the geological interpretation of gravity surveys.

Cores

Density measurements on core samples can give reliable values, especially in consolidated sediments, but are rarely available for any but very limited segments of the total geologic column, commonly only in an oil- or gas-producing section.

Cuttings from Drilling

Density measurements can be made with a Jolly balance (Bible, 1964) on the cutting samples which are routinely taken during drilling and saved for lithologic or paleontologic analysis. The differences in density probably are indicated better than the absolute values. There is a tendency for the sampling to be selective in that the harder and more resistive parts of the sections cut by the drill are those which may be selectively preserved in chips large enough for measurement. Also, some pore space is lost. Both these effects tend to make the values from cutting measurements too high.

At least one geophysical operator (Bible, 1964) made routine Jolly-balance measurements of drill cuttings as an aid to gravity interpretation. When they are made at frequent depth intervals, so that individual variations can be averaged out, such measurements may indicate major density discontinuities (Fig. 8-6), which can be useful in gravity interpretation. This procedure is not widely used, apparently because of the very considerable amount of work involved and some skepticism of its value, but in favorable circumstances it can be quite helpful.

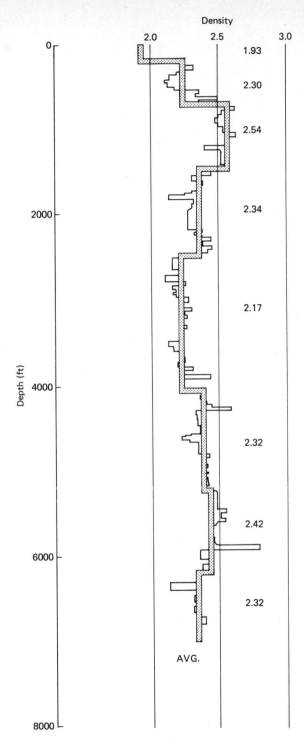

FIGURE 8-6
Density log, Galveston County, Texas, from drill cuttings. (*From Bible, 1964.*)

Gamma-Gamma Density Logs

This well-logging tool (Tittman and Wahl, 1965) measures gamma rays generated within the formation by Compton scattering of a collimated primary gamma-ray beam from a high-intensity artificially radioactive source such as irradiated cobalt. The intensity of the secondary radiation is approximately inversely proportional to the density of electrons in the rock material.

There are corrections because the relation of electron density to neutron density, and therefore to the bulk density of the material, depends, to a small degree, on the nature of the material (Zanes, 1960). For most of the common rock-forming minerals, the corrections for this effect are quite small, mostly less than about 0.02 g/cm^3. The important exception is salt, for which the apparent density is too low and must be corrected by about 0.13 g/cm^3.

The gamma-gamma log is run quite frequently in wells, not primarily for the density values as such but because it can contribute another factor of information in addition to the many electrical and other measurements often made to determine the nature of the rocks penetrated and their fluids.

The gamma-gamma log is run in an open hole because the gamma rays would be greatly reduced by penetrating steel casing. The effective depth of penetration (and hence the radius of the measurement) is quite small—of the order of 1 ft. In some special cases, this leads to a marked difference in the density value from a gamma-gamma log and that measured by a borehole gravity meter (see Fig. 8-11).

Density recordings are made in great detail (Fig. 8-7), i.e., at intervals of 1 ft or less. For use in connection with gravity surveys, the great detail of these records is not needed. A convenient way of using the logs is to estimate an average value for each 100 ft of depth and plot them on a relatively small vertical scale, such as 1000 ft to the inch. A smooth curve may be drawn through these points when changes are continuous, or steps or discontinuities may be shown where any abrupt changes in lithology occur. Such a plot gives the general pattern of density changes with depth which are relevant to gravity interpretation. An example from such a plot is Fig. 8-8.* These density logs may give quite satisfactory, regular, and reliable values in consolidated sediments. They often are very ragged and unreliable in shallow, unconsolidated sediments, as in the upper several thousand feet of section in the Gulf Coast. Part of the irregularity is due to the sensitivity of the measurement to variations in the size of the hole. It is common practice to run a caliper log to record the hole diameter. Corrections are then made for changes in size and also for mud thickness, but these corrections are not accurate in ragged holes. Such variations, of course, are much greater in unconsolidated sediments.

Densities from Seismic Velocities

There is an approximately regular relation between seismic wave speeds and densities. In general, rocks with high wave speeds are of high density. Seismic wave

---

* The minimum in this curve is not typical of the Gulf Coast sedimentary section. This well is close to a salt-dome flank and may reflect influence of the dome on the gravity or on the sediments.

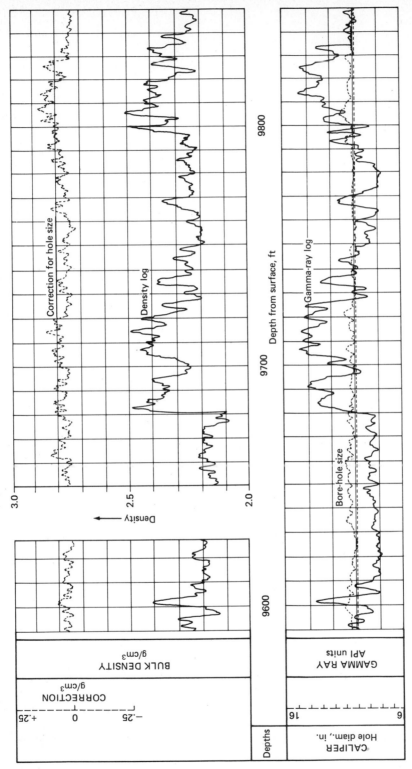

FIGURE 8-7
Sample of radioactivity (gamma-gamma) density log.

250

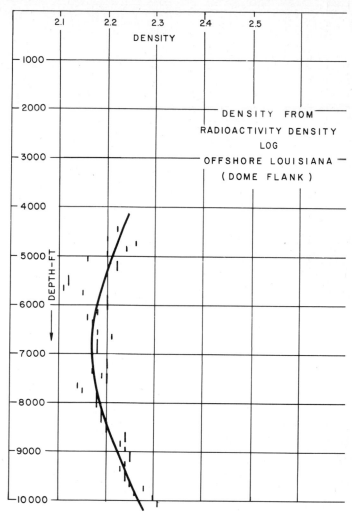

FIGURE 8-8
Densities from radioactivity (gamma-gamma) density log, offshore Louisiana.

speeds in rocks are quite commonly available from continuous velocity logs
(CVL), which are usually expressed in the reciprocal of wave speed or as microseconds per foot. Seismic wave speeds also may be available from refraction-
seismograph results and also from reflection-seismic results as velocity "gathers"
as a by-product of computer processing of digitally recorded seismic data.

Figure 8-9 shows relations between seismic velocity and density from different
sources. The curves by Woollard (1959) and Nafe and Drake (see Grant and West,
1965, p. 200) have been replotted as microseconds per foot to be directly comparable with the usual CVL recording. The curve between these two was deter-

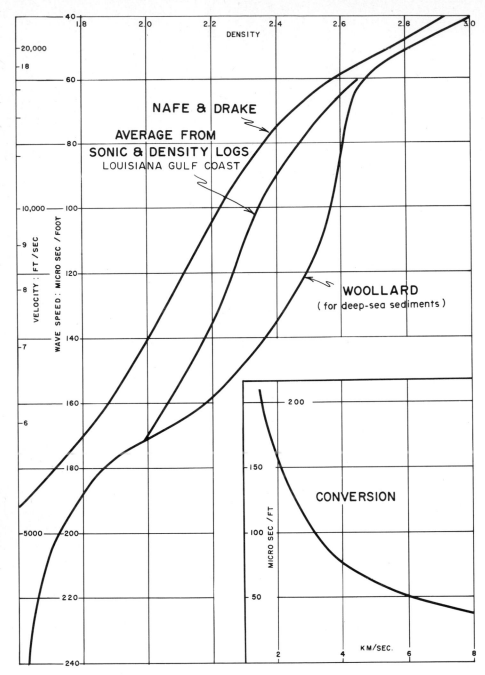

FIGURE 8-9
Empirical relations of density to seismic wave speed.

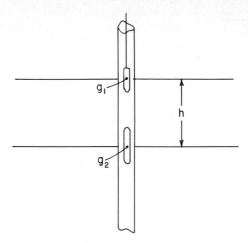

FIGURE 8-10
Principle of borehole gravity meter for
density determination.

mined by the writer from data from a number of wells in which both CVL and gamma-gamma density logs were run.

There is considerable variation in the different sets of data, but on the whole there is a fairly regular relation between wave speed and density. Therefore when wave-speed data from either seismic observations or CVLs are available, at least some approximation of densities may be determinable as a factor in gravity interpretation.

Densities by Borehole Gravity Meter

Measurement of gravity differences within a borehole gives a direct measure of the average density of the section between the points of measurement (Smith, 1950; Howell, Heintz, and Barry, 1966; Jones, 1972). An example of an instrument for such use is described on page 41.)

The principle of such measurements is indicated in Fig. 8-10. Let g_1 and g_2 represent two gravity measurements in a drill hole separated by a vertical interval h. The measurement at g_1 is increased by the downward attraction of the sheet of material of thickness h or by the quantity $2\pi G\sigma h = 0.01276\sigma h$ mgal/ft, corresponding to the Bouguer coefficient (see page 21). At g_2 the measurement is decreased by the upward attraction of the same amount. Also, of course, the lower measurement is increased by the normal vertical gradient of 0.09406 mgal/ft (see page 20). Thus, the difference in gravity between the two points is

$$g_2 - g_1 = \Delta g = 0.09406h - 2 \times 0.01276\sigma h$$

and the density σ is

$$\sigma = \frac{0.09406h - \Delta g}{0.02552h} = \begin{cases} 3.686 - 39.18 \dfrac{\Delta g}{h} & \text{for } h \text{ in ft} \\[2ex] 3.686 - 11.95 \dfrac{\Delta g}{h} & \text{for } h \text{ in m} \end{cases}$$

It is convenient to determine the density directly from the measured vertical gradient $\Delta g/h$ in the borehole. A chart showing the relation of density to gradient, expressed in milligals per foot and milligals per meter, is given in Fig. 8-11.

For accurate density measurements over small depth intervals (less than 25 to 50 ft), the gravity values may need to be corrected for tidal, topographic, and other disturbances (Jones, 1972).

An example of density measurements from a borehole gravity meter is given by the solid line of Fig. 8-12 (Jones, 1972). The measurements are mostly over intervals of about 1000 ft except for some smaller intervals in the lower part of the hole. The measurement, of course, gives the average density over the interval between the points of gravity measurement. It is this average value which is of direct concern in evaluation of gravity anomalies.

Figure 8-12 also includes the results from a gamma-gamma density log run in the same hole. The dotted line shows these measurements averaged over the same interval as the gravity measurements. Above a depth of 7500 ft the two densities are in quite good agreement. Below this depth there is a departure reaching nearly 0.15 g/cm^3 in the 8600 to 9600-ft interval. The 7500-ft depth is the top of a shale zone, part of which (below 8600 ft) is overpressured.*

For application to gravity interpretation (see page 273), unfortunately, instruments for gravity measurements in boreholes (see pages 43 to 45) are difficult to make, and their operation is slow and expensive, so that the results of such measurements are relatively rare. This situation is improving, as at least one geophysical company has provided borehole gravity measurements as a routine logging service.

Surface Sampling by Gravity Meter

An "elevation-factor profile" (Nettleton, 1942) for determining average density of surface material is often observed in connection with the field gravity observations (page 89). This gives a measure of the density of surface material within the range

---

* In the Gulf Coast, tight, i.e., very low permeability, shales are encountered quite frequently in which the fluid pressure is much above normal. In extreme cases it approaches pressures corresponding to the rock overburden, whereas the normal pressure corresponds approximately to that of a fluid column to the surface. In many cases, as in this example, the high-pressure zones are reported to have low densities, as if the shales were so tight that the pore-space fluids could not escape and keep the pore space open to maintain the low density. While the low-density indication in some cases is derived from measurements on cores or cuttings, many are derived from gamma-gamma logs. It may be that the phenomenon indicated by Fig. 8-12 is more common than realized, which would mean that the high pressures are not accompanied by low densities as often as thought.

We have no clear explanation of this phenomenon. The obvious difference in the two types of measurement is that the density determined from the borehole gravity averages over a horizontal distance several times the distance between the two points of measurement while the gamma-gamma measurement is influenced by the material within only 1 ft or less from the wall of the hole. The subject is complex and not well understood. Further discussion is outside the field of this book.

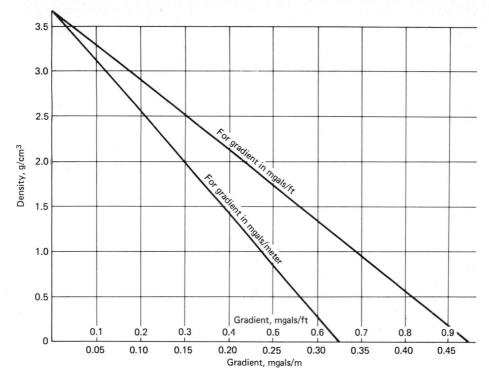

FIGURE 8-11
Relation of density to vertical gravity gradient in a borehole.

of the topographic relief which is useful for determining the elevation factor for data reduction and, occasionally, for estimating near-surface densities for interpretation of shallow anomalies.

RANGES OF ROCK DENSITIES

Despite all the possible ways of determining densities and the density contrasts which may be responsible for gravity anomalies, completely satisfactory data on densities are rarely, if ever, available. Approximate values and general ranges of densities are given in Table 8-1; much more detailed density data are given, for instance, by Heiland (1940), Birch (1942), and Dobrin (1960).

It is only differences in these densities, where rocks of different character are brought into horizontal juxtaposition, that can cause gravity anomalies. From Table 8-1 it is evident that the density contrasts involved in usual sedimentary sections will rarely be more than about 0.5.

With only limited geological data available such general values may be all that can be used to serve as a basis for estimating structural relief corresponding

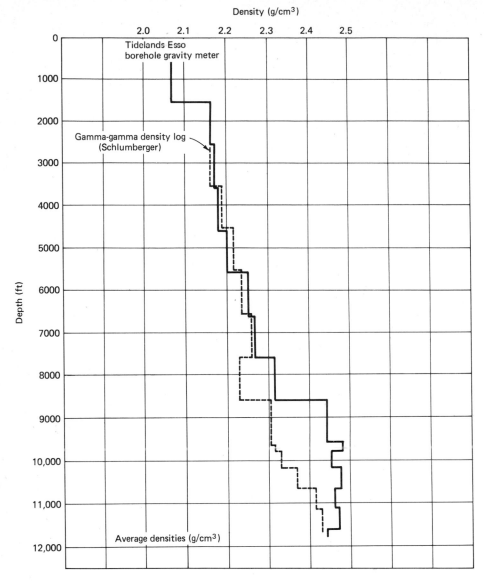

FIGURE 8-12
Example of density determination by borehole gravity meter.

to observed gravity anomalies. While the extreme range of possible densities is from about 1.0 (diatomite) to 2.9 or 3.0 (anhydrite, dolomite, or basic igneous rocks), the actual range likely to be encountered in any one local area is very much less.

Unconsolidated sands and shales are likely to have densities in the range of about 1.7 to 2.3, increasing rapidly with depth (Figs. 8-8 and 8-12). Classic studies of changes of density and porosity with depth have been made by Hedberg (1926) and Athy (1930). Figure 8-13 shows the variation of density of a single formation (Garber shale) from measurements on samples from a wide range of depths as the same formation dips into the Anadarko Basin. Hedburg's porosity data, based on measurements on cuttings from a well in Kansas (Hedberg, 1926, table IV), converted to density by the curve of Fig. 8-5 for grain density 2.8, plotted against depth, and included in the same figure show a quite similar increase of density with depth.

For the most part where there are large thicknesses of geologically recent (Upper Tertiary) sediments, as in the Gulf Coast, they increase in density rather regularly (page 258 and Fig. 8-14).

Compact sands and shales which are older (e.g., Paleozoic, as in Fig. 8-13) are likely to have densities of around 2.5 to 2.6. Heavy massive limestones, e.g., the Ellenburger of Texas or Arbuckle of Oklahoma, may have densities of around 2.7. Within the area of a single gravity survey the range of density contrast commonly is about 0.25, and often such a general figure can be used for estimating anomaly magnitudes when no definite information is available. However, any really quantitative interpretation of gravity anomalies in terms of geologic structure involves a reasonably reliable determination of the magnitudes of the density contrasts involved.

SALT-DOME DENSITIES AND GRAVITY ANOMALIES

Salt domes are of special interest in the application of the gravity method to petroleum exploration. They are economically important primarily because they control the deformation of the surrounding sediments in which oil accumulations occur. A very large fraction of the oil in the Gulf Coast zone from the Mississippi River westward to about 75 miles west of Houston and inland approximately 100

Table 8-1

Type	Density
Unconsolidated sediments	1.7–2.3, usually increasing with depth on account of compaction of the shales
Sandstones	2.0–2.5, varying principally with porosity
Salt, pure halite	2.16
In salt domes	2.20, including disseminated anhydrite
Limestone	2.5–2.7
Granite	2.6–2.8
Basalt (sial)	3.0
Dunite (sima)	3.3
Average density of crustal rocks above sea level	2.67

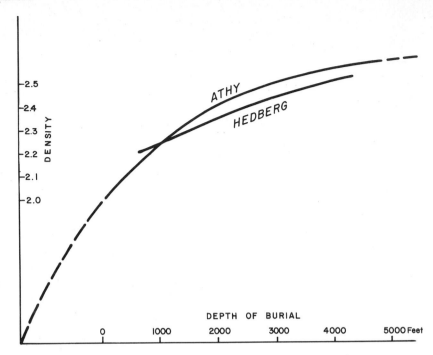

FIGURE 8-13
Variation of densities of shales with depth. (*Modified from Athy, 1930, and Hedberg, 1926.*)

miles from the coast is over or around salt domes. An even larger fraction of the oil fields now reaching over 100 miles offshore is in salt-dome structures. Salt domes are economically important also for the occurrence of free sulfur in some of the porous limestone cap rocks. The salt itself is mined in a few domes, either through conventional shafts or by solution. Also domes are important in a few other oil-producing areas. Because of their somewhat unusual geologic nature, resulting in near vertical juxtaposition of materials with strong density contrasts, they are particularly susceptible to gravity exploration.

The very beginning of gravity exploration by torsion balances in the Gulf Coast was in the search for domes. Even earlier torsion-balance surveys were made over salt domes in Germany and in Mexico. Throughout the history of gravity exploration up to the present time the gravity method has been very effectively applied to determining the presence and the depth of salt domes and delineating their approximate boundaries. This is because salt domes are relatively simple geologic disturbances and their flanks are usually steeply dipping to vertical or even overhanging. As is evident from the calculation of gravity effects from geometric models, the gravity method is inherently better suited to the detection or delineation of steeply dipping density contrasts, of which a vertical fault is the

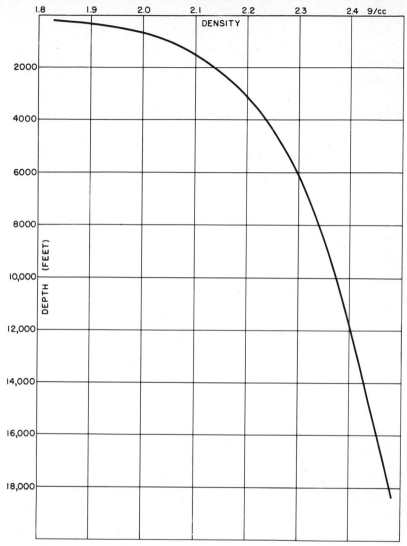

FIGURE 8-14
Density-depth curve, Gulf Coast. (*After Dickinson, 1953.*)

extreme example; a vertical-sided dome may be considered, geometrically, to be a curved fault which is closed on itself.

Salt-Dome Geology

Salt domes, particularly as they occur in the Gulf Coast region of the United States, are intrusions of rock salt (halite) through the sediments. The salt column rises from a very thick salt layer of unknown but very large depth (probably of the

order of 30,000 to 40,000 ft) and extends upward to the surface or to some intermediate depth. In plan, the salt column is commonly circular to elliptical, with
horizontal dimensions of the order of 1 to 5 miles. On shallow domes the salt is
often overlain by cap rock, of which the lower part is anhydrite and the upper part
is limestone.

Salt-Dome Densities

The density situation around salt domes is complex. The sediments increase in
density with depth from values as low as 1.6 near the surface to probably 2.5 to 2.6
at great depth. The density increase with depth is due primarily to compaction and
reduction of porosity in the shales. For pure rock salt (halite) the density is 2.16,
but in most domes there is some disseminated anhydrite in the salt, making the
resultant density about 2.20. There is practically no porosity, so there can be little
or no increase of density with depth, and density 2.20 generally is used for the
entire salt column.

Because of the increase with depth, the density of the sediments is less than
that of the salt at shallow depth and is greater than the salt below a depth at which
the increasing density of the sediments becomes equal to that of the salt. In the
Gulf Coast this "crossover depth" varies from around 2500 ft in the inland and
westerly part of the onshore dome area to as much as 5000 ft in the central Louisiana
offshore area. An example of a density-depth curve for the onshore Louisiana Gulf
Coast is given in Fig. 8-14. This curve, with a crossover depth of 3100 ft, was
compiled for quite another purpose, i.e., an investigation of overpressured shales,
but gives values which are generally applicable to inland and some near offshore
domes. The maximum crossover depths are in those areas where the rate of sedimentary deposition has been very rapid and there has been less time for compaction
and consolidation, particularly of the shales. The maximum thicknesses of Quater-
nary, Pliocene, and Miocene deposition are offshore from central and southeastern
Louisiana (Clark and Rouse, 1971, figs. 3 and 4), and the crossover depths there
are substantially greater than in those updip areas to the north and northwest,
where older sediments are present at much shallower depths.

The cap rock minerals over the salt are of substantially higher density than
the salt itself. The anhydrite (calcium sulfate, $CaSO_4$), which commonly forms the
lower part of the cap rock, usually has a high density, approaching 2.9 if nonporous. The limestone, which commonly overlies the anhydrite, may have a
density of around 2.4 to 2.7, depending on its porosity. In shallow domes which
apparently have been subjected to solution by circulating water, the limestone may
be absent and the anhydrite hydrated to gypsum so that the cap-rock density may
be only about 2.3.

Figure 8-15 shows how the gravity effects change with depth to the top of the
dome. Approximate calculations were made for assumed circular domes with
depth to base of salt of 30,000 ft and diameter of 15,000 ft. The several curves
correspond to the different depths to the top indicated. The density-depth curve
used has the crossover depth at 3000 ft and is approximately typical for a Texas
onshore dome near the coast.

SALT DOME GRAVITY ANOMALIES

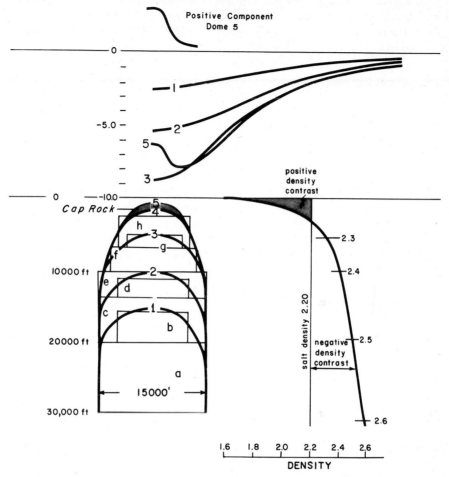

FIGURE 8-15
Variation with depth of the nature of the gravity anomaly over a salt dome.

The variations of depth to the top of the dome and of salt and cap rock distribution can lead to the several types of gravity anomalies illustrated by the different outlines and gravity profiles of Fig. 8-15.

1 The dome which has its top at 15,000 ft, all its salt far below the crossover depth, and no cap rock has a simple broad gravity minimum of relatively low relief (2 mgals).

2 A shallower dome, with depth of 10,000 ft, still has all its salt well below the crossover depth and has a similar anomaly profile but with much greater relief (over 5 mgals).

3   A still shallower dome, with its top near the crossover depth, has the maximum negative anomaly (about 9 mgals).

4   A dome with its top a moderate distance, say 1500 ft, above the crossover depth and with no cap rock should have a positive gravity component, but this may be recognizable only as a flattening in the bottom of the minimum to give a profile not very different from curve 3. The positive gravity effect of salt immediately above the crossover depth is, to a degree, canceled by the negative effect of the salt below this depth, so that there is a "blind zone" in the vicinity of the crossover depth. In this zone, the form of the dome (for instance, an overhang) is not directly determinable but has to be inferred from a reasonable connection between the upper and lower parts which may be delineated because of the stronger density contrasts.

5   If the dome has a shallow cap rock, it has a definite positive gravity effect which is more or less central within the surrounding negative anomaly. By estimating the magnitude of the minimum above the cap rock, effects can be separated out as shown by the "positive component." The magnitude of such positive components varies widely, depending on the depth, thickness, and breadth of the cap rock. Around 2 mgals is a common magnitude, but those in the Gulf Coast range from barely detectable to about 5 mgals.

The examples shown with their corresponding gravity effects are only a small sample of the very wide range of form and depth which salt domes may have.

In an area where the crossover depth is relatively large (4000 ft or more) a shallow dome without cap rock may have a strong positive effect. For example, the Ship Shoal Block 113 area, approximately 10 miles offshore from central Louisiana (Musgrave and Hicks, 1966), has a nearly circular positive anomaly of about 2 mgals over a dome with top of salt shallower than 1000 ft and with no cap rock.

A dome with a relatively slender salt column and a shallow, thick cap rock may have a small and rather broad negative gravity effect, within which there is a strong positive anomaly.

The negative effects of salt domes commonly are in the range of about 1 to 5 mgals, but a few are much larger and can range to over 15 mgals in the Gulf Coast and much more for some of the very large offshore domes. Positive effects for shallow salt known to the writer have a maximum of about 6 mgals but are commonly much smaller, and most are less than about 2 mgals. Thus the range of gravity expression can include large, very strong minima, e.g., Timbalier Bay; smaller minima with, e.g., Long Point, or without, e.g., Darrow, accompanying positive effects; minima with flat bottoms indicating small positive effects; large or small minima with weak to very strong central positive components, or very strong positive anomalies with the minimum so small as to be barely recognizable, e.g., Spindletop.

This wide range of gravity expression led to confusion in the early days of gravity prospecting in the Gulf Coast with the torsion balance. At one stage the search was directed toward finding maxima after a test survey, made over the Spindletop Dome near Beaumont, Texas, which was a spectacular oil field, found the positive anomaly there; this dome has shallow cap rock on a relatively small vertical-sided salt column, making the gravity picture almost entirely positive. The Nash dome, about 40 miles southwest of Houston, has a similar profile, and its discovery by a torsion-balance survey was the first American oil field clearly attributable to a geophysical operation of any kind. Later surveys over other domes showed the negative effects, and finally the complex nature of salt-dome gravity expressions was recognized.

EXAMPLES OF GRAVITY INTERPRETATION

Over the 40 years in which gravity surveys have been carried out, there are, of course, many examples of their geological interpretations. For the most part these have been done for commercial purposes and are proprietary with the companies for which they were carried out. A small number of them have been published, and the geophysical literature offers numerous interpretational examples in a rather wide range of geological circumstances. Several of these are included in the Society of Exploration Geophysicists Geophysical Case Histories volumes (Nettleton, 1948; Lyons, 1956). Several are given by Dobrin (1960, pp. 398–431). A few other selected examples are given in the following pages.

Quantitative Analysis of a Simple Salt Dome

The gravity map of the Humble dome, near Houston,* is given in Fig. 8-16. Figure 8-17 is a profile on line AA' of the map. A regional is removed to give the residual gravity curve shown, and the analysis is carried out on this residual. By application of the simple parameters of the normalized curve for the sphere (Fig. 7A-1) we can determine the general features of the dome. We designate the half-amplitude width of the residual anomaly $x_{1/2}$ as its width (from its center) where the amplitude is half the maximum (that is, $g/g_0 = \frac{1}{2}$). If we measure the total width of the curve at its half-amplitude, this horizontal distance is the double half-width or $2x_{1/2}$, and, from the normalized curve, the depth is $0.652(2x_{1/2})$. The residual gravity curve of Fig. 8-17 gives a double half-width of about 25,000 ft, from which the indicated depth to the center of the equivalent sphere is $25,000 \times 0.652 \doteq 16,300 = 16.3$ kft. The total amplitude of the anomaly is about 13.9 mgals, from which we can estimate the radius of the equivalent sphere. From page 190, the total amplitude factor is $g_0 = 8.52\sigma R^3/z^2$, and $R^3 = g_0 z^2/8.52\sigma$. If we assume an average density contrast of 0.3 and put in values of $g_0 = 13.9$ mgals and $z = 16.3$ kft, we have

$$R^3 = \frac{13.9 \times 16.3^2}{8.52 \times 0.3} = \frac{13.9 \times 266}{2.556} = 1446 \text{ kft}$$

and

$$R = \sqrt[3]{1446} = 11.3 \text{ kft} = 11,300 \text{ ft}$$

* The Humble Oil Field, over and on the flanks of this dome, was discovered in 1904 and is the one which gave the Humble Oil Company its name.

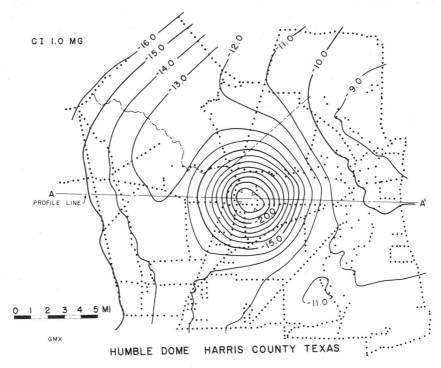

CI 1.0 MG

HUMBLE DOME HARRIS COUNTY TEXAS

FIGURE 8-16
Bouguer gravity map, Humble salt dome, Harris County, Texas. Contour interval 1.0 mgal. Control shown by dots representing gravity-station locations. (*From Nettleton, 1962.*)

The depth of the top of the dome is $T = z - R = 16{,}300 - 11{,}300 = 5000$ ft. The gravity anomaly from the equivalent sphere can be calculated very simply as shown by Table 8-2, with all the numerical operations being done quickly with a slide rule or one of the new minicomputers. For $z = 16.3$ kft, $g_0 = 13.9$ mgals and reading $f(x/z)$ from curve 1 of Fig. 7A-1, we have the results shown in Table 8-2 to give the calculated points shown along the residual curve.

A calculation like that shown by the table can be carried out in 5 to 10 min by anyone reasonably skilled in the use of the slide rule. The calculated values of Fig. 8-17 are as close to the residual-gravity curve as the reliability with which that curve is determined. Without more accurate information on densities and on the depth of the dome at some point, such as a drilling contact or seismograph data, the simple calculation with the spherical approximation gives (1) the position quite accurately, (2) a fair approximation of the depth, and (3) a rough measure of the diameter and therefore of the deeper flanks of the dome. For relatively shallow domes such as this, where the top of the dome is known to be about 1000 ft, the depth to the calculated top is too great because the density contrasts of the shallow sediments with respect to the salt are less than the average contrast used for the calculation (see page 260).

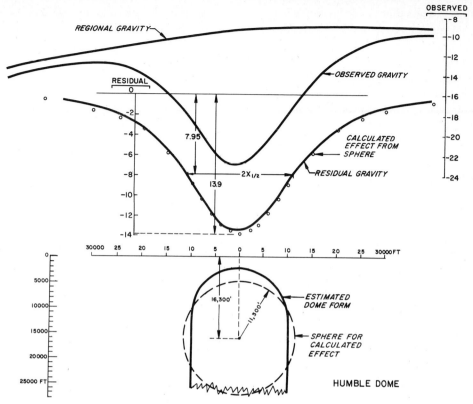

FIGURE 8-17
Gravity profiles and dome cross section on line AA' of Fig. 8-16.

Table 8-2

x	$\dfrac{x}{z}$	$f\left(\dfrac{x}{z}\right)$	g
0	0	1.000	13.90
2	0.123	0.975	13.70
4	0.246	0.912	12.70
6	0.369	0.822	11.42
8	0.492	0.718	9.98
10	0.614	0.615	8.55
15	0.920	0.410	5.70
20	1.228	0.256	3.56
25	1.536	0.162	2.25
30	1.840	0.110	1.53
40	2.429	0.050	0.69

The point of this example is to illustrate that very simple calculations may be adequate as control on geological assumptions and that more elaborate calculations, such as may be made with dot charts or computers, are justified only when additional information is available. For more complex gravity anomalies, e.g., those from shallow domes with cap rock, where the density and geological situation are much more complex, the gravity picture may have components which can be separated from the total anomaly and used for other interpretive calculations (see Fig. 8-15).

A corollary of the use of relatively simple calculations is that the degree of detail of the gravity data which can be useful depends very much on the objective of the survey. The gravity contour map of Fig. 8-16 is based on stations approximately $\frac{1}{2}$ mile apart on the principal roads and trails in the area, and there are some wide gaps between lines. The control is quite adequate for locating and determining the general nature of this dome. On the other hand, to determine the details of a shallow dome with an associated shallow cap rock would require much closer control over the central part of the dome to map the relatively sharp changes there and to separate the shallow-dome anomaly from the total gravity field. As a crude approximation, a gravity-station spacing of less than about one-half the minimum depth to the feature or density contrast to be mapped is about as close as can contribute to determination of significant geological details. Closer spacing may be desirable when shallow inhomogeneities produce sharper effects which must be filtered out in some way to accurately determine the anomaly to be analyzed.

Moderately Deep Dome, Offshore Louisiana

This example is from an underwater gravity-meter survey over a relatively small part of the salt-dome area of offshore Louisiana. The survey was made for the specific purpose of delineating a dome in an area which had been put up for bids in a lease sale. The Bouguer gravity map of the area over and around the particular dome (Fig. 8-18) shows a closed minimum (D-1) centering within the outlined 3-mile-square survey block. Parts of several partially overlapping minima from other domes are indicated as D-2, D-3, and D-4. Also, there is a general or regional gravity increase which was established from the gravity pattern over a much larger area.

The overlapping effects of the other domes complicate the isolation of the local minimum. Calculations were made using simple spherical approximations to remove these disturbances. These overlapping effects and how they were removed to leave the isolated gravity anomaly of the dome in the central block are illustrated by the profile in Fig. 8-19. The resulting isolated minimum is shown in Fig. 8-20.

A calculation of a dome form to account for the minimum was carried out by using the chart for solid angles of circles (Fig. 7A-9). The dome model was divided into several thick disks, with density contrasts increasing with depth. After some trial-and-error manipulation of the circularly symmetrical model, the dome with the cross section shown on the profile sheet (Fig. 8-19) gave a calculated effect, shown by small circles, which is quite close to that of the isolated dome

FIGURE 8-18
Bouguer gravity, with salt-dome anomalies, offshore Louisiana. Offshore Block
208 outlined. Dots show gravity-station locations.

minimum over the central part of the anomaly (below about -2.0 mgals). The
wide flanks of the calculated effect have values greater than those of the anomaly;
this is a frequently observed relation and comes from incomplete removal of
regional and other extraneous influences but does not seriously detract from the
effectiveness of the calculation.

At the time the analysis was carried out there was no drilling in the area.

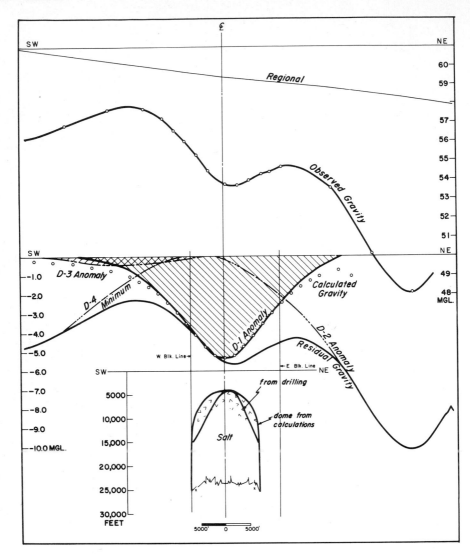

FIGURE 8-19
Observed, regional, and residual gravity on northeast-southwest line of Fig. 8-18.
Estimated contributions from adjacent anomalies D-2, D-3, and D-4 are shown.
Subtraction of these effects leaves the D-1 anomaly. The circles on and near this
curve show the anomaly calculated by solid angles of circles.

Several years later an oil field had developed, and many wells were drilled into salt
to give the structure contours on the dome as shown in Fig. 8-21. A cross section
over the dome from this drilling is shown for comparison with the section derived
from the gravity analysis in Fig. 8-19.

On the whole, the agreement of the dome form from the interpretation with

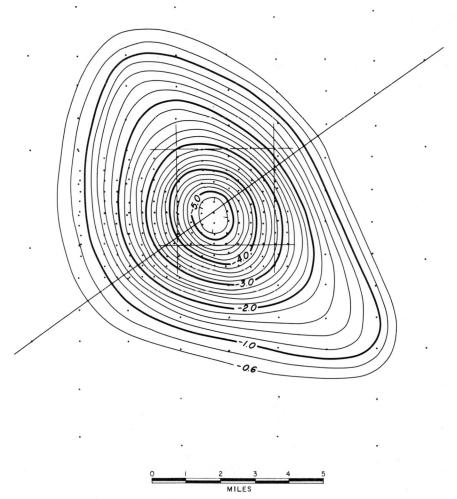

FIGURE 8-20
Isolated gravity for central anomaly of Fig. 8-18.

that actually found later from drilling is better than can be expected from a purely analytical standpoint. In particular the close agreement of the depth of the top of the dome is partly accidental because the depth to the top (top of salt in the shallowest well was at 3948 ft) is very near the crossover depth, where the density of salt and sediments is nearly the same. In this zone there is very little density contrast to affect the gravity anomaly, and therefore the depth is indeterminate, from gravity alone, over a range of around 2000 ft. Thus the form derived is to

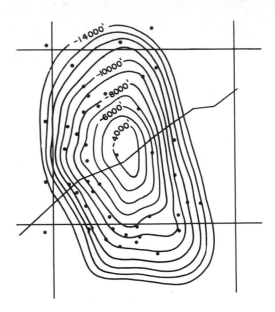

FIGURE 8-21
Contours on top of dome from drilling
in oil field. Dots show well locations.

some extent intuitive and based on general experience with the analysis of salt-dome gravity anomalies. From the standpoint of the purpose for which the survey and analysis were made, i.e., the location of the flanks and the general depth of the dome for control of bidding for leases, the result was entirely satisfactory.

The analysis made in 1955 was carried out by indirect methods, i.e., by trial-and-error calculations, until a satisfactory fit with the isolated anomaly was achieved. With modern computer techniques, the much more sophisticated iterative approach could be used. Ideally this would be applied to a much broader area so that the several nearby domes would be included in the calculation. Then the overlapping effects onto the area of primary interest would be included in those calculations so that the several anomalies would be isolated from one another and the forms of each of the causative salt masses would be determined.

As a more limited example of the use of the iterative technique, it has been applied to the isolated minimum (Fig. 8-20). The gravity minimum was digitized on a 4000-ft grid over an area of 20 by 20 grid units centered over the anomaly. Depth to the bottom of the salt column was taken to be 25,000 ft (after earlier trials with 35,000 and 30,000 ft). The calculation was constrained to a minimum depth of 4000 ft, as the program does not provide for the density reversal at the 4000-ft crossover depth.

The resulting depths to the tops of the 4000-ft-square prisms, within the area where they are significantly above the 25,000-ft bottom, and contours on these depths at 5000-ft intervals are shown in Fig. 8-22. The numbers on the horizontal and vertical lines designate the rows and columns of the central part of the 20 by 20 grid at 4000-ft spacing on which the gravity values from the residual map (Fig.

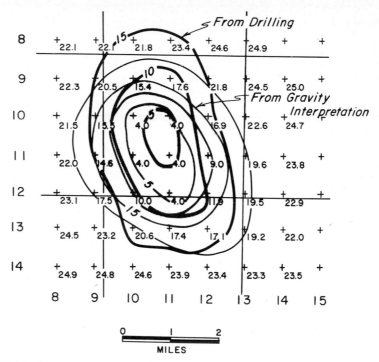

FIGURE 8-22
Comparison of dome form from interpretation by iterative calculation with that from drilling.

8-20) were written. The heavy lines show 5000-ft contours on the dome from drilling, from Fig. 8-21.

Profiles and a cross section on row 11 (Fig. 8-22) are shown in Fig. 8-23. The upper part shows curves for the observed (input) values and that for the calculated values after 10 iterations. The fit is quite good over the central part of the anomaly, with the calculated values too large on the flanks; the same effect was noted above with respect to the calculation from a geometric model.

The lower part of the figure shows the heights of the 4000-ft-square prisms corresponding to the calculated gravity curve. The density-depth curve used is shown at the left; it is almost parallel with the Dickinson curve (Fig. 8-15) but with values about 0.04 g/cm³ lower. The dome cross section from drilling (Fig. 8-22) is shown for comparison.

On the whole, the comparison of the calculated with the actual dome is very good, especially when one considers the complex field from which the anomaly was isolated. The difference at the top is due partly to the lack of density contrast immediately below the crossover depth and partly to the forced minimum depth of 4000 ft. The elongation of the dome outline to the south is in quite good agreement. The calculation does not closely follow the northward elongation of the

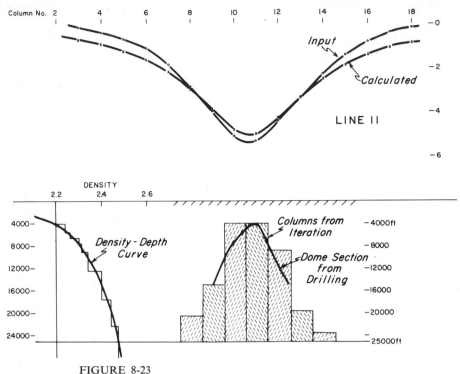

FIGURE 8-23

Cross section and gravity profile on row 12 of Fig. 8-22, showing section from iteration and comparison of input and calculated gravity.

actual dome because of the form of the original anomaly isolation; minor changes in the details of the peripheral effects removed could account for this deficiency.

An iterative operation, applied to a wide area and including the several peripheral anomalies such as mentioned above, might give a somewhat better result but is not available. Also, use of a smaller grid interval (say 2000 ft) would give some improvement but would increase the number of points calculated by a factor of 4 and the computer time by a factor of 4 also.

Gravity Stripping

An interpretive process for improving the gravity expression of a deep source by removing, or "stripping," the effects of a shallower source has been described by Hammer (1963). The process is applicable where the general structure is known and the anomaly results from two primary sources at widely different depths. When the effects of the shallow source are removed, the anomaly expression of the deeper source is enhanced. Because of the structural and density data required, the process is applicable only where considerable subsurface information is available.

The example given is the analysis of a gravity survey around and over the Citronelle Field in southwestern Alabama. A residual gravity map shows a sprawling irregular minimum of only 1 mgal relief and dimensions of about 10 by 20 miles. The structure on which the oil field is located is also quite broad, with relief of some 600 ft at a depth of 7000 ft. The general area is one in which structures due to salt flow are known.

The geological factor which made the stripping process applicable is that gamma-gamma density logs (page 249) had shown a sharp increase in density, from about 2.1 to 2.3 g/cm^3, at a depth of about 4000 ft. Thus, the positive gravity effect of the known structure could be calculated. Removing this effect from the residual anomaly increased the negative relief from about 1.0 to 3.0 mgals and also greatly reduced the area and sharpened its definition. The oil field is well within the stripped anomaly.

The process may be useful where the general geology is well known so that the information is available for estimating the shallow gravity effects and separating them from the deeper contributions. In such cases the anomaly expression from deep sources may be clarified. If density contrasts at the two levels are of opposite sign, the amplitude of the deep-source anomaly is increased; if they are of the same sign, it is decreased but may be more sharply defined.

SUBSURFACE GRAVIMETRY

The use of gravity measurements in a borehole for the measurement of density was described briefly in a previous section (page 253). Possible applications of subsurface gravity measurements for other purposes have been described in the literature, particularly in an extensive article by Smith (1950). Applications to the possible determination of the nature of fluids in porous sands have been described by McCulloh (1967) and by McCulloh, Schoellhamer, and Pampeyan (1967), and for determination of porosity by Jones (1972). Measurements in a mine shaft for the evaluation of ore bodies are discussed by Rogers (1952) and for density determinations by Hammer (1950) and by Allen, Callouet, and Stanley (1955).

Of particular interest is the use of borehole measurements to give, in effect, a third dimension to gravity field measurements. This is a very interesting possibility, but is not as simple as it may seem. The primary difficulty comes about because any anomaly due to a mass irregularity lateral to the borehole must be determined as a departure from other changes in gravity. This means that any departures which may be indicative of anomalous mass lateral to the hole are small deviations from a large regional effect which is strongly dependent on the details of the density-depth relations in the general environment of the drill hole.

For instance, let us consider those regional effects in a vertical hole in a normal Gulf Coast sedimentary section with density-depth changes as represented by the Dickinson curve of Fig. 8-14. From the surface to, say, 10,000 ft, the total gravity change is made up of (1) the normal linear vertical gradient of 0.09406 mgal/ft, or +941 mgals, and (2) the double Bouguer effect of -0.02552σ mgal/ft, which, for the densities of the Dickinson curves from 0 to 10,000 ft, integrates to − 581 mgals, to give a net change of 360 mgals. Other density-depth curves of course would give somewhat different values. The precision with which the details of a base curve from which anomalous departures are determined depends entirely on the precision with which the normal density-depth relations are known. Local unknown variations would have effects which could be confused with those caused by lateral mass anomalies. The situation can be greatly improved if a formation-density log from gamma-gamma radioactivity measurements is available through the normal section in the same hole or one or more others in the general area of the suspected anomaly.

Application to a Salt Dome

The analysis of vertical gravity measurements made in connection with the exploitation of a salt-dome oil field is given below. The gravity data are from measurements in a borehole drilled near the dome with the objective of trying to determine the proximity of the well to the dome flank. The analysis is based on the principle that a formation-density log in the same hole gives correct density values and that any differences between the gravity calculated from these densities and that observed in the well are due to the effect of the nearby dome.

In the available example the formation-density log was run from 2900 to 6500 ft. Gravity measurements were made at various intervals over the depth range 1000 to 6500 ft. The comparison can be made, of course, only over the common interval beginning at 2900 ft. The results are shown in Fig. 8-24.

Curve 1 shows, for each depth at which a gravity observation was made, the difference (accumulated from an arbitrary starting point) between a gravity change computed from the densities of the formation-density log and those observed by the borehole meter. This curve has a slope to the right. This is qualitatively the effect to be expected from a nearby dome; the effect of the dome would be to decrease gravity because of its negative density contrast and therefore the "calculated-minus-observed" values should be positive if the dome has any influence.

To evaluate the effect quantitatively, calculations were made of the lateral effect of a simple cylindrical dome with vertical flanks and with dimensions similar to those of the actual dome, which was partially defined by drilling and with a density-depth curve appropriate for the area. These calculations were made for a well tangent to the model dome and for another 2000 ft from the flank. The resulting theoretical effects are shown by curve 2 for the tangent well and by curve 3 for the well 2000 ft from the flank. The smaller effect of the more distant well is shown by its flatter slope.

The comparison of these calculated effects with the differences observed can be made by matching the difference curves at some arbitrary starting point, which has to be within the range where both types of measurements, i.e., gravity and formation density, are available. The curves are shown with the match made at the 3600-ft level, which is just below the relatively large and apparently anomalous irregularity between the 3200- and 3600-ft depths.

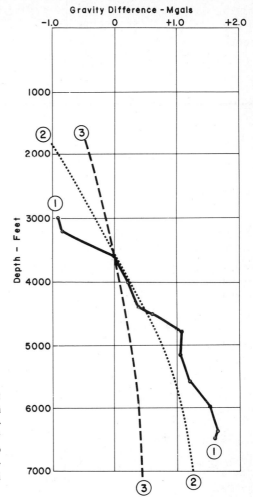

FIGURE 8-24
Borehole gravity on flank of a salt dome:
curve 1: gravity variation calculated from
densities measured by a formation density (gamma-gamma) log; *curve* 2: gravity
variation calculated for a well tangent to
the model dome; *curve* 3: gravity variation calculated for a well 2000 ft from
the flank of the model dome.

Curve 2 for the tangent well is generally parallel with, but shows a somewhat smaller anomaly than, that of the calculated difference. The calculated well at 2000 ft from the flank shows a clearly smaller slope and poorer agreement with the calculated difference. From this we conclude that since its curve is close to that of the tangent well, the actual well in which the measurements were made is very close to the dome. Actually, from later information, the test well drilled into salt a few hundred feet below the lowest gravity measurement shown.

This example illustrates these points:

1   The significant gravity differences are very small within large changes; the anomaly in the depth range 3600 to 6500 ft is about 1.5 mgals; the total gravity change in this interval is 105 mgals.

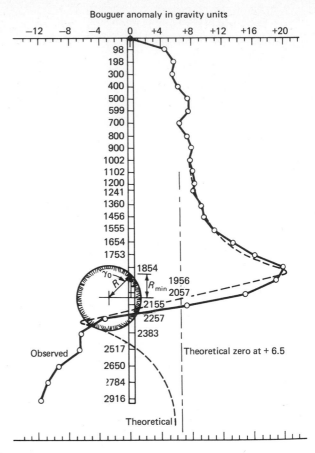

FIGURE 8-25
Comparison of observed gravity in a drill hole with the theoretical gravity calculated from a sphere derived from the gravity data. (*From Rogers, 1952; by permission of the Society of Exploration Geophysicists.*)

2　Determination of the significant anomaly is completely dependent on an independent means, i.e., the formation-density log, to establish a base from which the anomaly can be measured.

Other Vertical Gravity Measurements

Measurements by Hammer (1950) were made in a limestone shaft primarily for the evaluation of the gravity meter as a means of measuring densities. The measurements were generally in agreement with density values determined from measurements on rock samples from the shaft, although there were some unexplained inconsistencies.

The measurements by Allen, Callouet, and Stanley (1955) were made in a vertical shaft in a salt mine in Harris County, Texas. These measurements were made primarily as a test of the gravity meter to determine densities. This makes a good example because the distinctly different densities of the material penetrated make obvious changes in the slope of the gravity-depth curve.

From the slopes measured by the author on fig. 2 of Allen, Callouet, and Stanley (1955) the densities are approximately 1.96 for the sediments, 2.30 for the upper cap rock with some gypsum, 2.75 for the lower limestone and anhydrite cap, 2.83 for the solid anhydrite, and 2.22 for the salt.

Measurements by Rogers (1952) were made with the objective of estimating the mass anomaly of an ore body. The gravity measurements were in a vertical shaft which penetrates a sulfide ore body. Figure 8-25 shows the anomaly calculated as the departure from the theoretical vertical gravity for density 2.67. Rogers also shows the theoretical anomalous gravity for a well penetrating a spherical mass through its center and at various distances from the center. A comparison of this theoretical curve with the observed vertical gravity anomaly gave a mass estimate quite consistent with that made from drilling in the mine.

The examples by Hammer and by Rogers paid careful attention to the effects of local topography on the meter penetrating the shaft. The terrain effects are considerably larger than those for gravity measurements on the surface because the vertical angle between the instrument and a given piece of topography increases as the meter goes to greater depths.

On the whole, the studies and examples above indicate that there are definite possibilities of using vertical gravity measurements to get a third dimension to gravity surveys as well as their application to accurate measurement of densities. Care and understanding must be used, together with considerable knowledge of the density-depth relations of the normal rock environment, aided by model calculations in using such measurements.

9

GRAVITY, ISOSTASY, AND THE EARTH'S CRUST

In Chap. 2 we discussed the variation of gravity with latitude because it contributes one of the basic corrections to gravity measurements. There it was shown that this variation has been expressed at different times in the history of geodesy by several slightly different formulas and that the one now generally used is

$$\gamma_0 = 978.049(1 + 0.0052884 \sin^2 \varphi - 0.0000059 \sin^2 2\varphi)$$

where the first term is average gravity at the equator and φ is the latitude. Also, it was pointed out (page 59) that recent measurements show that the first term is about 14 mgals too high, but since practically all gravity evaluations with which we are concerned are based on relative gravity differences, they are not affected by this correction.* The quantitative relations between the shape of the earth and the variation of gravity from equator to pole are discussed above, beginning on page 15. The geodetic formula gives the value of gravity on an ideal spheroid of

---

* The absolute value of gravity is a factor in the precise definition of certain primary physical constants such as the ampere and barometric pressure and therefore the correction by over 1 part in 10^5 could be of some significance.

revolution. Gravity measurements and their anomalies determine variations from this ideal.

In this chapter we shall deal with broad-scale gravity variations over the earth's surface and their interpretations in terms of the nature of the crust.

THE DIFFERENT GRAVITY ANOMALIES

In previous sections we have considered an anomaly to be any local deviation of gravity from a more regular or smooth trend, which is defined by a group of several to many stations. Such anomalies are usually the primary data from which a geological interpretation of their source is attempted. In the geodetic sense the anomaly is a single numerical value for any individual observation and is the difference of the observed value from a theoretical or calculated value based on certain assumptions about the form of the gravity field over the earth as a whole. The observed gravity value is determined by relative gravity measurements, made by pendulums or gravity meters, referred to certain primary gravity base stations, where absolute measurements have been made by special instruments such as the Kater reversible pendulum, by very long pendulums, or, more recently, by falling-body techniques. The anomaly is positive if the measured value is higher than that calculated and negative if lower.

The Free-Air Anomaly

The vertical gradient of gravity (page 19) is approximately the simple effect of a station at higher elevation being farther away from the center of the earth. To a close approximation the effect is linear and independent of latitude so that the vertical gradient is simply $-Kh$, where K, the free-air coefficient, is 0.3086 mgal/m or 0.09406 mgal/ft. The theoretical free-air gravity is $g_f = \gamma_0 - Kh$, where γ_0 is the gravity at sea level at the latitude of the station, calculated from the gravity formula used. It is called free-air because the theoretical anomaly is calculated as if the gravity measurement were made at the elevation of the station but without taking into account the attraction of material between that elevation and sea level, i.e., as if the gravity-measuring instrument were suspended free in the air.

The free-air anomaly is the difference between the observed value g_0 and the calculated free-air gravity.

The Bouguer Anomaly

Consider gravity observations made at points 1, 2, and 3 of Fig. 9-1. Station 1 is at an elevation h above sea level, station 2 is at sea level, and station 3 is on the ocean surface, where the water depth is d (where gravity is measured by a pendulum in a submarine or by a shipborne gravity meter). The measurement at point 1 is subject to the attraction of the slab of earth material with a thickness h between it and sea level (which is the same as the Bouguer effect removed in making the

Bouguer correction in routine mapping of gravity field measurements). A station at point 3 has a deficiency in gravity because the material between the ocean bottom and the surface is seawater with density 1.03 rather than the material of the earth's crust. The Bouguer effect is a calculation to take these attractions into account.*

The simple Bouguer effect is calculated as if the material under the station were horizontal (except when calculations include effects from great horizontal distances and the curvature of the earth's surface is taken into account). For a simple horizontal slab of material the attraction is $B = 2\pi G\sigma h$, where G is the gravitational constant, σ the density of rock, and h the elevation of the station or the thickness of the slab of material between the station and sea level. When we put in the numerical values for 2π and the gravitational constant (6.6732×10^{-8}), the effect becomes $0.04193\sigma h$ mgal/m or $0.01278\sigma h$ mgal/ft. The calculated theoretical gravity at the station, including free-air and Bouguer effects, is $g_B = \gamma_0 - 0.09406h + 0.01278\sigma h$ (for h in feet).

For station 3 the Bouguer effect is that calculated as if the water were replaced by rock of density σ and becomes (for a measurement made at the sea surface)

$$(g_B)_w = \begin{cases} [0.04193(\sigma - 1.03)]d & \text{mgals/m} \\ [0.01278(\sigma - 1.03)]d & \text{mgals/ft} \end{cases}$$

The Bouguer anomaly is the difference between the measured value at the point of observation and the theoretical value calculated for that elevation or water depth and the appropriate density of the earth's materials. An average value for the density of the crustal material within the range of visible topography and which has been used worldwide for calculation of general Bouguer effects is 2.67 g/cm³.† For this density the Bouguer correction coefficient is 0.05994 mgal/ft or 0.1966 mgal/m for land stations and 0.02095 mgal/ft or 0.06876 mgal/m for sea-surface measurements with water density 1.03.

---

* The Bouguer effect at sea is somewhat fictitious. On land, the Bouguer correction corresponds to the attraction of the material between sea level and the station site. If the elevation is zero, the Bouguer correction is zero, the free-air correction is also zero, and the Bouguer and free-air anomalies are the same. A gravity measurement on the sea surface also is made at sea level, and, again, strictly speaking, the free-air and Bouguer anomalies are the same. As a practical matter a Bouguer correction is made which is a first step of the interpretation in that the calculation introduces the effect of a part of the earth's crust, i.e., the water, for which the density and thickness are known. This distinction between the meaning of a Bouguer correction on land and at sea is not generally recognized. To those who do recognize this difference, any interpretation is preferably based on the free-air anomaly, and the water becomes one of the layers whose density and thickness are included in the geologic section accounting for the gravity observations. On the other hand, the effects of high-relief local topography on the ocean bottom can seriously distort the free-air map (see Fig. 9-8). The Bouguer correction, i.e., the removal of the effect of the density contrast between seawater and the underlying rock, produces a map on which the gravity effect from sources below the sea bottom are more apparent.

† The value 2.67 is generally applicable to rocks within about 2 km of the earth's surface. For deeper materials, where cracks and fissures are closed and porosity is reduced by overburden pressure, a value of around 2.85 may be more realistic.

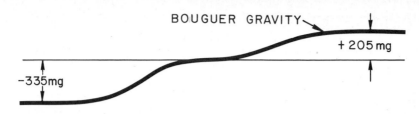

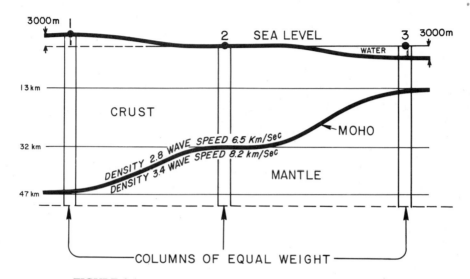

FIGURE 9-1

Idealized diagram of crustal changes and gravity effects from deep ocean to continental areas. The magnitudes shown for gravity anomalies are for heights and depths of 3000 m, calculated for rock density 2.67 and water density 1.03 and assuming ideal isostatic compensation. (*From Nettleton, 1971.*)

For gravity measurements made in rugged topography, terrain effects may be calculated. These, of course, are the same as those previously discussed for reducing gravity measurements on land (page 42) and at sea (page 124).

THE CONCEPT OF ISOSTASY

The Bouguer effect on gravity measurements is a very obvious one, and it is necessary to make such a correction if the effects of the visible topography or variations in ocean depth are to be taken into consideration. However, it was found very early in the history of gravity measurements that Bouguer anomalies have a strong correlation with topography over broad areas. In general, areas with high

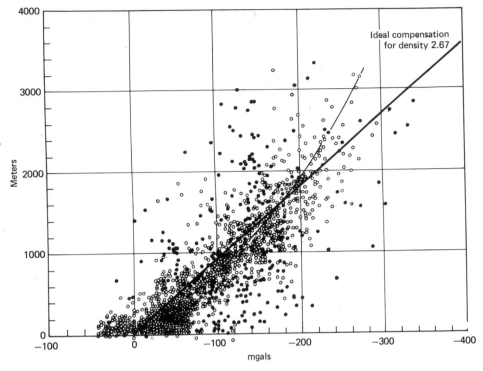

FIGURE 9-2
Relation of Bouguer gravity anomaly to surface elevation (land stations). Straight line (added to original diagram) shows the Bouguer effect for density 2.67. If all locations were completely and ideally isostatically compensated for 2.67, they would fall on this line. (*Modified from Woollard, 1959, p. 1523; by permission of the American Geophysical Union.*)

topography have negative Bouguer anomalies, as shown by Fig. 9-2; also (but not included in the figure) those with low topography, i.e., the ocean basins, usually have positive Bouguer anomalies. For broad areas of large relief these anomalies can be several hundred milligals.

The existence of such anomalies points to the existence of some other component of the earth's crust which is not obvious at the surface. In the first place, it is evident from general knowledge of the earth's crust and its physical properties that the crustal material is not strong enough to support the weight of a very broad mountainous area. This weight corresponds to the magnitude of the Bouguer attraction averaged over the area. Therefore there must be some other property of the earth's crust which supports the topography. In the early history of such measurements two competing theories were put forth to provide a principle which would hold up the mountains:

1   The Pratt theory assumes that the density of the earth's material below sea level under areas of high topography is lower than average and that this low-density material extends to a fixed "depth of compensation."

2   The Airy theory assumes that there is a discontinuity in density at some depth within the earth's crust and that this discontinuity is at variable depths, being deeper under high topography and shallower under low topography. Since the difference in density would be much smaller than the actual density of the topography, the vertical relief of such a surface would have to be several times greater than the relief of the topography and thus the mountains would have deep roots in a manner similar to that in which an iceberg projects down into the water to a depth several times its visible height. Thus the Airy proposal became known as the "roots of mountains" type of compensation.

The system or principle by which topography is compensated by variations of the earth's crust was called "isostasy" by C. E. Dutton in 1889.

Calculations of isostatic effects can be made on the basis of either principle to determine the magnitude of the required deficiency in density under areas of high topography, and systematic "isostatic corrections" can be carried out and corresponding isostatic anomaly maps can be produced. In general it is found that these maps show very much smaller anomalies than the Bouguer maps. The foregoing statements have been made with respect to elevations of the topography and negative Bouguer anomalies, but exactly the same considerations apply to depressions, i.e., ocean basins, and corresponding positive anomalies. The difference between the two is that in one case the crustal surface is bounded by air and in the other by seawater, so that the effective density contrast in ocean areas is reduced by the density of seawater (usually taken as 1.03).

In areas of flat but elevated topography, such as a high plateau or in broad, deep ocean basins, the isostatic anomaly is approximately the same as the free-air anomaly. This is because the mass deficiency which compensates for any mass above sea level (or deficiency for mass below sea level) is nearly equal to the Bouguer mass and the Bouguer and isostatic corrections are approximately equal but of opposite sign. This is not true where boundaries are steep, as in areas of grossly rugged topography or at continental margins.

THE MOHO

The above general relations were the basis of much of the geodetic analysis of worldwide gravity surveys through approximately the first quarter of the present century. Isostatic compensation was calculated, usually on the basis of the Pratt theory because it is easier to handle analytically, for early gravity measurements over the continental United States, other continental areas, and the relatively few gravity measurements at sea made in submarines.

In more recent years the Mohorovičić discontinuity (commonly abbreviated as "Moho") was first recognized as a level at which a sharp increase in seismic velocity occurs, from an average of approximately 6.5 km/sec above to a little over 8.2 km/sec below the boundary. From general relations of velocity and density and also from considerations of the gravity requirements for the variation of density with depth within the earth's crust, there are strong indications that this surface is also one where a sharp density contrast occurs, with probable density values of roughly 2.8 above to 3.4 below the Moho (Ludwig, Nafe, and Drake, 1971) (Fig. 9-1).* The difference in density at the Moho serves as a natural surface to be the upper boundary of the high-density material proposed by the Airy roots-of-mountains system of isostasy. The material of the upper mantle must be viscous, corresponding to Airy's asthenosphere, to permit adjustment to changes in composition or topography of the crust. The basic difference between the Airy-Moho roots, on the one hand, and the constant depth of compensation of the Pratt theory, on the other, is that in the former case the contact between crust and mantle is at variable depth and in the latter it is constant. The center of gravity of the zone of compensation is at about the same depth for either system.

From these general considerations we have the following expectations:

1 Under areas of high topography there should be strong negative Bouguer anomalies, and the Moho should be relatively deep.
2 Under areas of low topography or coastal areas the Bouguer anomalies should average near zero, and the Moho should be at an intermediate depth.
3 Over deep ocean areas the Bouguer anomalies should be strongly positive, and the Moho should be relatively shallow.

This means that in all three types of areas the weight of a column of material from the surface to some uniform depth below the Moho would be the same (Fig. 9-1). In mountain areas the greater thickness of low-density material is compensated by a smaller thickness of high-density material below the Moho. In ocean areas the smaller thickness of low-density crustal material is compensated by a greater thickness of high-density material below the Moho.

Depths to the Moho have been obtained from refraction-seismograph measurements in many areas both on continents and under the oceans. When these depths are related to the topography and to the Bouguer anomalies (Fig. 9-3), the general relations are as indicated above (Fig. 9-1). While the details vary considerably, there is no question from these measurements that the general condition

---

* These values, of course, are inferred from indirect evidence. No drilling or core from the Moho is available. It was because of an intense interest in the nature and properties of the mantle, i.e., the part of the crust below the Moho, that the multimillion dollar "Moho project" to drill a hole into the mantle was conceived and started but never completed. The origin of the concept and the first stages of the attempt to carry it out are described in Bascom (1961).

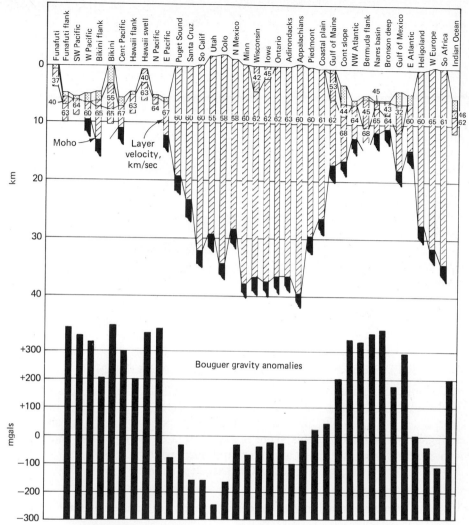

FIGURE 9-3

Relation of crustal thickness obtained from seismic measurements to Bouguer gravity anomalies. (*From Woollard, 1959, p. 1526; by permission of the American Geophysical Union.*)

of isostasy is very real and that it is adequately explained, in broad terms, by variations in the depth of the Moho. Very extensive studies of the relations of gravity to topography have been carried out since about 1960 as worldwide gravity data have become available, particularly over the oceans. They have shown that there are many departures from ideal compensation.

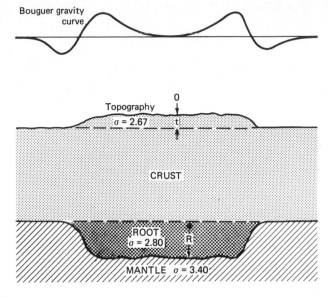

FIGURE 9-4
Idealized topography and compensating root. For the densities shown, the depth R of the root is 6 times the height t of the topography.

DEPARTURES FROM IDEAL COMPENSATION

If isostatic compensation were complete in all details and the densities of topography and mantle were constant, it might be expected that the values in Fig. 9-2 would fall on a straight line with its slope corresponding to the density of the topography. Actually, as this figure shows, there are many positive and negative departures of over 100 mgals and a few approaching 200 mgals.

There are two general reasons for these departures:

1   The shape effect
2   Uncompensated masses supported by strength or stresses in the crust

The Shape Effect

This comes about because a gravity measurement on the surface is affected by the attraction of material at different depths. A measurement at point O, Fig. 9-4, is attracted by:

1   The positive effect of the slab of near-surface material of thickness t, with the assumed density 2.67
2   The nominally equal negative effect of the balancing slab of root material with a thickness for the assumed densities shown, of

$$R = t\,\frac{2.67}{3.40 - 2.80} = 4.45t$$

The general shape of the gravity curve for the balanced masses is shown in the upper part of Fig. 9-4. At point O, at the center of the plateau, the attractions of the two slabs are closely equal, the Bouguer anomaly is balanced by the compensation, and the free-air anomaly and isostatic anomaly both are near zero. Toward the edges of the plateau, the effect of the surface slab does not decrease much until the edge is reached, but that of the slab of root material begins to decrease much farther from the edge because of its great depth. The combined effects give a positive anomaly inside the plateau area and a negative anomaly near the plateau on the outside, as shown by the gravity curves of Fig. 9-4. Approximate quantitative effects can be estimated from the characteristic curves for faults (Fig. 7A-4). For the slab of topography, a point at, say, 10 km back from the edge would have a large x/t ratio, and the gravity effect would be near the total Bouguer magnitude. For the slab of root material, with a depth of, say, 30 km, the x/t ratio would be around 1:3 and the gravity effect about two-thirds of the Bouguer magnitude. Thus, even with ideal mass compensation there would be an anomaly of about one-third the ideal Bouguer effect, which would be in the direction to make the Bouguer anomaly more positive, i.e., decrease numerically. Similar effects are to be expected in actual situations near the edges of masses of high topography or within more complex topographic features.

Uncompensated Masses

The gross worldwide departures from isostasy indicate uncompensated masses. Relations of gravity to topography and departures from compensation have been studied in great detail. A review by Woollard (1969) is a study of the relation of Bouguer gravity to elevation, taking blocks of topography of various sizes (from 1 by 1° to 3 by 3° squares). Results are shown by a series of figures similar to Fig. 9-2. For data from the United States, the diagrams are quite regular, with some decrease in scatter as larger units are averaged. Also, each diagram includes an average slope calculation for the relation of Bouguer gravity to elevation. For ideal compensation this slope would be a measure of the average density of the topography (corresponding to the 2.67 line of Fig. 9-2, with a slope of 0.112 mgal/m). For the blocks of 1, 2, and 3° squares, the slopes (for the lines including the highest elevations) are 0.109, 0.0995, and 0.1005 mgal/m. Dividing these by the Bouguer coefficient of 0.04185 gives indicated densities of 2.61, 2.38, and 2.40, respectively, which suggests a value appreciably lower than the 2.67 commonly used as an average for rocks above sea level. These values are not to be considered as indications of real departures in the average density of the topography above sea level but as somewhat systematic departures from ideal isostasy. The edge effects discussed above would be in the direction to decrease the slope and the indicated density.

There are many other studies of the relation of the Moho to gravity in many parts of the world. Several of them are reviewed in Hart (1969, sec. 4), and many of the sources are listed in the bibliographies accompanying the several separate reviews in that section. All this material is of interest in any study of the earth's

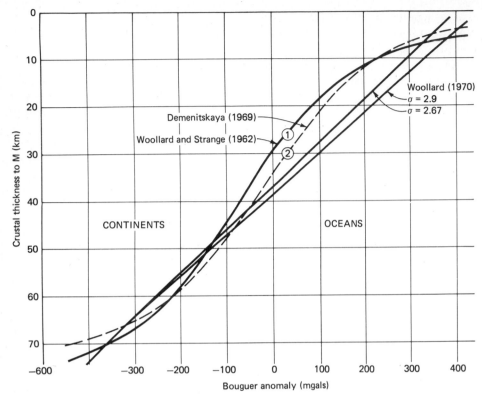

FIGURE 9-5
Average worldwide relations between surface elevation and depth to the Moho.
(*Curves 1 and 2 from Woollard and Strange and from Demenitskaya; straight lines from Woollard, 1970.*)

crust and the broad relations of gravity to the depth of the Moho and other features of the crust.

These general worldwide relations are not very pertinent to the application of gravity to petroleum exploration, except at the continental margins, particularly the continental-shelf areas, discussed in the next section. For average conditions, applicable to such areas, the relation of depth to the Moho (corresponding to those shown schematically by Fig. 9-1) is summarized in Fig. 9-5. Curves 1 and 2 give general relations of Moho depth, i.e., crustal thickness, to Bouguer anomalies from worldwide gravity measurements. The straight lines are from another review of gravity data and their relation to the Moho. It is pointed out that the fact that these lines are flatter than expected from simple compensation is to be expected from modification of gravity effects due to the curvature of the earth.

The data, such as given by Fig. 9-5, may be somewhat useful in continental-shelf exploration for indicating general expectations in depth to the Moho and rough approximations of basement depth in shelf areas where thick sedimentary deposits may be expected. The rather wide range in these general or average relations, as shown by Fig. 9-5, and the much wider range in any particular series of measurements, as shown, for instance, by Fig. 9-2, point to the caution which must be exercised in any application to a particular area.

GRAVITY AT CONTINENTAL MARGINS

The changes in gravity from continents across continental shelves and slopes to the deep oceans have been measured in many parts of the world. The older measurements, by pendulums in submarines, were assembled by Worzel (1965). Many similar traverses made later by shipborne gravity meters have shown similar results.

The measurements show great variation in detail, but the gross characteristics are similar to the idealized relations illustrated by Fig. 9-1. The Bouguer anomaly tends to increase seaward, beginning somewhat back of the outer edge of the continental shelf, partly as would be expected from the shape effect, mentioned above.

A typical example is shown by Fig. 9-6. Several such sections are included in Worzel (1965). His original figure has been modified to include the Bouguer gravity. The calculated free-air gravity is based on the section shown at the bottom of the diagram. The densities and density boundaries used to approximate the section for the calculation are shown in the central part of the figure. Some of the details of the layering of the model used in the calculations are controlled by refraction-seismograph observations, which give depths and velocities for the major discontinuities. The calculated free-air curve is in quite close agreement with the observed free-air gravity values.

The Bouguer curve shows the rapid outward rise caused by the increase in water depth and the corresponding rise in the Moho. When, in making the Bouguer correction, the water is replaced by rock of density 2.67, the gravity effect is added to the free-air anomaly. Actually the density deficiency due to the deep water is largely balanced by the rise in the Moho. This is an example of the advantage of using the free-air rather than the Bouguer anomaly for the interpretation of marine gravity.

The details in the free-air curve, such as the sharp maximum in the central part of the section at the beginning of the continental slope, are accounted for by details in the form of the density layering. They are much too sharp to be compensated by variation in depth to the Moho and are attributed to the variations in sediment thickness and a local basement uplift, as shown in the interpretive cross section.

The pendulum data off the east coast of the United States include seven profiles from Nova Scotia to Cape Hatteras. All show strong (50- to 120-mgal) free-air anomalies near the outer edge of the continental shelf, which are generally similar to that of Fig. 9-6. They are similarly attributed to local variations in thickness of sediments and depth to basement along with the relatively sharp

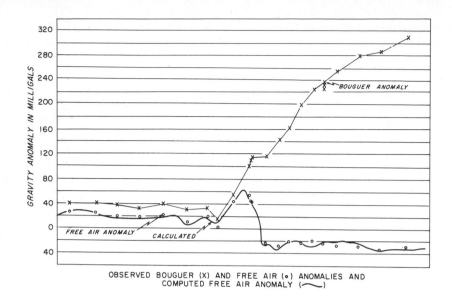

OBSERVED BOUGUER (X) AND FREE AIR (•) ANOMALIES AND
COMPUTED FREE AIR ANOMALY (⌒)

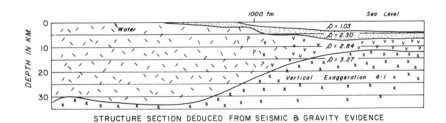

DENSITIES AND SECTION USED FOR COMPUTED ANOMALY

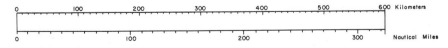

STRUCTURE SECTION DEDUCED FROM SEISMIC & GRAVITY EVIDENCE

FIGURE 9-6
Gravity anomalies, calculated gravity, and interpreted crustal section, western
Atlantic (southeast from central Maine coast). (*From Worzel, 1965.*)

increase in water depth. These features also are too sharp to have their origin in details of compensation at or below the Moho.

ANOMALIES SUPPORTED BY THE EARTH'S CRUST

Numerous quite large anomalies (up to 100 mgals or more) are much too local to be compensated at the depth of the Moho. Therefore they must be supported either by strength of the crust or by unknown variations in density in the upper 30 to 50 km of the crust. Such features are not of immediate interest in oil prospecting but a few examples may be mentioned as typical of large anomalies which may appear on regional gravity maps.

On the island of Cyprus, a large gravity disturbance, covering much of the area of the island, has a positive Bouguer anomaly of over 200 mgals and local relief of about 120 mgals (Harrison, 1955; Garland, 1971, p. 195), which is attributed to an intrusion of ultrabasic rocks of unusually high density.

A nearly circular anomaly with some 120 mgals relief in northern Canada (Stacey, 1971) is attributed to an inverted cone of gabbroic material with a diameter, near the surface, of approximately 75 km and with its apex at a depth of around 40 km.

A very strong gravity maximum, the Mid-Continent Gravity High, extends from the west end of Lake Superior for some 700 miles south-southwest across Minnesota, Iowa, Nebraska, and Kansas (Coons, Woollard, and Hershey, 1967). Total gravity relief from the central maximum to the flanking minima is about 150 mgals, and its source appears to be intrusions of gabbro and basalt. These must be supported by the crust to produce this large uncompensated feature. A syncline in near-surface sediments has been attributed to loading of the crust, but its relief is relatively small and it could account for gravity effects of only a small fraction of the total relief of the anomaly. (Gravity maps of portions of the anomaly are shown by Figs. 6-24 and 6-25, where they are used to illustrate the least-squares anomaly-filtering process.)

These and many other large anomalies demonstrate that the crust of the earth has a certain strength. On the other hand, there are some indications that the process of isostasy is measurably active in the geological present. This evidence comes from changes in land height, particularly in Fennoscandia, i.e., the Finland-Scandinavian part of northern Europe (Heiskanen and Vening Meinesz, 1958, p. 365; Stacey, 1969, p. 199), where the rate of height changes has been measured by differences in elevations of a series of raised beaches with the time intervals determined from glacial varves. Movement has been confirmed by repeated precise leveling. The uplift of the surface is attributed to the unloading resulting from the removal of the thick ice sheet of the last glacial epoch. From the rate of uplift and estimates of the ice thickness, a value has been derived for the viscosity of the upper mantle. Similar indications have been obtained from studies in formerly glaciated areas of the Canadian Shield.

Very simple and clear evidence for the continued operation of the isostatic process during geologic time is given by the erosion of mountains. As succintly stated by Clark (1971, p. 112):

The theory of isostasy offers a ready explanation of the deep erosion that has affected ancient mountain systems. Erosion removes mass from above the geoid; consequently, the mountains become overcompensated. The compensating mass is lighter than its surroundings and therefore is buoyant, so it tends to rise. The total vertical movement may be many times the height of the mountains. The common exposure of the deeply eroded cores of ancient mountains is a direct consequence of this process of isostatic uplift.

An inverse corollary is the deposition of sediments in basin areas. As sediments are deposited, the basins sink from the loading to attract water flowage, to bring in more sediments, and the process continues until great thicknesses of shallow water sediments are accumulated.

GROSS FEATURES OF THE EARTH'S CRUST

The extension, largely since the end of World War II, of gravity, magnetic, seismic, and other geophysical observations over the water-covered areas of the globe has made a great body of new information available. The study and correlation of these new data have revolutionized the concepts of the evolution of the earth to its present form. Only a few special examples relating to gravity are mentioned here.

The Great Ocean Trenches

Under various parts of the oceans, there are great deeps, or trenches, which are the sources of some of the largest gravity anomalies known. The first to be studied gravimetrically is in the East Indies, just south of the island arc of Sumatra and Java. Pendulum gravity measurements in submarines by Vening Meinesz showed an arcuate anomaly some 2000 miles long, extending south of Sumatra, Java, and Timor with negative isostatic anomalies to -130 mgals and positive anomalies of 50 mgals (Heiskanen and Vening Meinesz, 1958, p. 388). There are similar ocean-trench anomalies under many other parts of the world's oceans.

A profile across the Tonga Trench, east of the Fiji Islands in the South Pacific, is given by Fig. 9-7. Here, the maximum water depth is over 9000 m, the largest negative free-air anomaly is -224 mgals, and the largest positive one is 153 mgals, to give a total range, among the submarine pendulum stations, of 377 mgals. The Bouguer anomaly is strongly positive because of the rise in the Moho under the deep ocean (over 6000 m) under the easterly part of the profile. These values are typical of the Tonga-Kermadic Trench, which extends some 2000 miles from Samoa to New Zealand.

The Mid-Ocean Ridges

In the geology of plate tectonics, a corollary to the ocean trenches is the globe-encircling system of mid-ocean ridges. These are the presumed origin of the ocean-floor material which spreads out to terminate in the subduction zones of the

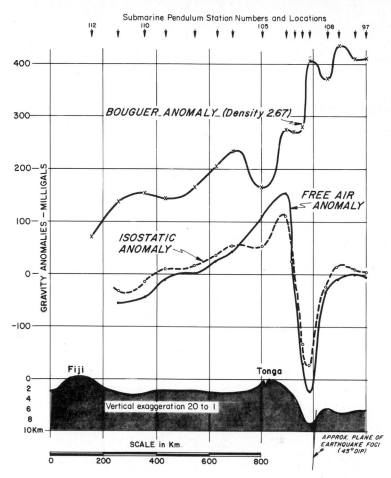

FIGURE 9-7

Ocean depths and gravity anomalies across the Tonga Trench, South Pacific. (*Replotted from data by Talwani, Worzel, and Ewing, 1961.*)

trenches. The ridges are not as spectacular, gravitationally, as the trenches, largely because they are more nearly compensated, but gravity studies have contributed to their understanding.

An example is given by Fig. 9-8. The free-air and Bouguer gravity profiles are from a Graf shipborne-gravity-meter traverse some 2200 km long running southwesterly and perpendicular to the central part of the mid-Atlantic ridge (at about 30° north latitude). The free-air anomaly averages near zero but has large local departures (±50 mgals) caused by local topography. Smooth curves added to the original chart show the overall topographic rise of the ridge of about 3600 m and a corresponding decrease in the Bouguer anomaly of about 160 mgals. This means nearly complete overall compensation for a density of about 2.5 (neglecting

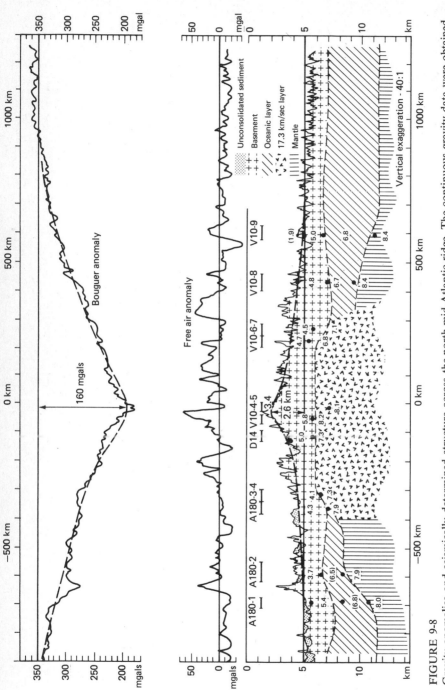

FIGURE 9-8

Gravity anomalies and seismically determined structure across the north mid-Atlantic ridge. The continuous gravity data were obtained on *Vema* cruise 17. Bouguer anomalies were calculated assuming two-dimensionality and a density of 2.60 g/cm³ for the basement layer. A correction was also made for the sediment layer. The seismic section is obtained by projecting the structure at seismic stations along the gravity profile. Seismic horizon locations are represented by dots with corresponding values of compressional-wave velocities in kilometers per second. Numbers in parentheses denote assumed seismic velocities. Smooth dashed lines through Bouguer-anomaly and bottom-topography curves have been added to indicate the fairly complete broad or average compensation corresponding to the near-zero average free-air anomaly. (*Modified from Talwani, LePichon, and Ewing, 1965.*)

earth-curvature effects). From this and other examples it appears that the dynamic processes of the ridges are more nearly compensated than those of the trenches, suggesting much slower movement.

Plate Tectonics

The mid-ocean ridges and the trenches are only the most conspicuous aspect of the whole "new geology" of plate tectonics and continental drift. The concept of ocean floor being formed continuously at the ocean ridges and disappearing into subduction zones at the trenches has developed from many lines of geophysical evidence. For instance, the relatively rapid movement into the subduction zones, as indicated by their lack of isostatic compensation, apparently is the cause of frequent earthquakes along them. Also, when depths of sources are plotted against horizontal distance, they tend to fall on a plane dipping at around 45°, which may be the plane of the descending crustal sheet. This plane, from measured earthquake foci in the Tonga Trench (Stacey, 1969, p. 67), extends to a depth of 600 km; its upper part is shown by the near-vertical line (because of the vertical exaggeration) in relation to the topography in Fig. 9-7. It occurs at the steep dip on the deep ocean side of the trench, as to be expected if this is part of a subduction zone.

The examples of gravity related to the ocean ridges and trenches are only small samples from the great mass of geophysical data and literature bearing on plate tectonics. For the most part this new geology is not directly related to the search for oil, but, in a broader sense, worldwide oil accumulation is controlled by worldwide geology. The recent expansion of the search for oil into the very large shelf areas of the continental margins is approaching more closely those zones where better understanding of the geology (largely dependent on geophysics in some form) can contribute to the broad application of geological principles for the guidance of petroleum exploration.

REFERENCES FOR PART ONE

ADLER, JOSEPH L.: Simplification of Tidal Corrections for Gravity Meter Surveys, *Geophysics*, vol. 7, pp. 35–44, 1942.

AFFLECK, JAMES: Magnetic Anomaly Trend and Spacing Patterns, *Geophysics*, vol. 28, pp. 329–395, 1963.

ALLEN, W. E., H. J. CALLOUET, and L. STANLEY: Gravity Investigations in the Hockley Salt Dome, Harris County, Texas, *Geophysics*, vol. 20, pp. 829–849, 1955.

ASKANIA: The New Askania Seagravity Meter Gss 3, Askania Werke, Berlin, 1970.

ATHY, L. F.: Density, Porosity and Compaction of Sedimentary Rocks, *Bull. Am. Assoc. Pet. Geol.*, vol. 14, pp. 1–14, 1930.

BASCOM, WILLARD: "A Hole in the Bottom of the Sea," Doubleday & Company, Inc., Garden City, New York, 1961.

BELL AEROSYSTEMS: Stabilized Gravity Measurement Systems, Bell Aerosystems, Buffalo, N.Y., 1970.

BERROTH, A.: Referenzpendelmessungen am Salzhorst Oldau-Hambühren (Hann.), *Z. Geophys.*, vol. 3, pp. 1–16, 1927.

BEYER, L. A., R. E. VON HUENE, T. H. MCCULLOH, and J. R. LOVETT: Measuring Gravity on the Sea Floor in Deep Water, *J. Geophys. Res.*, vol. 71, no. 8, pp. 2091–2100, 1966.

BIBLE, JOHN L.: Terrain Correction Table for Gravity, *Geophysics*, vol. 27, pp. 716–718, 1962.

——: Gravity for the Geologist, *World Oil* (Gulf Publishing Co.), October–November 1964.

BICKEL, H. C.: A Note on Terrain Corrections, *Geophysics*, vol. 13, pp. 255–258, 1948.

BIRCH, FRANCIS: Handbook of Physical Constants, *Geol. Soc. Am. Spec. Pap.* 36, 1942.

BLAU, L. W.: Black Magic in Geophysical Prospecting, *Geophysics*, vol. 1, no. 1, p. 1, January 1936.

BOTT, M. H. P.: The Use of Electronic Digital Computers for the Evaluation of Gravimetric Terrain Corrections, *Geophys. Prospect.*, vol. 7, no. 1, pp. 45–54, 1959.

——: Solution of the Linear Inverse Problem in Magnetic Interpretation with Application to Oceanic Magnetic Anomalies, *Geophys. J. R. Astron. Soc.*, vol. 13, pp. 313–323, 1967.

——: Inverse Methods in the Interpretation of Magnetic and Gravity Anomalies, pp. 133–162 in Bruce A. Bolt (ed.), "Methods in Computational Physics," vol. 13, Academic Press, Inc., New York, 1973.

BOUCHER, F. G.: U.S. Patents 2,318,665, May 11, 1943, and 2,327,697, Aug. 24, 1943.

BOWIE, WILLIAM: Investigations of Gravity and Isostasy, *U.S. Coast Geod. Surv. Spec. Pub.* 40, 1917.

BOWIN, CARL, THOMAS C. ALDRICH, and R. A. FOLINSBEE: VSA Gravity Meter System: Tests and Recent Developments, *J. Geophys. Res.*, vol. 77, no. 11, pp. 2018–2033. 1972.

BOYS, CHARLES VERNON: The Mean Density of the Earth, "A Dictionary of Physics," vol. 3, pp. 270–285, Macmillan & Company, Ltd., London, 1923.

BROWN, DONALD A.: Instrumentation on the *Gulfrex*, *Undersea Technol.*, pp. 16–24, 1970.

BROWN, HART: U.S. Patent 2,125,282, Aug. 2, 1938.

BRYAN, A. B.: Gravimeter Design and Operation, *Geophysics*, vol. 2, pp. 301–308, 1937.

BULLARD, E. C., M. N. HILL, and C. S. MASON: The Observation of Gravity by Means of Invariable Pendulums, *Proc. R. Soc.*, vol. 141, pp. 233–258, 1933.

————, ————, and ————: Chart of the Total Force of the Earth's Magnetic Field for the Northeastern Atlantic Ocean, Geomagnetica, Serviço Meteorologico Nacional, Lisbon, 1962.

BYERLY, P. E.: Convolution Filtering of Gravity and Magnetic Maps, *Geophysics*, vol. 30, pp. 281–283, 1965.

CAPUTO, MICHELE: "The Gravity Field of the Earth from Classical and Modern Methods," Academic Press, Inc., New York, 1967.

CASSINIS, A., P. DORÉ, and S. VALLARIN: "Fundamental Tables for Reducing Gravity Observed Values," Royal Italian Geodetic Committee, Milan, 1937.

CHAMPION, F. C., and N. DAVY: "Properties of Matter," Prentice-Hall, Inc., New York, 1937.

CLARK, R. H., and J. T. ROUSE: A Closed System for Generation and Entrapment of Hydrocarbons in Cenozoic Deltas, Louisiana Gulf Coast, *Bull. Am. Assoc. Pet. Geol.*, vol. 55, pp. 1170–1178, 1971.

CLARK, SYDNEY, JR.: "Structure of the Earth," Prentice-Hall, Inc., Englewood Cliffs, N.J., 1971.

COOK, A. G.: A New Determination of the Acceleration Due to Gravity at the National Physical Laboratory, England, *Phil. Trans. R. Soc., Lond.*, vol. A261, pp. 211–252, 1967.

————: Report on Absolute Measurements of Gravity, *Trav. Assoc. Int. Geod.*, tome 23, p. 271, 1968.

————: "Gravity and the Earth," The Wykeham Science Series, Wykeham Publications, Ltd., London, 1969.

COONS, R. L., J. W. MACK, and W. STRANGE: Least-Squares Polynomial Fitting of Gravity Data and Case Histories, Computers in the Minerals Industries, pt. 2, *Stanford Univ. Pub. Geo. Sci.*, vol. 9, no. 2, pp. 498–519, 1964.

———— and M. SMALET: Surface Ship Gravity in Oil Exploration, *Soc. of Explor. Geophys.*, *37th Meet.*, 1967.

————, G. P. WOOLLARD, and G. HERSHEY: Structural Significance of the Mid-Continent Gravity High, *Bull. Am. Assoc. Pet. Geol.*, vol. 51, no. 12, pp. 2381–2399, 1967.

CORDELL, L., and R. G. HENDERSON: Iterative Three-dimensional Solution of Gravity Anomaly Data Using a Digital Computer, *Geophysics*, vol. 33, pp. 596–601, 1968.

DAMPNEY, C. N. G.: Three Criteria for the Judgment of Vertical Continuation and Derivative Methods of Geophysical Interpretation, *Geoexploration*, vol. 4, pp. 3–24, 1966.

————: The Equivalent Source Technique, *Geophysics*, vol. 34, pp. 39–53, 1969.

DANES, Z. FRANKENBERGER: On a Successive Approximation Method for Interpreting Gravity Anomalies, *Geophysics*, vol. 25, pp. 1215–1228, 1960.

DARBY, E. K., and E. B. DAVIES: The Analysis and Design of Two-dimensional Filters for Two-dimensional Data, *Geophys. Prospect.*, vol. 15, pp. 383–406, 1967.

————, E. J. MERCADO, R. M. ZOLL, and J. R. EMANUEL: Computer Systems for Real-Time Marine Exploration, *Geophysics*, vol. 38, no. 2, pp. 301–309, 1973.

DARWIN, SIR GEORGE H.: "Scientific Papers," vol. 3, p. 78, Cambridge University Press, Cambridge, 1910.

DEAN, W. C.: Frequency Analysis for Gravity and Magnetic Interpretation, *Geophysics*, vol. 23, pp. 97–127, 1958.

DEGOLYER, E. L.: The Seductive Influence of the Closed Contour, Editorial, *Econ. Geol.*, vol. 23, pp. 681–682, 1928.

DICKINSON, GEORGE: Geological Aspects of Abnormal Reservoir Pressures in Gulf Coast Louisiana, *Bull. Am. Assoc. Pet. Geol.*, vol. 37, pp. 410–432, 1953.

DOBRIN, MILTON B.: "Introduction to Geophysical Prospecting," McGraw-Hill Book Company, New York, 1952; 2d ed., 1960.

ELKINS, T. A.: The Second Vertical Derivative Method of Gravity Interpretation, *Geophysics*, vol. 16, pp. 39–56, 1951.

———— and SIGMUND HAMMER: The Resolution of Combined Effects and Applications to Gravitational and Magnetic Data, *Geophysics*, vol. 3, no. 4, pp. 315–331, 1938.

EÖTVÖS, ROLAND V.: Untersuchungen über Gravitation und Erdmagnetismus, *Ann. Phys.*, vol. 59, pp. 354–400, 1896.

————: Geodätische Arbeiten in Ungarn, besonders über Beobachtungen mit der Drehwage, *Verh. XVI Allg. Konf. Int. Erdmessung Lond. Camb., 1909.*

————: Nachweis der Schwereänderung, die ein auf normal geformter Erdoberfläche in östlicher oder westlicher Richtung bewegter Körper durch diese Bewegung erleidet, *Ann. Phys.*, ser. 4, vol. 59, pp. 743–752, 1919.

EVJEN, H. M.: The Place of the Vertical Gradient in Gravitational Interpretations, *Geophysics*, vol. 1, pp. 127–136, 1936.

FAJKLEWICZ, Z. J.: Fictitious Anomalies of Higher Vertical Derivatives of Gravity, *Geophysics*, vol. 30, pp. 1094–1107, 1965.

FALLER, JAMES E.: Results of an Absolute Determination of the Acceleration of Gravity, *J. Geophys. Res.*, vol. 70, no. 16, pp. 4035–4038, 1965.

FORWARD, ROBERT L., and CURTIS C. BELL: Simulated Terrain Mapping with the Rotating Gravity Gradiometer, "Advances in Dynamic Gravimetry," pp. 115–129, Instrument Society of America, Pittsburgh, 1970.

FROWE, EUGENE W.: Diving Bell in Underwater Gravimeter Operations, *Geophysics*, vol. 12, pp. 1–12, 1947.

FULLER, BRENT D.: Two-dimensional Frequency Analysis and Design of Grid Operators, *Min. Geophys.*, vol. 2, Theory (Society of Exploration Geophysics), pp. 658–708, 1967.

GARLAND, G. D.: "The Earth's Shape and Gravity," Pergamon Press, Ltd., Oxford, 1965.

————: "Introduction to Geophysics," W. B. Saunders Co., Philadelphia, 1971.

GAY, MALCOLM W.: Relative Gravity Measurements Using Precision Pendulum Equipment, *Geophysics*, vol. 5, pp. 176–191, 1940.

GILBERT, R. L. G.: A Dynamic Gravimeter of Novel Design, *Proc. Phys. Soc. Lond.*, vol. 62B, pp. 445–454, 1949.

————: Gravity Observations in a Borehole, *Nature*, vol. 170, pp. 424–425, 1952.

GLICKEN, MILTON: Eötvös Correction for a Moving Gravity Meter, *Geophysics*, vol. 27, pp. 531–533, 1962.

GOGUEL, JEAN: A Universal Table for the Prediction of the Lunar-Solar Correction in Gravimetry (Tidal Gravity Correction), *Geophys. Prospect.*, vol. 2, suppl. March 1954.

GOODELL, R. R., and C. H. FAY: Borehole Gravity Meter and Its Application, *Geophysics*, vol. 29, pp. 774–782, 1964.

GRAF, ANTON: Ein neuer statischer Schweremesser zur Messung und Registrierung lokaler und zeitlicher Schwereänderungen, *Z. Geophys.*, vol. 14, pp. 152–172, 1938.

————: Das Seegravimeter, *Z. Instrumentenkd.*, pp. 151–162, 1958.

————: Improvements on Sea Gravimeter Gss 2, *J. Geophys. Res.*, vol. 66, no. 6, pp. 1813–1821, 1961.

GRANT, F. S.: Review of Data Processing and Interpretation Methods in Gravity and Magnetics, 1964–1971, *Geophysics*, vol. 37, pp. 647–661, 1972.

———— and G. F. WEST: A Problem in the Analysis of Geophysical Data, *Geophysics*, vol. 22, pp. 309–344, 1957.

—— and ——: "Interpretation Theory in Applied Geophysics," McGraw-Hill Book Company, New York, 1965.

GRIFFIN, W. RAYMOND: Residual Gravity in Theory and Practice, *Geophysics*, vol. 14, pp. 39–58, 1949.

GUDMUNDSSON, G.: Interpretation of One-dimensional Magnetic Anomalies by Use of the Fourier Transform, *Geophys. J. R. Astron. Soc.*, vol. 12, pp. 87–97, 1966.

——: Spectral Analysis of Magnetic Surveys, *Geophys. J. R. Astron. Soc.*, vol. 13, pp. 325–337, 1967.

HAALCK, H.: Lehrbuch der angewandten Geophysik, Verlag von Gebruder Borntraeger, Berlin, 1934.

——: Der statische (barometrische) Schwermesser für Messungen auf festen Lande und auf See, *Beitr. angew Geophysik*, vol. 7, pp. 285–316, 1938.

HAMMER, SIGMUND: Investigation of the Vertical Gradient of Gravity, *Trans. Am. Geophys. Union*, 1938.

——: Terrain Corrections for Gravimeter Stations, *Geophysics*, vol. 4, pp. 184–194, 1939.

——: Note on the Variation from Equator to Pole of the Earth's Gravity, *Geophysics*, vol. 8, p. 57, 1943.

——: Estimating Ore Masses in Gravity Prospecting, *Geophysics*, vol. 10, pp. 50–62, 1945.

——: A New Calculation Technique for Quantitative Interpretation in Gravity Prospecting (abstract only), *Geophysics*, vol. 12, p. 498, 1947.

——: Density Determination by Underground Gravity Measurements, *Geophysics*, vol. 15, pp. 637–652, 1950.

——: Usefulness of High-Quality Gravity Surveys, *Oil Gas J.*, pp. 106–110, 1953.

——: Modern Methods of Gravity and Magnetic Interpretation, *Proc. 4th World Pet. Congr.*, sec. I/E, 1955.

——: Deep Gravity Interpretation by Stripping, *Geophysics*, vol. 28, pp. 369–378, 1963.

——: Approximation in Gravity Calculations, *Geophysics*, vol. 39, pp. 205–222, 1974.

HAMMOND, JAMES A.: A Laser-Interferometer System for the Absolute Determination of the Acceleration of Gravity, *Univ. Colo. Joint. Inst. Lab. Astrophys. Rep.* 103, Boulder, Colo., Feb. 26, 1970.

—— and J. E. FALLER: 12.5 Laser Interferometer System for the Determination of the Acceleration of Gravity, *IEEE Jour. Quantum Electron.*, QE-3, 597 (1967), fig. 1.

HARRISON, J. C.: An Interpretation of Gravity Anomalies in the Eastern Mediterranean, *Phil. Trans. R. Soc., Lond.*, vol. A248, pp. 283–324, 1955.

HART, PEMBROKE J. (ed.): The Earth's Crust and Upper Mantle, *Am. Geophys. Union Geophys. Monogr.* 13, Washington, D.C., 1969.

HARTLEY, K. A.: A New Instrument for Measuring Very Small Differences in Gravity, *Physics*, vol. 2, pp. 123–130, 1932.

HASTINGS, W. K.: Gravimeter Operations in Foothills of Alberta, *Geophysics*, vol. 10, pp. 526–534, 1945.

HEDBERG, H. D.: The Effect of Gravitational Compaction on the Structure of Sedimentary Rocks, *Bull. Am. Assoc. Pet. Geol.*, vol. 10, pp. 1035–1072, 1926.

HEDSTROM, HELMER: A New Gravimeter for Ore Prospecting, *Am. Inst. Min. Met. Eng. Tech. Publ.* 953, 1938.

HEILAND, C. A.: Gravimeters: Their Relation to Seisometers, Astatization, and Calibration, *Am. Inst. Min. Met. Eng. Tech. Publ.* 1049, 1939.

——: "Geophysical Exploration," Prentice-Hall, Inc., New York, 1940.

HEISKANEN, W. A., and F. A. VENING MEINESZ: "The Earth and Its Gravity Field," McGraw-Hill Book Company, New York, 1958.

HELMERT, F. R.: Die Schwerkraft und die Massenverteilung der Erde, pp. 85–177, in "Encyklopädie der mathematischen Wissenschaften," Band VI, Teil 1, Heft 7, B. B. Teubner, Leipzig, 1910.

HENDERSON, GARRY C., WILLIAM E. STRANGE, and ROBERT M. IVERSON: Dynamic Gravimetry in Marine Exploration, preprint, Offshore Technology Conf., Paper No. 1287, 1970.

HENDERSON, ROLAND G.: A Comprehensive System of Automatic Computation in Magnetic and Gravity Interpretation, Geophysics, vol. 25, pp. 569–585, 1960.

———: Field Continuation and the Step Model in Aeromagnetic Interpretation, Geophys. Prospect., vol. 14, no. 4, pp. 528–546, 1966.

——— and ISADORE ZIETZ: Computation of Second Vertical Derivatives of Geomagnetic Fields, Geophysics, vol. 14, pp. 517–534, 1949.

HEYL, PAUL R.: A Redetermination of the Constant of Gravitation, Bur. Stand. J. Res., vol. 5, pp. 1243–1290, 1930.

HOLWECK, F., and P. LEJAY: Un Instrument transportable pour las mesure rapide de la gravité, Compt. Rend., vol. 190, pp. 1387–1388, 1930.

HORSFIELD, W., and E. C. BULLARD: Gravity Measurements in Tanganyika Territory Carried Out by the Survey Division, Department of Lands and Mines, Mon. Not. R. Astron. Soc., Geophys. Suppl., vol. 4, pp. 43–113, 1937.

HORTON, C. W., W. B. HEMPKINS, and A. A. J. HOFFMAN: A Statistical Analysis of Some Aeromagnetic Maps from the Northern Canadian Shield, Geophysics, vol. 29, pp. 582–601, 1964.

HOSKINSON, ALBERT J.: Recent Developments in Gravity Instruments, Trans. Am. Geophys. Union, 17th Annu. meet., 1936, pp. 44–45.

HOWELL, LYNN G., K. O. HEINTZ, and H. BARRY: The Development of a High Precision Downhole Gravity Meter, Geophysics, vol. 31, pp. 764–772, 1962.

HOYT, ARCHER: Gravimeter, U.S. Patent 2,131,737, Oct. 4, 1938a.

———: Helical Ribbon Spring Measuring Apparatus, U.S. Patent 2,131,739, Oct. 4, 1938b.

———: Optical System, U.S. Patent 2,131,738, Oct. 4, 1938c.

HUBBERT, M. KING: Line-Integral Method of Computing Gravity, Geophysics, vol. 13, pp. 215–225, 1948a.

———: Gravitational Terrain Effects of Two-dimensional Topographic Features, Geophysics, vol. 13, pp. 226–254, 1948b.

ISING, GUSTAF: Use of Astatized Pendulums for Gravity Measurements, Am. Inst. Min. Met. Eng. Tech. Publ. 828, 1937.

———: U.S. Patent 2,221,480, Nov. 12, 1940.

JAKOSKY, J. J.: "Exploration Geophysics," 1st ed., Times Mirror Press, Los Angeles, 1940.

———: "Exploration Geophysics," 2d ed., Trija Publishing Co., Los Angeles, 1950.

JEFFREYS, SIR HAROLD: Who Named the Milligal?, Geophys. J. R. Astron. Soc., vol. 6, p. 553, 1962.

JONES, BILL R.: Downhole Gravity Tool Spots Distant Porosity, World Oil, vol. 175, no. 2, pp. 56–59, 1972.

KANASEWICH, E. R., and R. G. AGARAWAL: Analysis of Gravity and Magnetic Fields in Wave Number Domain, J. Geophys. Res., vol. 75, no. 29, pp. 439–447, 1970.

KANE, M. F.: A Comprehensive System of Terrain Corrections Using a Digital Computer, Geophysics, vol. 27, pp. 455–462, 1962.

KOLOGINCZAK, JOHN B.: Marine Geophysical Surveys, Oil Gas J., pt. 1, Nov. 16, 1970, p. 204; pt. 2, Nov. 22, 1970, p. 106.

KUNARATNAM, K.: An Iterative Method for the Solution of a Non-linear Inverse Problem in Magnetic Interpretation, *Geophys. Prospect.*, vol. 20, pp. 439–447, 1972.

KUO, J. T., R. C. JACKSON, M. EWING, and G. WHITE: Transcontinental Tidal Profile across the United States, *Science*, vol. 168, pp. 964–971, 1970.

————, MARIO OTTAVIANI, and SHIRA K. SINGH: Variation in Vertical Gravity in New York City and Alpine, New Jersey, *Geophysics*, vol. 34, pp. 235–248, 1969.

LACOSTE, LUCIEN J. B.: A New Type Long Period Vertical Seismograph, *Physics*, vol. 5, no. 7, 1934.

————: A Simplification in the Condition for the Zero-Length Spring Seismograph, *Bull. Seism. Soc. Am.*, vol. 25, no. 2, 1935.

————: Measurement of Gravity at Sea and in the Air, *Rev. Geophys.*, vol. 5, pp. 477–526, 1967.

————: Crosscorrelation Method for Evaluating and Correcting Shipboard Gravity Data, *Geophysics*, vol. 38, no. 4, pp. 701–709, August 1973.

————, NEAL CLARKSON, and GEORGE HAMILTON: LaCoste and Romberg Stabilized Platform Shipboard Gravity Meter, *Geophysics*, vol. 32, no. 1, pp. 99–190, 1967.

LAFEHR, T. R.: The Estimation of the Total Amount of Anomalous Mass by Gauss's Theorem, *J. Geophys. Res.*, vol. 70, pp. 1911–1919, 1965.

———— and L. L. NETTLETON: Quantitative Evaluation of a Stabilized Platform Shipboard Gravity Meter, *Geophysics*, vol. 32, pp. 110–118, 1967.

LAMBERT, WALTER D.: The Figure of the Earth from Gravity Observations, *J. Wash. Acad. Sci.*, vol. 26, pp. 491–506, 1936.

LANCASTER-JONES, E.: Principles and Practice of the Gravity Gradiometer, *J. Sci. Instrum.*, vol. 9, pp. 341–353, 373–380, 1932.

LONGMAN, I. M.: Formulas for Computing the Tidal Accelerations Due to the Moon and the Sun, *J. Geophys. Res.*, vol. 64, pp. 2351–2355, 1959.

LUDWIG, WILLIAM J., JOHN E. NAFE, and CHARLES L. DRAKE: Seismic Refraction, pp. 53–84 in "The Sea," vol. 4, pt. 1, Wiley-Interscience, New York, 1971.

LYONS, PAUL L. (ed.): "Geophysical Case Histories," vol. II, The Society of Exploration Geophysicists, 1956.

MARKOWITZ, WILLIAM: SI, the International System of Units, *Geophys. Surv.*, vol. 1, pp. 217–241, 1973.

MASKET, A. V. N., et al.: Tables of Solid Angles, *USAC Div. Tech. Inf.* TID 14975, 1962.

MCCULLOH, THANE H.: Mass Properties of Petroleum Rocks as Related to Petroleum Exploration, *U.S. Geol. Surv. Prof. Pap.* 528-A, 1967.

————, J. E. SCHOELLHAMER, and E. H. PAMPEYAN: The U.S. Geological Survey, LaCoste and Romberg Precise Borehole Gravimeter System, Test Results, *U.S. Geol. Surv. Prof. Pap.* 575-D, pp. D101–D112, 1967.

MECHTLY, E. A.: International System of Units, Physical Constants and Conversion Factors Revised, *NASA Sci. Tech. Inf. Div., Off. Technol. Util.* NASA SP-7012, 1969.

MEISSER, O.: Ein neuer Vierpendelapparat für relative Schweremessungen, *Z. Geophys.*, vol. 6, pp. 1–12, 1930.

MELCHIOR, PAUL: "The Earth Tides," Pergamon Press, New York, 1966.

MESKÓ, C. A.: Two-dimensional Filtering and the Second Derivative Method, *Geophysics*, vol. 31, pp. 606–617, 1966.

MILLETT, FRANK B., JR.: A Dot Chart for Calculation of Gravitational and Magnetic Attraction of Two-dimensional Bodies, *Min. Geophys.*, vol. II (Society of Exploration Geophysicists), pp. 642–657, 1967.

MOTT-SMITH, LEWIS M.: Torsion Gravimeter, U.S. Patent 2,130,648, Sept. 20, 1938.

MUFTI, IRSHAD R.: Design of Small Operators for the Continuation of Potential Field Data, *Geophysics*, vol. 37, no. 3, pp. 488–506, 1972.

MUSKAT, M.: "The Flow of Homogeneous Fluids Through Porous Media," McGraw-Hill Book Company, New York, 1937.

MUSGRAVE, A. W., and W. G. HICKS: Outlining of Shale Masses by Geophysical Methods, *Geophysics*, vol. 31, pp. 711–725, 1966.

NAGI, J. G.: Convergence and Divergence in Downward Continuation, *Geophysics*, vol. 32, pp. 867–871, 1967.

NAIDU, P. S.: Extraction of Potential Field Signal from a Background of Random Noise by Strakhov's Method, *J. Geophys. Res.*, vol. 71, pp. 5987–5995, 1966.

———: Statistical Properties of Potential Fields over a Random Medium, *Geophysics*, vol. 32, pp. 88–98, 1967.

NAUDY, H., and H. DREYER: Essai de filtrage nonlinéaire appliqué aux profile aéromagnetique, *Geophys. Prospect.*, vol. 16, pp. 171–178, 1968.

NETTLETON, L. L.: Determination of Density for Reduction of Gravimeter Observations, *Geophysics*, vol. 4, pp. 176–183, 1939.

———: "Geophysical Prospecting for Oil," McGraw-Hill Book Company, New York, 1940.

———: Gravity and Magnetic Calculations, *Geophysics*, vol. 7, pp. 293–310, 1942.

——— (ed.): "Geophysical Case Histories," vol. I, Society of Exploration Geophysicists, Tulsa, 1948.

———: Regionals, Residuals and Structures, *Geophysics*, vol. 19, pp. 1–22, 1954.

———: Gravity and Magnetics for Geologists and Seismologists, *Bull. Am. Assoc. Pet. Geol.*, vol. 46, pp. 1815–1838, 1962.

———: Elementary Gravity and Magnetics for Geologists and Seismologists, *Soc. Explor. Geophys. Monogr. Ser.* 1, 1971.

———, L. J. B. LACOSTE, and MILTON GLICKEN: Quantitative Evaluation of Precision of Airborne Gravity Meter, *J. Geophys. Res.*, vol. 67, pp. 4395–4410, 1962.

———, ———, and J. B. HARRISON: Tests of an Airborne Gravity Meter, *Geophysics*, vol. 25, pp. 181–202, 1960.

NEUMAN, LAWRENCE D., and MANIK TALWANI: Accelerations and Errors in Gravity Measurements in Surface Ships, *J. Geophys. Res.*, vol. 77, pp. 4330–4338, 1972.

NISKANEN, E.: Gravity Formulas Derived by the Aid of the Level Land Stations, *Publ. Isostatic Inst., IAG, Helsinki* 16, 1945.

NORGAARD, G.: *Dan. Geodet. Inst. Publ.* 10, Copenhagen, 1939.

———: Danish Geodetic Institute, 3d ser., vol. 7, Copenhagen, 1945.

ODEGARD, M. E., and J. W. BERG, JR.: Gravity Interpretation Using the Fourier Integral, *Geophysics*, vol. 30, p. 424, 1965.

PEPPER, T. B.: The Gulf Underwater Gravimeter, *Geophysics*, vol. 6, pp. 34–44, 1941.

PETERS, L. J.: The Direct Approach to Magnetic Interpretation and Its Practical Application, *Geophysics*, vol. 14, pp. 290–320, 1949.

POYNTING, J. H., and J. J. THOMPSON: "A Textbook of Physics," vol. 1, "Properties of Matter," 11th ed., Charles Griffin and Co., Ltd., London, 1927.

PRESTON-THOMAS, H., L. G. TURNBULL, E. GREEN, T. M. DAUPHINÉE, and S. N. KALRA: An Absolute Measurement of the Acceleration of Gravity at Ottawa, *Can. J. Phys.*, vol. 38, pp. 824–852, 1960.

ROGERS, GEORGE R.: Subsurface Gravity Measurements, *Geophysics*, vol. 17, no. 2, pp. 365–377, 1952.

ROSENBACH, OTTO: A Contribution to the Computation of "Second Derivative" from Gravity Data, *Geophysics*, vol. 18, pp. 894–912, 1953.

ROY, A.: Convergence in Downward Continuation for Some Simple Geometries, *Geophysics*, vol. 32, pp. 853–866, 1967.

RUDMAN, ALBERT J., JUDSON MEAD, JOSEPH F. WHALEY, and ROBERT F. BLAKELY: Geophysical Analysis in Central Indiana Using Potential Field Continuation, *Geophysics*, vol. 36, pp. 878–890, 1971.

SAKUMA, A.: Etat actuel de la nouvelle détermination absolue de la pesanteur au Bureau International des Poids et Measures, *Bull. Geod.*, vol. 69, pp. 249–260, 1963.

———, J-M CHARTIER, and M. DUHAMEL: Gravimetrie: determination absolute de g, *P-V Com. Int. Poids Mes.*, 2e ser., tome 35, pp. 45–53, 1967.

SKEELS, D. C.: Ambiguity in Gravity Interpretation, *Geophysics*, vol. 12, pp. 43–56, 1947.

———: What Is Residual Gravity?, *Geophysics*, vol. 32, pp. 872–876, 1967.

SMITH, NEAL J.: The Case for Gravity from Boreholes, *Geophysics*, vol. 15, pp. 605–636, 1950.

SPECTOR, A., and F. S. GRANT: Statistical Models for Interpreting Aeromagnetic Data, *Geophysics*, vol. 35, pp. 293–302, 1970.

STACEY, FRANK D.: "Physics of the Earth," John Wiley & Sons, Inc., New York, 1969.

STACEY, R. A.: An Interpretation of the Gravity Anomaly at Darnley Bay, N.W.T., *Can. J. Earth Sci.*, vol. 8, no. 8, 1971.

STEENLAND, N. C.: Review of "Regional Correction of Gravity Data," by Vajk, *Geophysics*, vol. 17, p. 409, 1952.

STRAKHOV, V. N.: The Smoothing of Observed Strengths of Potential Fields I, II, *Izv. Akad. Nauk SSR, Ser. Geofiz.*, nos. 7–12, pp. 897–995, 1964.

SWARTZ, C. A.: Some Geometrical Properties of Residual Maps, *Geophysics*, vol. 19, pp. 46–70, 1954.

TALWANI, MANIK: Developments in Navigation and Measurement of Gravity at Sea, *Geoexploration* (Elsevier Publishing Co.), no. 8, pp. 151–183, 1970.

———: Computer Usage in the Computation of Gravity Anomalies, pp. 344–389 in Bruce A. Bolt (ed.), "Methods in Computational Physics," vol. 13, Academic Press, Inc., New York, 1973.

——— and MAURICE EWING: Rapid Computation of Gravitational Attraction of Three-dimensional Bodies of Any Shape, *Geophysics*, vol. 25, pp. 203–225, 1960.

———, XAVIER LEPICHON, and MAURICE EWING: Crustal Structure of the Mid-ocean Ridges, 2: Computed Model from Gravity and Seismic Data, *J. Geophys. Res.*, vol. 70, pp. 341–352, 1965.

———, G. H. SUTTON, and J. L. WORZEL: A Crusted Section across the Puerto Rico Trench, *J. Geophys. Res.*, vol. 64, pp. 1545–1555, 1959.

———, J. LAMAR WORZEL, and MAURICE EWING: Gravity Anomalies and Crustal Section across the Tonga Trench, *J. Geophys. Res.*, vol. 6, pp. 1265–1278, 1961.

———, ———, and MARK LANDISMAN: Rapid Gravity Computations for Two-dimensional Bodies with Application to the Mendocino Submarine Fracture Zone, *J. Geophys. Res.*, vol. 64, no. 1, pp. 49–59, 1959.

TANNER, J. G.: An Automated Method of Gravity Interpretation, *Geophys. J. R. Astron. Soc.*, vol. 13, pp. 339–347, 1967.

TATE, D. R.: Absolute Value of g at the National Bureau of Standards, *J. Res., Natl. Bur. Stand. (U.S.)*, vol. 70c, p. 149, 1966.

THOMPSON, L. G. D., and L. J. B. LACOSTE: Aerial Gravity Measurements, *J. Geophys. Res.*, vol. 65, pp. 305–322, 1960.

THULIN, A.: *Trav. Mem. Bur. Int. Poids Mes.*, vol. 22, p. A1, 1961.

THYSSEN-BORNEMISZA, STEPHEN V.: Vergleiche zwischen direkten und Schiefenmessungen mit dem Thyssen-Gravimeter, *Beitr. Angew. Geophys.*, vol. 7, pp. 212–229, 1938a.

———: U.S. Patents 2,108,421, Feb. 15, 1938b, and 2,132,865, Oct. 11, 1938c.

TITTMAN, J., and J. S. WAHL: The Physical Foundations of Formation Density Logging (Gamma-Gamma), *Geophysics*, vol. 30, pp. 284–296, 1965.

VACQUIER, VICTOR: Ultimate Precision of Borometric Surveying, *Bull. Am. Assoc. Pet. Geol.*, vol. 21, pp. 1168–1181, 1937.

VAN VOORHIS, G. D., and T. M. DAVIS: Magnetic Anomalies North of Puerto Rico: Trend Removal with Orthogonal Polynomials, *J. Geophys. Res.*, vol. 69, pp. 5363–5371, 1964.

VENING MEINESZ, F. A.: "Theory and Practice of Pendulum Observations at Sea," pt. II, Waltman, Delft, 1941.

WING, CHARLES G.: An Experimental Deep-Sea-Bottom Gravimeter, *J. Geophys. Res.*, vol. 72, pp. 1249–1257, 1967.

———: MIT Vibrating String Surface Ship Gravimeter, *J. Geophys. Res.*, vol. 74, pp. 5882–5894, 1969.

WINKLER, H. A.: Simplified Gravity Terrain Corrections, *Geophys. Prospect.*, vol. 10, pp. 19–34, 1962.

WOODSIDE, J. M.: The Mid-Atlantic Ridge near 45°N: XX, The Gravity Field, *Can. J. Earth Sci.*, vol. 9, pp. 942–959, 1972.

WOOLLARD, G. P.: Crustal Structure from Gravity and Seismic Measurements, *J. Geophys. Res.*, vol. 64, pp. 1521–1544, 1959.

———: Regional Variations in Gravity, in Pembroke J. Hart (ed.), "The Earth's Crust and Upper Mantle," *Am. Geophys. Union Geophys. Monogr.* 13, 1969.

———: Evaluation of the Isostatic Mechanism and Role of Mineralogic Transformations from Gravity and Seismic Data, "Physics Earth Planet Interiors," vol. 3, pp. 484–498, North-Holland Publishing Co., Amsterdam, reprinted as *Univ. Hawaii Contrib.* 273, Honolulu, 1970.

——— and JOHN C. ROSE: "International Gravity Measurements," Geophysical and Polar Research Center, University of Wisconsin, Madison, 1963.

WORZEL, J. L.: "Pendulum Gravity Measurements at Sea, 1936–1959," John Wiley & Sons, Inc., New York, 1965.

——— and G. L. SHURBET: Gravity Anomalies at Continental Margins, *Proc. Natl. Acad. Sci. (U.S.)*, vol. 41, pp. 458–469, 1955a.

——— and ———: Gravity Interpretation from Standard Oceanic and Continental Crustal Sections, in "The Crust of the Earth," *Geol. Soc. Am. Spec. Pap.* 62, pp. 87–100, 1955b.

WRIGHT, F. E., and J. L. ENGLAND: An Improved Torsion Gravity Meter, *Am. J. Sci.*, 5th ser., vol. 35A, pp. 373–383, 1938.

WYCKOFF, R. D.: The Gulf Gravimeter, *Geophysics*, vol. 6, pp. 13–33, 1941.

ZANES, Z. F.: A Chemical Correction Factor in Gamma-Gamma Density Logging, *J. Geophys. Res.*, vol. 65, pp. 2149–2153, 1960.

ZENOR, H. M.: U.S. Patent 2,322,681, June 22, 1943.

ZURFLUEH, E. G.: Application of Two-dimensional Linear Wavelength Filtering (Residual Regional Bandpass), *Geophysics*, vol. 32, pp. 1015–1035, 1967.

PART TWO

The Magnetic Method

10

FUNDAMENTAL PRINCIPLES AND UNITS

INTRODUCTION

The physical background of the magnetic method of geophysical prospecting has much in common with that of the gravitational method. Both are "potential" methods, having their fundamentals in the mathematical theory of the potential. Just as the gravitational force in a given direction is the derivative, or rate of change, in that direction of the gravitational potential, so also the magnetic force in a given direction is the derivative in that direction of the magnetic potential.

In the gravitational case we may consider the external gravitational effect of a body as the sum of the effects of the mass particles constituting the body. In the magnetic case we may consider the external magnetic effects of a body as the sum of the effects of the magnetic particles or dipoles that give the body its magnetic state. An essential difference is that the magnetic case is inherently more complicated because the magnetic dipole is a vector with an associated sense of direction. Since the orientations of the dipoles which determine magnetization may be in any direction, definition of the magnetic state or magnetization of a body requires both a magnitude and a direction rather than a single magnitude (mass), as in the case of gravity.

While the gravity and magnetic methods are both based on potential theory and therefore have certain elements in common, their applications in oil exploration are quite different. Measured gravity effects are caused by sources which may vary in depth from the grass roots down. On the other hand, the sedimentary rocks, which are the ones in which oil may occur, nearly always are so much less magnetic than the underlying basement, usually igneous or metamorphic rocks, that the magnetic effects are almost the same as if the sediments were not present. This is especially true with airborne measurements, which are made high enough above the surface to reduce or eliminate the effects of small, near-surface inclusions of magnetic material or of cultural objects (buildings, pipelines, railroads, etc.). Therefore the sources of airborne magnetic anomalies are nearly always from local relief on the basement surface or from inhomogeneities in the magnetic rocks below that surface.

PHYSICAL PROPERTIES

The magnetic properties of materials vary over very wide limits. This variation may be considered as a result of the variation in the volume density of the elementary magnets, the ease with which they can be oriented, and the persistence with which they maintain a given orientation once it has been acquired.

The properties of magnetized bodies and their fields are closely related to the properties of the magnetic field of an electric current and may be quantitatively defined in such terms. The relationship of an electric current to a magnetic field is expressed by physicists as Ampere's law (also called the Biot-Savart law). Both electric currents and magnetic fields are vector quantities and have direction as well as magnitude associated with them. Figure 10-1 illustrates the interrelationship between them.

The magnetic field about a straight electric current (Fig. 10-1a) has the shape of circles perpendicular to the current direction and concentric about the current. The strength of the magnetic field is proportional to the electric current and decreases as the distance from the current.

A series of circular electric currents such as would be produced by an electric current flowing through a helical wire (solenoid) (Fig. 10-1b) produces a uniform magnetic field within the helix. The strength of the magnetic field is proportional to the electric current (which is measured in amperes) and also to the number of turns per meter of length. This provides a basis for units for measuring magnetic field strength as ampere-turns per meter. The strength of the magnetic field may depend also on the nature of the medium; thus the center of a helical coil of wire (Fig. 10-1c) might contain a material (such as soft iron) which produces a stronger magnetic field. This is accommodated in Ampere's law by including a proportionality constant μ, which is called the "relative permeability."

It is convenient to think of the magnetic state of a body as resulting from elementary magnets or dipoles, each of which consists of a positive and a negative

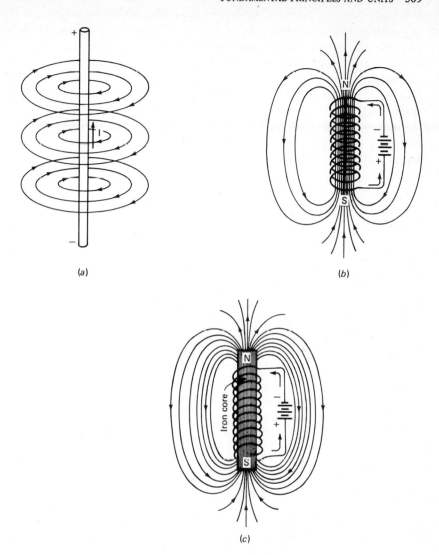

(a)

(b)

(c)

FIGURE 10-1

Relations between electric-current flow and magnetic fields. The convention for current direction I is in the direction of voltage drop; electrons actually flow in the direction opposite to I. (a) The strength of the magnetic field about a long straight wire is $B = \mu\mu_0\frac{1}{2}\pi r$, where r is the perpendicular distance to the wire, μ is the relative permeability ($\mu = 1$ for empty space), and μ_0 is a constant (which involves the units of measurement). (b) The magnetic field strength inside a helical current is $B = \mu\mu_0 NI/l$, where N is the number of turns per unit of length. (c) The helix contains a material with large permeability so that the magnetic field is stronger than in Fig. 10-1b. (In some references B is called the magnetic induction and $B/\mu\mu_0 = H$ is called the magnetic field.)

magnetic pole separated by a small distance. The dipoles react to the influence of an external magnetic field by aligning themselves with that field.

In the demagnetized state the elementary dipoles have a random distribution and orientation. The fields of the different dipoles neutralize each other so that there is no resultant field and no magnetic influence outside the body. However, if the body is placed in a magnetic field, the elementary dipoles tend to align themselves parallel with the field. The stronger the field, the more completely they are aligned. After the body has become magnetized, it has a field of its own in the space around it. If the external magnetizing field is removed, the alignment will largely disappear for most "soft," easily magnetized materials and they will no longer have their own magnetic field. However, in magnetically "hard" materials, the alignment of the elementary magnets will persist, and the magnetization will be "permanent."

UNITS IN MAGNETIC PROSPECTING

A system of magnetic units can be built up from various points of view. Changes in systems and terminology over the years have produced some confusion in definitions and in numerical values. Earlier definitions were in terms of the cgs-emu system of units whereas the more recent set of SI units is in terms of the meter, kilogram, second, and ampere. While this book is not the place for detailed definitions and history of the several units, a brief review will help clarify the relations.

The recent changes and present confusion in units are in part due to questions of fundamental definitions and in part to the adoption of the International System (SI) in physics and in engineering. Conceptually the SI regards magnetism as an effect of electric currents, whereas the cgs system begins with the forces between magnetic poles. The Committee on Units and Nomenclature of the European Association of Exploration Geophysicists (EAEG) proposed adopting the International System (Kolfoed, 1967), but as of 1974 it has not been done officially.

Certain changes in details have been proposed by Green (1968) and by Parasnis (1968), and a consistent set of recommendations is reviewed by Reilly (1972). An extensive review of the history and present status of the SI is given by Markowitz (1973).

The various magnetic units in the SI (mksa) and cgs systems are summarized in Table 10-1.

Magnetic Poles and Dipoles

The force between magnetic poles can be measured experimentally. Coulomb showed that this force varies as the strength of the poles and inversely as the square of the distance between them. The force is one of attraction if they are of opposite sign and of repulsion if of like sign. By convention, the north-seeking pole (corresponding to that at the north end of a compass needle) is called positive and the south, or south-seeking, pole is called negative.

An isolated magnetic pole cannot exist, but a positive pole can be separated by a small distance from its companion negative pole to constitute a "dipole." The strength of a dipole is given by the product of the strength of the magnetic poles involved and the distance by which they are separated. A circular electric current is equivalent to a magnetic dipole.

Magnetic Field Strength B

The force between magnetic poles may be considered as the reaction of one pole to the magnetic field of the other. A magnetic field can be produced either by a magnet or by an electric current. The magnetic field B is a vector quantity with a magnitude and direction which can be defined as the force which would be felt by a unit positive magnetic pole. Thus a unit field strength exerts a unit force on a unit pole. (In some references B is called the "magnetic induction" and $\beta/\mu\mu_0$ is called the "magnetic field.") The magnetic field strength is the quantity measured by the instruments of magnetic prospecting. In the SI the unit is the tesla, which is the magnetic field such that a force of 1 newton is exerted on a pole with a strength of 1 ampere-meter. The tesla is equal to 10^4 gauss or 10^9 gammas (γ) (see Table 10-1). Most magnetic work uses the gamma; $1 \gamma = 10^{-5}$ gauss $= 10^{-9}$ tesla $= 1$ nanotesla.

Significant magnetic anomalies, after removal of the earth's normal field, range from 1γ or less up to several hundred, or, in exceptional cases, a few thousand γ's. Thus, to detect the smallest anomalies of interest in geophysical prospecting, differences in magnetic intensity must be measured to an accuracy of 1γ or less. Contours, or lines of equal magnetic intensity, are commonly called "isogams."

The magnetic field can be represented by "lines of force," often designated by the symbol ϕ. The local direction of lines of force is the direction of the field at that point, and the density of the lines of force (number of lines per unit area in a plane perpendicular to their direction) is a measure of field strength.

Intensity of Magnetization or Polarization (M or I)

The magnetic field strength within a body may be thought of as consisting of two parts, a magnetizing force H and the resulting intensity of magnetization of the body. At any point within a magnetized body the intensity of magnetization depends on the number and degree of orientation of the elementary magnetic dipoles within the body. The intensity of magnetization is the same as the magnetic dipole moment per unit volume. It is a vector quantity which usually has a direction parallel to the direction of the magnetizing force. A uniformly magnetized body has the same intensity of magnetization and the same direction throughout.

Susceptibility k and permeability μ

Placing a magnetizable body in the influence of a magnetizing force tends to align the dipole moments within the body in the direction of the magnetizing force. The

Table 10-1 MAGNETIC QUANTITIES AND UNITS

Term	SI (mksa system)		cgs system	
	Unit*	Symbol	Unit*	Symbol
Magnetic field strength, magnetic intensity, magnetic induction, magnetic flux density	1 tesla = 1 weber/meter^2 = 1 newton/ampere-meter	$B = \mu_0(H + M)$ $= \dfrac{\Phi}{\text{area}}$	= 10^4 gauss = 10^9 γ	$B = H + 4\pi I$
Dipole moment	1 ampere-meter^2	md	= 10^{10} pole centimeters	md
Magnetic pole	1 ampere-meter	m	= 10^8 unit poles	m
Magnetic flux, magnetic line of force	1 weber	Φ	= 10^8 maxwells	Φ
Magnetomotive force, magnetic potential	1 ampere-turn	mmf	= 0.4π gilbert = 1.26 gilberts	mmf
Magnetizing force	1 ampere-turn/meter	$H = \dfrac{B}{\mu\mu_0}$	= $4\pi 10^{-3}$ oersted = 0.0126 oersted = $4\pi 10^{-3}$ gilbert/ centimeter	$H = \dfrac{B}{\mu'}$

Quantity	SI units	SI	cgs units	cgs
Magnetization, magnetic dipole moment per unit volume, magnetic polarization, magnetization intensity	1 ampere-meter^2/meter^3 = ampere/meter	$M = kH$ $= \dfrac{md}{V}$		$I = k'H$
Magnetic permeability†			1 gauss/oersted	$\mu' = \dfrac{|B|}{|H|}$
Relative permeability	Dimensionless	$\mu = \dfrac{|B|}{\mu_0|H|}$	Dimensionless	$\mu = |\mu'|$
Magnetic susceptibility	Dimensionless k_{SI}	$k = \dfrac{|M|}{|H|}$ $= \mu - 1$	Dimensionless $= 4\pi k_{cgs}$	$k' = \dfrac{|I|}{|H|}$
Reluctance	1 ampere-turn/weber	$R = \dfrac{mmf}{\Phi}$	$= 4\pi 10^{-7}$ gilbert/maxwell	$R = \dfrac{mmf}{\Phi}$
Inductance	1 henry = 1 weber/ampere	L		

* The magnitudes of the SI and cgs units given here are equal.

† μ_0 = permeability of free space = $4\pi 10^{-7}$ weber/ampere-meter
$= 12.57 \times 10^{-7}$ weber/ampere-meter
$= 12.57 \times 10^{-7}$ henry/meter
$= 1$ gauss/oersted

Modified from Sheriff (1973).

body thus takes on a degree of magnetization which is proportional to the magnetizing force and also depends on the ease of magnetization of the body. The susceptibility is a measure of the ease of magnetization and may be considered as a measure of the number of elementary magnets per unit volume of the material and of their mobility, or the ease with which they can be oriented.

The susceptibility is a dimensionless ratio, but its magnitude depends on the units of measurement, the susceptibility in SI units being larger than the susceptibility in cgs units by the factor 4π. Most sedimentary rocks are relatively nonmagnetic, with susceptibilities in the range 0.00001 to 0.00005 for averages of groups of measurements (or roughly 0.0001 to 0.0005 in SI units). Igneous rocks are much more magnetic, ranging in susceptibility from about 0.003 to 0.03 in cgs units (or 0.03 to 0.3 in SI units). Many individual values fall far outside both the low and the high side of these averages for both rock types (see Table 14-2). Susceptibilities in cgs units often are listed in terms of 10^{-6} cgs unit.

The magnetizing force H and resulting magnetic induction B are usually parallel and proportional. The proportionality factor is called the "relative magnetic permeability"

$$\mu = \frac{|B|}{\mu_0 |H|}$$

From the definition of B it is evident that

$$\mu = 1 + k \qquad \text{for } k \text{ in SI units}$$

$$\mu' = 1 + 4\pi k' \qquad \text{for } k' \text{ in cgs units}$$

While we have defined μ as dimensionless, $\mu\mu_0$ is sometimes called the permeability and it then has dimensions. The quantity μ_0 is the permeability of free space. The relative permeability is the same in either system and is also numerically equal to the permeability in cgs units (without the gauss per oersted dimensions).

DISTORTION OF MAGNETIC FIELD BY A MAGNETIZABLE BODY

The distortion or modification of the magnetic field by a magnetizable body may be pictured (Fig. 10-2) as a concentration of lines of induction within the body (increase of magnetic induction at A) and a rarefaction (decrease of magnetic induction) at B in the space immediately outside the body. This concentration may be thought of as the resultant of the magnetizing force H plus the field due to the magnetized body M or I. Similarly, the spreading out of the magnetic lines in the region just outside the magnetic body may be considered as the resultant of the magnetizing force H minus the effect of the field due to the polarized body; the lines of force due to the polarized body have a direction opposite to those of the original field in this region. When the effect is considered in this way, we do not have to distinguish between a magnetization induced by an external magnetizing force and that which may be due to permanent magnetization of the material. If a

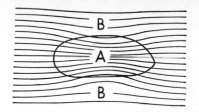

FIGURE 10-2
Distortion of magnetic field by a
magnetizable body.

material is magnetized for any reason, it has a field around it that causes a distortion or an anomaly in the normal field in its vicinity.

From these considerations it is evident that the intensity of magnetization *M* or *I* is the most important property of material with which we are concerned in magnetic prospecting. It corresponds in importance to the density in gravitational prospecting. Just as the gravitational effect of a homogeneous body at a distance depends on size, shape, and density of the body, so the magnetic effect of a homogeneously magnetized body depends on its size, shape, and the intensity and direction of its magnetization.

ELEMENTS OF THE MAGNETIC FIELD

The direction and magnitude of the geomagnetic field at any point on the earth's surface are represented by a vector or arrow parallel to the direction of the field, pointing in the direction of force on a positive pole, and having a length proportional to the strength of the field at that point. We shall refer this vector to a set of mutually perpendicular axes directed astronomically north and east and vertically downward. Let us consider the vector as passing diagonally from the origin to the far corner of a rectangular box (Fig. 10-3). The various magnetic elements then correspond to certain sides and angles of this box (or parallelepiped) as follows:

T = total intensity or total length of field vector

X, Y, V = north, east, and vertical components, respectively, of magnetic field

h = total horizontal component (different from magnetizing force H)

d = angle of declination or angle between direction of horizontal component (direction taken by a compass needle; magnetic north) and true (astronomic) north

i = angle of inclination or dip of total intensity vector below horizontal plane

The following relations between the quantities are obvious.

$$T = \frac{h}{\cos i} \qquad T^2 = h^2 + V^2$$

$$X = h \cos d \qquad Y = h \sin d \qquad \frac{T}{h} = \tan i$$

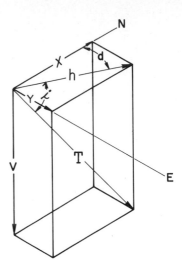

FIGURE 10-3
Notation for components of the magnetic field vector.

For many years, magnetic surveying involved mainly ground measurements of the vertical component V and (in rather limited applications) the horizontal component h. The components of greatest interest in airborne and ship-towed magnetic prospecting are the total intensity T, which ranges in value over the earth's surface (Fig. 11-1) from about 0.25 to 0.60 gauss, and the inclination i, which ranges (Fig. 11-2) from vertically downward at the north magnetic pole ($+90°$) through zero at the magnetic equator to vertically upward at the south magnetic pole ($-90°$). The operating principles of the various magnetic instruments will be discussed in Chap. 12.

FUNDAMENTALS OF MAGNETIC PROPERTIES

Modern physics has developed theories to explain the magnetic properties of physical materials in terms of their atomic structure. The following section briefly describes the atomic basis for magnetism, particularly the very special conditions which make ferromagnetism the property of the magnetic elements in rocks—on which the entire system of magnetic prospecting depends.

Atoms can be thought of as nuclei about which electrons revolve in circular orbits. An electron in a circular orbit constitutes a circular electric current (as in Fig. 10-4); consequently it has an associated magnetic field just as if it were a small magnet or magnetic dipole. Usually the small atomic magnets in a material are oriented randomly, so that their magnetic effects nearly cancel.

The presence of an external magnetic field modifies the electron orbits slightly, so that their net effect is to oppose the external field. Consequently the magnetic field is weakened very slightly. This phenomenon, called "diamagnetism," characterizes all substances, although some substances also display other phenomena

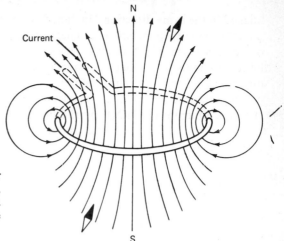

FIGURE 10-4
The magnetic field strength at the center of a single current loop is $B = \mu\mu_0 \frac{1}{2}r$, where r is the radius of the loop. Such a single current loop is called a "simple dipole."

which override and obscure the very weak diamagnetic effects. In many common minerals such as quartz, feldspar, and salt, the diamagnetism is dominant; this can be expressed by saying that their relative permeability μ is slightly less than unity or that they have negative susceptibility.

A small magnetic effect (magnetic moment) is also associated with the spin of electrons (as opposed to their orbital motion). Orbital electrons tend to be arranged in pairs with their spins so oriented that the magnetic moments cancel. Hence atoms or molecules which contain an even number of orbital electrons do not have any net magnetic moment because of electron spin. In those atoms or molecules which have a net magnetic moment, the moment tends to line up with an external magnetic field and thereby increase the field strength. This phenomenon is called "paramagnetism." The effect is at its greatest for transitional elements, for which several electrons may be in unpaired states. Paramagnetic substances have relative permeability slightly greater than 1.

Usually the net magnetic moments of atoms tend to be randomly oriented because of thermal agitation. However, in some crystals the bonding produces a systematic alignment of the magnetic moments resulting in macroscopic magnetization. This phenomenon is called "ferromagnetism." As the temperature increases, the thermal agitation increases. For ferromagnetic substances above a certain Curie temperature (page 322n) the macroscopic magnetization disappears.

The atomic magnetic moments in ferromagnetic substances within a certain region, called a "domain," tend to align themselves in the same direction, even without the influence of an external magnetic field. The tendency to produce such systematic alignments is opposed by the natural tendency of unlike magnetic poles to attract and of like poles to repel. Energy is associated with the formation of domains, and domains have definite domain walls.

Domains of different substances have different orientation patterns. In ferromagnetic materials the magnetic moments of adjacent domains tend to be

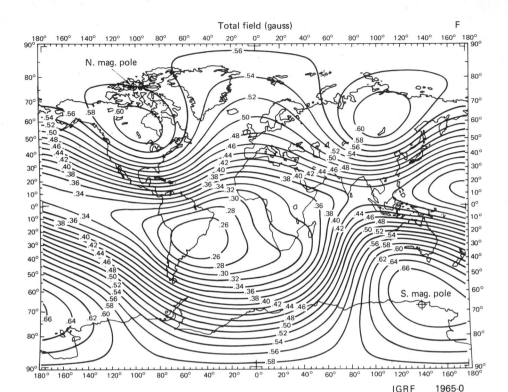

FIGURE 11-1
Contours of the total intensity of the earth's magnetic field in oersteds. (*From Leaton, 1971.*)

2 A relatively small "diurnal" part, which changes somewhat erratically with time but is repeated approximately in daily cycles and may have relatively strong short-time disturbances due to magnetic storms (see page 326).
3 Minor variations in the field from place to place which are caused by magnetic inhomogeneities of the earth's crust. This is the part which is the chief interest of magnetic prospecting, and we shall call it the anomaly part.

The secular and diurnal parts must be removed to isolate the anomaly part, for a geomagnetic observation necessarily is affected by the sum of all three parts.

The Inner Field of the Earth and Secular Variation

To a first approximation the general form of the magnetic field at the surface of the earth is that of a polarized sphere, with one magnetic pole near the north geographic pole and one near the south pole. We have defined (page 310) a positive magnetic pole as that corresponding to the north-seeking pole of a compass. Since the com-

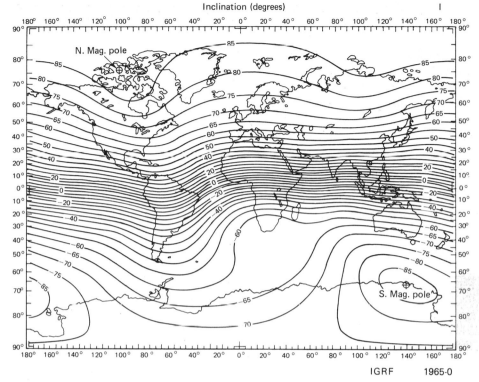

FIGURE 11-2
Contours of the inclination of the earth's magnetic field in degrees from the horizontal. (*From Leaton, 1971.*)

pass tends to point toward an opposite pole, it is evident that the magnetic pole near the north geographic pole is negative; that near the south pole is positive.

The source of the earth's inner magnetic field is one of the great unsolved problems of geophysics. Measurement of the field and its changes have been made for over 400 years in London and for nearly as long in Paris. In that time the declination has changed from a little over 10° east to a little over 20° west. Changes of similar magnitude have been recorded in Sicily since 1600. Also, there have been substantial changes in the total intensity of the field.

It was suggested as early as 1600 that the gross form of the magnetic field of the earth is that of a polarized sphere. This was shown by William Gilbert, who cut a natural lodestone (magnetite) into the form of a sphere and measured its magnetic field. However, from later and more detailed studies of the actual nature of the magnetic field and consideration of the physical and magnetic properties of materials, particularly the loss of magnetism at higher temperatures, it is quite clear that the source of the earth's magnetism is much more complex.

The long-wavelength components are attributed to electric currents in the highly conducting hot metallic core. The change in their general pattern with time is accompanied by a slow shift of the locations of the actual magnetic poles, i.e., the two points of vertical dip. These slow changes are the secular variation of the field mentioned above, and their accumulation over long periods of time is expressed by the gross changes in direction and magnitude over hundreds of years.

In ways which are not well understood, there must be a relation between the magnetization of the earth and its rotation. The technical field of "magneto-hydrodynamics" has attempted to work out a "dynamo" theory which compares the buildup of the magnetic field, its relation to the rotation of the earth, and the conductivity of the high-temperature materials of the core with the process by which an ordinary electric dynamo becomes self-polarized once it has an initial magnetic impulse. These studies also have proposed mechanisms by which the polarization of the earth can reverse itself.*

The gross elements of the field can be appreciated by picturing the earth as a uniformly polarized sphere, as in Fig. 11-4. The magnetic lines of force are some-

* While not a direct part of the application to oil prospecting, the study of reversals of the earth's magnetic field and the resulting reversed polarization of rocks is a very important factor in the evidence for continental drift and plate tectonics. While the matter is still somewhat controversial, there is a large body of evidence indicating moving continents. A major component of this evidence has been obtained by airborne magnetic surveys over the oceans. These surveys have shown extensive parallel bands of relatively high and low field strength, and these bands have similar patterns on the right and left sides of mid-continent ridges. Similar sequences are found in widely different parts of the ocean.

The widely but not universally accepted explanation of the magnetic bands is that they are related to reversals of the polarity of the earth's magnetic field while its general direction is maintained roughly parallel with the axis of rotation. It is assumed that as material from a molten layer rises and spreads laterally from the mid-ocean ridges to form new ocean floors, it becomes magnetized on cooling in the earth's field. As this spreading continues through a reversal in the direction of the earth's field, a new piece of the floor becomes magnetized in the opposite direction. Thus the alternating bands of relatively high and low field intensity represent times of normal (as at present) and reversed directions of the field, and their worldwide correlation (if accepted) makes a universal geologic calendar. There are now recognized some 160 reversals from the present back about 76 million years into Cretaceous time. The periods between reversals are widely variable, in the range of 50,000 to 3,000,000 years.

The timing of a given geologic section in the magnetic calendar is not simple. Measuring the direction of polarization of magnetized rocks determines only whether they are normal or reversed. It may be necessary to find a considerable series of reversals to establish a pattern which can be fitted reliably into the complex column of normal and reversed fields.

All this is in a stage of rapid change and development at this time. (For summaries of recent ideas on the internal magnetic field the reader is referred to Stacey, 1969, chap. 5; Jacobs, 1963, chap. 5; and Strangway, 1970, chaps. 4 and 5.) A series of readings ("Continents Adrift," 1971) is a collection of 15 well-illustrated papers from *Scientific American*, which give many details of data and their interpretation in terms of the new global tectonics.

All the above is not very directly related to the applications of the magnetic method of prospecting for oil. However, the techniques of the various airborne magnetometers as developed by the petroleum industry have made this body of evidence possible. In a broad way, the resulting concepts have contributed to explanations of the time and mechanics of deformation at continental margins and other areas where their understanding is an important factor in petroleum exploration.

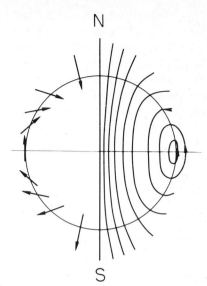

FIGURE 11-4
The earth as a magnet: (*right half*) magnetic lines of force through the earth and in the external space; (*left half*) direction and relative magnitude of total magnetic vector at the surface.

what as shown by the right half of the figure. Near the poles the lines are close together, giving a relatively strong field, pointing inward near the north geographic pole and outward near the south geographic pole. Near the equator (the zero contour of Fig. 11-2) the field has about half its intensity at the poles, is parallel to the surface, and points north. The arrows on the left half of the figure indicate the direction of the earth's field relative to the surface and, by their relative length, its approximate magnitude. As we go north or south from the magnetic equator, the angle with the surface, or the magnetic dip, increases rapidly (see Fig. 11-2) until it is vertical at the magnetic poles. The increase of dip and increase of strength both contribute to a pronounced increase in the vertical component of the earth's field as we go north or south from the magnetic equator, where, of course, its value is zero. In magnetic prospecting for oil a correction is always made for this normal northward or southward increase in the vertical component or in the total intensity.

The variation in the dip from horizontal at the magnetic equator to vertical at the magnetic poles makes gross changes in the nature of magnetic anomalies. Most anomalies are produced by induction by the earth's field and their form, for the same geometric body, will be greatly different depending on their magnetic latitude. These effects are treated in some detail in Chap. 15.

The External Field and Diurnal Variations

If accurate determinations of the earth's magnetic field are made continuously at a fixed point, it is found that the intensity changes by an appreciable amount over short time intervals. These changes have a more or less regular daily cycle, which is approximately the same at the same solar time at different points but differs materially in detail from place to place and from day to day.

The external field is probably due to electric currents high in the earth's atmosphere. It is well known that the upper atmosphere is ionized (the Kennelly-Heaviside layer, which is important in the transmission of radio signals). This ionization is caused by ultraviolet radiation from the sun and also perhaps by electrons shot out from the sun. Any variation in the number of ions or the velocity of their motion also causes a change in the magnetic field at the surface of the earth. The tidal circulation of the upper air, induced by the gravitational attraction of the sun and the moon, furnishes a fairly satisfactory explanation of the daily magnetic variations (Bartels, 1926, pp. 415–419). The amplitude of the daily change is greater in summer than in winter. The range of the daily variation in vertical or total intensity may be as great as 100γ but more commonly is 10 to 30 γ.*

At times the magnetic field may be disturbed by "magnetic storms." This term designates conditions when relatively violent and rapid changes in magnetic intensity occur. They are related to sunspots and the influence of the "solar wind" on ionization of the earth's upper atmosphere. The more severe storms are accompanied by other manifestations, such as unusually bright and extensive auroral displays and interruptions to radio communication systems. They are more common and more intense in high magnetic latitudes. Storms may be severe enough to interfere seriously with accurate magnetic surveying or even make it temporarily impossible. Such conditions usually last for only a few hours or occasionally for 1 or 2 days. There are a very few local mountainous areas where, for reasons yet unknown, the variations are so strong and continuous that precise magnetic surveying is virtually impossible.

The nature of the diurnal changes and the procedures used for correcting magnetic observations for their effects are discussed on pages 355 to 358.

The Anomaly Field of the Earth

This consists of that part of the field which is caused by irregularities in the distribution of magnetized material in the outer crust of the earth. If this crust were uniform to a depth of some tens of miles, there would be no anomaly field. The fact that variations in the magnetic field exist which, from their nature and extent, must have their source within the outer crust of the earth tells us that this crust is not magnetically homogeneous. The whole purpose of magnetic prospecting for petroleum is to measure the anomaly field and to attempt to interpret the magnetic inhomogeneities indicated in terms of geologic detail relevant to the occurrence and accumulation of petroleum. This interpretation is the subject of Chap. 15.

---

* For sample daily-variation curves, see Fig. 13-5a and b and also Vacquier (1937) and Soske (1933).

12

GEOMAGNETIC MEASURING INSTRUMENTS

HISTORICAL BACKGROUND

The instruments used for magnetic field measurements have changed greatly since the beginning of magnetic exploration. The earlier instruments were mechanical and depended on measuring the deflection of a bar magnet supported on a carefully made quartz knife-edge, which rested on cylindrical bearings. Such instruments, usually made with horizontal magnets, measured the vertical component of the field. Occasionally they were made with vertical magnets to measure the horizontal component. These mechanical magnetometers were used for oil prospecting from their earliest application until they were largely replaced by electronic devices beginning in the late 1940s.

During its period of application the vertical field magnetometer was used very extensively by the oil industry. Unlike the gravity and seismic methods, the magnetic method is not suited to the search for salt domes, and its early application was not on the Gulf Coast, where the seismic and gravity methods were used.

Very extensive surveys were made in the middle and late 1920s in West Texas and were applied on an intuitive basis in the selection of large lease blocks.*

The flux gate and other electronic magnetic measuring systems, which are without moving parts in their magnetic elements and therefore insensitive to motional acceleration, made possible a whole new field of magnetic surveying from airplanes and ships.† This means that the ground magnetometer is obsolete for petroleum prospecting except in special circumstances. Furthermore, even in such circumstances and in applications to mining geophysics the usual ground instruments are those based on the principles of electronic magnetometers.

The remainder of this chapter is devoted to descriptions of the magnetic field measuring instruments. Brief accounts of the mechanical field balances are included for their historical interest and because maps made with them may occasionally be encountered. Detailed accounts of their theory, mechanical construction, and operating practice are given in the older books on geophysical exploration such as Nettleton (1940), Heiland (1940), Jakosky (1940 or 1950), and Dobrin (1952).

THE VERTICAL MAGNETOMETER

The moving system of a vertical magnetometer (Fig. 12-1) consists essentially of a pair of bar magnets, carrying a mirror, balanced on a horizontal knife-edge. The center of gravity of the moving system is displaced horizontally and vertically with respect to the knife-edge so that its magnetic axis is substantially horizontal.

The magnetometer is always oriented with its magnetic axis in an east-west direction so that the horizontal component of the magnetic field produces no torque on the moving system. Therefore, we need consider only the effects of the vertical component.

In terms of the quantities shown in Fig. 12-1, the moving system is acted on by two torques: (1) the magnetic torque (tending to rotate it counterclockwise in the figure), resulting from the reaction of the vertical component of the earth's field V on the poles of the magnet with pole strength P, and (2) a gravitational torque (tending to rotate it clockwise in the figure), resulting from the weight m acting at the center of gravity c with displacement components a and d from the knife-edge. At the equilibrium position these two torques are equal, and

$$2PVL \cos \theta = mg(d \sin \theta + a \cos \theta)$$

---

* One major oil company established substantial reserves in the Permian Basin by simply leasing large ranches along linear features indicated by magnetic maps. Later it developed that these features were expressions of the flanks of the large Central Basin Platform, which localizes accumulation in several large West Texas oil fields.

† When airborne magnetometers began to be used, a question often asked was: If an airborne magnetometer can be operated, why not an airborne gravity meter? The answer, of course, is that any gravity sensor is inherently sensitive to the effects of motional acceleration because physically acceleration from gravitational or motional sources is the same. The nature of the magnetic field is entirely different, and therefore instruments can be made which are sensitive to magnetic fields and completely insensitive to gravitational or acceleration fields.

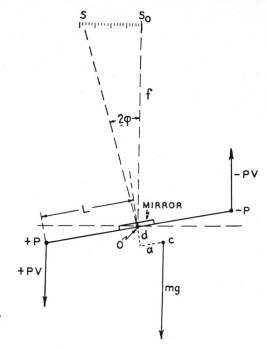

FIGURE 12-1
Diagram and notation for theory of
vertical magnetometer.

It can be shown (Nettleton, 1940) that, for small values of θ, the difference in the vertical intensity, $V - V'$, between any two points is

$$V - V' = K(s - s')$$

where K is a calibration constant depending on the physical constants and dimensions of the instrument and $s - s'$ is the difference in scale readings (Nettleton, 1940, pp. 171–173).

A large proportion of all the ground magnetic prospecting in the United States before about 1950 was done with the field balance designed by A. Schmidt and manufactured by the Askania Werke in Berlin.

THE HORIZONTAL MAGNETOMETER

The moving system of the horizontal magnetometer (Fig. 12-2) is essentially the same as that of the vertical-component instrument, except that the magnetic axis is substantially vertical. However, in the horizontal magnetometer the effect of the vertical component cannot be eliminated completely. The reference characters in the diagram (Fig. 12-2) represent the same quantities as those in Fig. 12-1.

From the equilibrium between the gravitational and magnetic torques,

$$2PHL \cos \varphi - 2PVL \sin \varphi = mg(a \cos \varphi + d \sin \varphi)$$

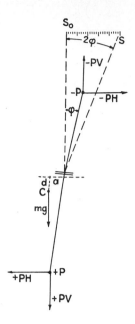

FIGURE 12-2
Diagram and notation for theory of
horizontal magnetometer.

It can be shown (see, for instance, Nettleton, 1940, p. 174) that

$$H_2 - H_1 = K_1'(s_2 - s_1) + \frac{\Delta V}{2f}(s_2 - s_0)$$

where ΔV is the difference in vertical field intensity between the two points of measurement.

The dependence of the horizontal component on the vertical component complicates the theory and practice of measurement of small differences in the horizontal component. Most applications have been in mining prospecting, where anomalies are strong, and over small areas within which the variation in the vertical component is not large and can be neglected (see Nettleton, 1940, p. 175).

Comparison of Measurements by Vertical and Horizontal Magnetometers

The vertical magnetometer measures the isolated vertical component of the earth's field, but the horizontal magnetometer measures differences in the total horizontal component H, as shown by Fig. 12-3. If the anomalous component is perpendicular to the total field, the measured difference is zero. This is analogous, in two dimensions, to the manner in which the several electronic magnetometers, discussed below, measure the variation in the total magnetic field intensity in three-dimensional space.

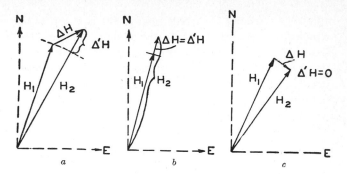

FIGURE 12-3
The variation of the effect of a given small increment of the horizontal component ΔH on the total horizontal component.

ELECTRONIC MAGNETOMETERS

Magnetic prospecting for petroleum, from its beginning early in the 1920s until about 1945, was carried out almost entirely on the ground with the mechanical magnetometers described previously and very largely with the vertical magnetometer. Magnetic prospecting for oil was completely revolutionized by the development of electronic magnetometers.

As mentioned briefly on page 328n, since electronic magnetometers are almost free from the effects of motional acceleration, they can be operated on airplanes and ships. Thus the development of electronic instruments lifted magnetic field measurements from point-by-point observations on the ground with a sensitivity of 5 to 10 γ to continuous observations in the air, at speeds of over 100 mph and with a sensitivity of the order of 1 γ or less.

Furthermore, measurements made at elevations of several hundred to 1000 ft or more above the ground are much less disturbed by cultural installations (cities, railroads, pipelines, etc.) or by near-surface natural magnetic material (such as in glacial drift).

While many new problems were encountered, to be discussed later, the speed and precision of magnetic surveying were increased so much that the magnetic method as applied to prospecting for petroleum entered a completely new phase. Many new areas were covered, the old areas of ground magnetic surveys were resurveyed from the air, and airborne magnetic surveying was extended offshore for considerable distances over the continental shelves. Also, magnetometers towed behind ships have mapped large areas of the deep oceans with results which have had a very important bearing on our understanding of the geological history of the earth as a whole (see footnote on page 324).

There were three stages in the development of electronic magnetometers. The so-called flux gate was the first electronic instrument and was used almost exclusively from 1945 until it began to be replaced by other instruments in about

1960. The second electronic instrument is the proton-precession magnetometer, which has a sensitivity about the same as that of the flux gate and depends on fundamental properties of an atom, usually hydrogen. The third instrument uses the spectra of "pumped" alkali vapors and has an inherent sensitivity two to three orders of magnitude greater than that of either the flux-gate or proton-precession instruments. The remainder of this chapter outlines the physical principles on which these various electronic magnetometers operate.

The Flux-Gate Magnetometer

The development of a magnetometer for airborne use was begun at the Gulf Research Laboratories shortly before the beginning of World War II.* After the war, development of the flux-gate instrument as an airborne magnetometer for petroleum exploration proceeded rapidly, and extensive surveys for petroleum exploration were begun.

The basic magnetically sensitive element of the flux-gate magnetometer is a small core of very highly permeable magnetic material such as Permalloy or Mumetal. The permeability is so high that the core approaches saturation in the earth's field and therefore has a nonlinear magnetization curve. This means that the reactance of a coil with a core of such material is also nonlinear and that the waveform of the magnetizing current of such a coil is distorted. With very highly permeable materials this distortion is quite appreciable within fields of the magnitude of that of the earth. With appropriate circuitry, magnetic field changes as small as 1 γ can be detected.

There are several systems of circuitry by which this distortion of the magnetizing current can be exploited to make continuous quantitative measurements of changes in the ambient field. The following brief description is based on Wycoff (1948) and is generally that of the Gulf magnetometer, widely used in routine aeromagnetic surveying.

* The development of the flux-gate magnetometer at Gulf was initiated and carried out largely by Victor Vacquier, with direction and guidance by R. D. Wycoff. Vacquier was interested in a magnetic test device which could be passed through the inside of a drill pipe to reveal incipient flaws. He found the flux-gate principle of a purely electronic magnetometer in some relatively old electrical engineering literature and attempted to use it for his original purpose. When a high sensitivity had been achieved, it was a natural step to suggest it for airborne applications. E. A. Eckhart, then director of the geophysical section at Gulf, was a member of one of the scientific advisory committees aiding in the war effort. He suggested the use of the airborne magnetometer as a submarine detector. After considerable resistance by the military, it was tested by the Navy and found sensitive enough to indicate submerged submarines. The development was then taken over by Bell Telephone Laboratories and Airborne Instruments Laboratory independently, who produced a practical "magnetic airborne device" (MAD) with a resolution of the order of 1 γ. This MAD was designed as an anomaly detector operating with a bandpass filter, with a pass frequency appropriate to the overflight of a submarine by a search plane. The military models therefore lacked both the stability and the ability to make continuous relative measurements which were required for application to prospecting.

After the war the development was resumed by Gulf and by others, and licenses under the Vacquier patents (Vacquier, 1946; Vacquier and Muffly, 1951) were offered to the exploration industry.

As outlined schematically in Fig. 12-4, the magnetically sensitive element consists of two high-permeability strips wound with separate coils and energized by a 1000-Hz current with the magnetization in opposite directions in the two coils. These two coils are enclosed together in another coil carrying direct current, by which the ambient magnetic field can be changed.

If there is no external magnetic field, i.e., if the normal earth's field is completely balanced by the field from the exterior dc winding, the distortion of the magnetizing winding of the two coils will be equal and opposite and there will be no output from the pulse transformer. If there is an ambient magnetic field, the hysteresis curves for the two strips are displaced in opposite directions, the pulses from the two windings no longer balance each other, and there is a net output pulse to the amplifier. This output is directional and changes sign if the ambient field changes sign. It has a sawtooth form, as indicated in the diagram, with a 1000-Hz frequency, and is filtered and rectified to produce a dc component. With no field at the detector these two components are equal, and their difference is zero. When there is a net field, the two components are not equal and their difference is a quantity which is proportional to the ambient field intensity at the magnetometer element. This difference activates a balancing circuit, which changes the current in the dc coil to bring the net field to zero. The change in the dc balancing current is recorded on a suitable analog strip chart or digitally or both.

In order to keep the recording pen on the chart for large changes of the magnetic field and with relatively high sensitivity, a range selector included in the detector unit is calibrated to provide full-scale deflection for various changes in the field, usually for 300 or 600 γ. With a 600-γ range on a 10-in. chart it is not difficult to read within 1 γ.

There is a fundamental difference between measurements made on the ground with mechanical magnetometers and those made in the air. The older mechanical magnetometers were leveled for each reading and therefore measured the vertical component. This is not feasible in the air because of the elaborate accessory equipment, with gyroscopes and inertial elements which would be required to maintain a level reference from which to determine the vertical during flight. Also, measurements of total intensity are just as useful as those of the vertical component. As a result, virtually all* airborne detectors and all those used for magnetic prospecting measure the total intensity of the magnetic field. This means that the magnetic-measuring elements must be continuously aligned with the direction, in space, of the total magnetic field vector. This is also the position in which any residual changes in intensity from misalignment are at a minimum. Therefore measurements in the direction of the total-intensity vector can be made with a very high degree of accuracy.

As the magnetometer is commonly used, the necessary orientation is achieved

---

* Special magnetometers with stable platforms and other special equipment can make airborne measurements of the vertical, horizontal, and total components of the field. This has made it possible to continue over the oceans the charts (see Figs. 11-1 and 11-2) which formerly were dependent on measurements on the ground or, to a limited extent, on special nonmagnetic ships.

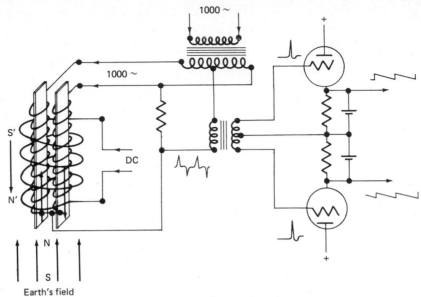

Magnetometer element schematic

FIGURE 12-4
Schematic diagram of flux-gate magnetometer detector circuit. Secondary voltage is here obtained by the pulse transformer from modification of primary circuits and is equivalent to a single secondary. Secondary winding is shown as Helmholtz-coil compensating means, producing a bucking north–south field. (*From Wycoff, 1948.*)

with two additional similar magnetically sensitive elements arranged so that the three are mutually perpendicular. The two orienting elements are in the plane of a gimbal-mounted plate, and the third, or measuring element, is perpendicular to this plate. Servomechanisms controlled by each of the two orienting elements operate small motors to keep the output signal zero for each of the two orienting elements. This means that the plane of the plate that carries them is perpendicular to the earth's field, and the third or measuring element is therefore parallel with the earth's field and measures the total component. In practice, the gimbal orienting mechanism is very fast and will keep the measuring element in the proper orientation in spite of rapid motions of the "bird" in which the instrument is mounted. The shell or bird, shaped like an aerial bomb, is some 6 to 8 in. in diameter and 3 or 4 ft long. The length is desirable aerodynamically but also is required to keep the motors of the servomechanism at some distance from the magnetically sensitive elements; the gimbal movements are accomplished through small, highly non-magnetic shafts, gears, strings, and pulleys.

The bird containing the magnetometer element may be carried on a cable suspended several hundred feet below the airplane in order to be free from magnetic disturbances caused by the plane itself. It is also possible to use a mounting in which the instrument is assembled in a short projection, or "stinger," on the tail of the airplane. This requires very careful compensation of the magnetic parts of the airplane itself so that at the instrument location there are no residual magnetic effects which change as the attitude of the plane changes with respect to the magnetic field. Wing-tip mountings have been used also; this requires even more complex magnetic compensation to counter the magnetic effects from eddy currents in the metal of the wing which result from its angular motion through the magnetic field of the earth and its flexing from the variable stresses in flight.

The Proton-Precession Magnetometer

This was the first new airborne electronic magnetometer after the development and extensive application of the flux-gate instrument. The fundamental principles disclosed by Packard and Varian (1954) were rapidly developed into a practical airborne or shipborne magnetometer. The device has been used rather extensively for measurements at sea, along with gravity and sometimes seismic measurements at the same time, because it is relatively simple to install with the sensitive device at the end of a long (approximately 500 ft) cable behind the ship and because the relatively simple electronic components are easy to purchase or rent off the shelf.

This magnetometer depends on certain fundamental properties of the atomic nucleus and the phenomenon of the Larmor precession. The protons (hydrogen nuclei) have a spin which makes each nucleus equivalent to a tiny magnet. Under normal conditions the spin axes are randomly oriented, so that their individual fields cancel each other and there is no external field. If a polarizing field is applied in a direction at an angle or perpendicular to the earth's field, the spin axes become aligned with the polarizing field; when that field is removed, the spinning nuclei behave like tiny tops and precess toward alignment with the earth's field at a frequency which is determined by the magnitude of that field. The frequency of precession is $f = \lambda T/2\pi$, and $T = 2\pi f/\lambda$, where T is the magnitude of the ambient total intensity and λ is the gyromagnetic ratio. This ratio is an invariant property of the nucleus and its value is

$$\lambda = 2.67513 \times 10^4 \text{ (oersted-second)}^{-1} \pm 0.00002 = 0.267513 \ (\gamma\text{-sec})^{-1}$$

and the total field strength in gammas is

$$T = \frac{2\pi}{0.267513} f = 23.4874f$$

The numerical coefficient is independent of the surroundings or chemical state of the atom involved. Therefore if the frequency of precession f can be measured, the total field intensity can be determined. In the earth's field of, say, 50,000 γ, the frequency is

$$f = \frac{T}{23.4874} = \frac{50,000}{23.4874} \approx 2000 \text{ Hz}$$

To measure changes in the field intensity to, say, 1 γ, the precession frequency must be measured to about 1 part in 10^5, or around 0.02 Hz. This measurement must

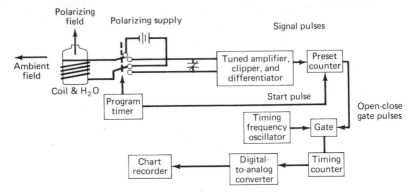

FIGURE 12-5

Block diagram of free-nuclear-induction proton magnetometer. (*From Duffus, 1966.*)

be made in a very short time because after the polarizing magnetic field is turned off, the precession dies out exponentially so that a precession field strength suitable for measurement persists only about 1 sec.

In the Varian instrument the frequency is measured by determining the time elapsed between the beginning and the end of a fixed number of whole cycles. An electronic gate (Fig. 12-5) opens a counting circuit, which then closes when a fixed number of cycles has passed. While the gate is open, a timing frequency of the order of 100 kHz is connected to another counter, so that the total number of cycles in the high-frequency timing circuit is counted during the passage of the fixed number of whole cycles from the precession frequency.

When this simple counting system is used, the number obtained is inverse to the magnetic field; i.e., the higher the precession frequency the shorter the time required for the given number of cycles to pass through the high-frequency counter circuit. In modern equipment, this inverse relation is rectified electronically so that the output is directly proportional to the field.

Because of the charge-up-and-then-count sequence, which is inherent in the above system, the output is discontinuous and appears as small steps on the record; see Fig. 12-6, which is a sample from a proton-precession record made in a routine marine geophysical survey.

In modern equipment, the charging and measuring cycle takes place in 1 sec or less. For 1-sec intervals at an airplane speed of, say, 200 mph the recording is equivalent to a spacing of about 300 ft on the ground. Since this distance is usually considerably less than the height of the airplane, a record obtained by smoothing or filtering between the individual times of measurement is quite satisfactory. At ship speeds of 7 to 10 knots or less, the measuring interval is only 15 ft or less.

There are other instrumental arrangements by which the proton-precession frequency is measured without the intermittent charge and measurement cycle

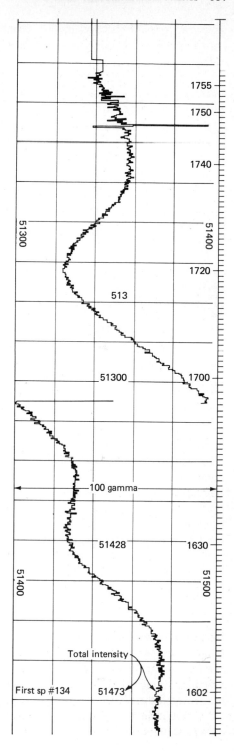

FIGURE 12-6

Example of routine record from proton-precession magnetometer from a combined magnetic-gravity survey at sea. (*Tidelands Exploration Co.*)

described previously. In one, a resonant circuit measures the Larmor frequency continuously and thereby gives a continuous measure of the earth's field (Duffus, 1966, p. 110).

Another system, developed in France, uses the Overhauser effect (Hood, 1967, p. 12). In this system the liquid containing the protons also contains a paramagnetic substance which makes it possible for the spin energy of the electrons to be coupled to the proton spin in the solvent. Then the polarized protons of the solvent precess continuously at the Larmor frequency, which in turn is generated by the ambient magnetic field and is detected by a suitable pickup. This makes continuous recording of the changes in the magnetic field possible.

In the proton-precession instruments the frequency of precession is determined by the total value of the magnetic field and is not affected by the orientation of the measuring system with respect to the field vector. However, the sensitivity is modified by orientation, the maximum sensitivity occurring when the axis of the magnetizing coil is perpendicular to the magnetic vector. At an angle of 45° the sensitivity decreases by half. This difficulty can be avoided by using two coils at right angles or by using a toroidal coil.

In general the mechanics of operating airborne magnetic surveys with the proton-precession magnetometers designed for that use is similar to the techniques developed in the application of the flux-gate instruments. Some developments in the electronic circuitry have shortened the time between individual readings to as little as 0.2 sec, with some sacrifices of sensitivity. One advantage of the proton magnetometer is that because the readings depend on a fixed constant (the gyromagnetic ratio) the values derived are in absolute units and do not need to be referred to a base. Any differences in values at flight-line intersections, therefore, must be attributed to time variations (diurnal changes), variations in elevation, or navigation errors so that the two measurements are not at the same spot.

Since the value depends on the precession frequency with respect to the earth's field, any rotation of the instrument itself would have the effect of modifying that frequency and therefore of changing the measured value. Because of its high sensitivity ($\frac{1}{23}$ Hz/γ) rotations of the instrument assembly, which might be caused by rough air (or by rough water in marine applications), will produce noise of 1 γ for 15° of rotation during the measuring cycle or proportionately for smaller rotations. Thus small-amplitude noise on the records may be caused by such motions.

Optically Pumped Vapor Magnetometers

This class of magnetometers also depends on the Larmor precession but applied to electrons rather than protons. Since the mass of the electron is approximately 2000 times smaller than that of the proton, the frequencies involved are much higher. Also, they all depend on the phenomenon of optical pumping. Practical instruments of this type have been based on the properties of cesium, rubidium, and metastable helium.

When atoms of a vapor are illuminated by electromagnetic radiation, they can absorb or emit energy. This process is highly restricted because absorption can take place only when the energy of photons of the incoming radiation matches a natural energy level of an atom and this energy is determined by its electron state. A photon is a definite unit of energy determined by the radiation frequency. Photons of light in the visible range have enough energy to cause a transition of electrons from one orbit to another at a higher level. This transition is reversible, and elections can fall back to the lower energy level; in doing so they emit radiation at the same characteristic frequency which corresponds to the difference in energy levels.

The electron spin may be either parallel with the ambient magnetic field in the upper level or antiparallel in the lower level. The transition can be confined to one level by passing the light through a circular polarizer. Then it is possible to fill a certain energy level selectively, depending on the frequency of the radiation, which is restricted, by a filter, to a single line of the spectrum of the rubidium (or other) vapor. As the selected energy level is filled (this is the process of optical pumping), the vapor can no longer absorb that particular frequency and the vapor cell becomes relatively transparent. If the cell is now swept with an alternating magnetic field of varying frequency, there is a certain frequency at which the cell becomes transparent; this is the Larmor precession frequency of the electrons. It is proportional to the magnetic field and is determined by the Zeeman splitting of the energy levels. Thus, if the frequency of the magnetic field around the vapor cell is continuously "tuned" to the frequency of the energy-level change, that frequency is proportional to the magnetic field. Since frequencies can be measured with high precision, a highly sensitive measurement of the magnetic field is possible.

One system for application of the principle of optical pumping to measuring the magnetic field requires the following components, shown schematically in Fig. 12-7:

1 A rubidium-vapor lamp, which is energized by an excitation oscillator and which is kept at a moderately elevated temperature (35°C) to produce the optimum vapor pressure in the lamp
2 A lens and collimator to direct the light through the other optical components
3 An interference filter which sharply limits the frequency of the transmitted light to the 7948-Å line (for rubidium)
4 A circular polarizer to limit the transmitted light to rotation in one direction, inducing electron spin to rotate about an axis parallel (or antiparallel) with the field to be measured
5 An absorption cell filled with rubidium vapor and surrounded by a radiofrequency coil
6 A final lens which focuses the light on a photocell
7 The photocell

In practice it is common to arrange components 2 through 7 on each side of a single vapor lamp to make the arrangement shown.

Because of filter 3 the light entering cell 5 is limited to a single line of the energy spectrum of the rubidium vapor, and nearby lines are eliminated.

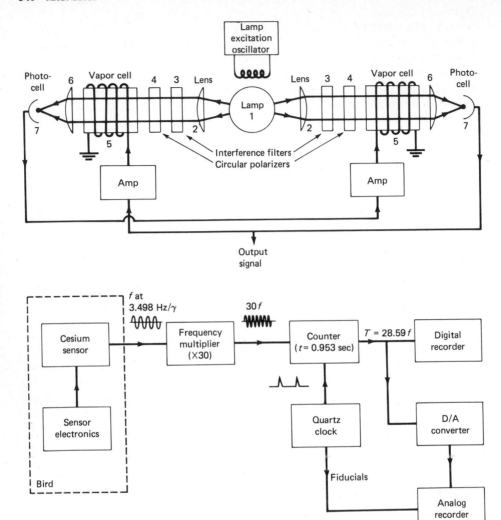

FIGURE 12-7
Block diagram of dual-cell self-oscillating pumped vapor magnetometer (*upper*) and of complete operating system for a cesium-vapor airborne magnetometer (*lower*). (*From Hood, 1967.*)

This has the result of concentrating the atoms into a single energy state. Then no further absorption can take place, the absorption cell becomes transparent, and the photocell current increases. If the cell is now swept with a weak magnetic field of variable frequency, a sharp absorption will occur when this frequency matches the Larmor precession frequency of the electrons.

In the instrument arrangement shown, the current from the photocell is amplified, shifted in phase by 90°, and used as a feedback to control the frequency of the alternating field in the coil around the vapor cell. This coupling of the light transmission and absorption with the vapor cell produces a continuous precession at the Larmor frequency, and this frequency is proportional to the total magnetic field intensity. The frequency is fixed by the atomic constant, which for rubidium 85 is 4.667 Hz/γ (3.498 Hz/γ for cesium); that is, the rubidium field strength is $f/4.667 = 0.2142\ \gamma/\text{Hz}$ (0.2859 γ/Hz for cesium).

The measurement of the magnetic field strength and its variations is reduced to the measurement of a continuous frequency of the order of 200 kHz, which is relatively straightforward. In laboratory installations variation of the field can be measured to 0.001 γ or less. In practical airborne equipment, measurements to 0.1 γ or somewhat less are feasible.

The accurate measurement of the frequency can be made by methods similar to those used for the proton-precession instrument but without the severe short, time limits. In one system (Jensen, 1965) the approximately 200-kHz signal is multiplied by 8 and is then digitally counted during a precise time interval of about 0.6 sec so that one count corresponds to about 0.05 γ. On the analog chart, recordings with full scale (10 in.) can be made for 10 and 100 γ simultaneously.

Vertical- and Horizontal-Gradient Measurement

The development of the very high-sensitivity optically pumped magnetometers has provided a basis for the measurement of the vertical gradient of the total magnetic field. Two separate magnetometers are flown with a vertical separation of the order of 200 ft. In some installations, the upper instrument is in a stinger mount at the tail of the airplane and the lower one is carried on a suspended cable. In other installations both instruments are on a cable. In either case the measurement is made by recording variations in the difference between the two instruments. With a separation of 100 to 200 ft, this variation can be measured to a precision of around 0.01 γ/ft. A continuous record is made of the gradient variation, and on the same recorder a trace of the total intensity is shown also. The vertical gradient then can be mapped from these variations.

It is quite possible to calculate the vertical gradient from an accurate map of the magnetic field, and there is some disagreement whether or not measuring the gradient adds value corresponding to the rather considerable increase in the cost of the operation. However, as mentioned below (page 358), and entirely aside from its usefulness in mapping the magnetic field, the vertical gradient is very useful in recognizing diurnal effects.

The vertical-gradient map is much more responsive to local influences than to broad or regional effects and therefore tends to give a considerably sharper picture than the map of the total field intensity. Thus the smaller anomalies are more readily apparent in an area of strong regional disturbances. The same general effect is obtained by a vertical-derivative operation, and therefore vertical-derivative maps calculated from the total-intensity field tend to be very similar to maps

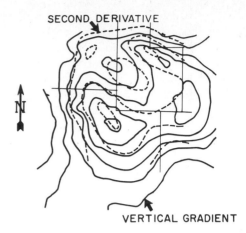

SECOND DERIVATIVE

N

VERTICAL GRADIENT

FIGURE 12-8
Comparison of measured vertical-gradient and calculated second-derivative
maps of the same area.

made from field measurements of the vertical gradient. In particular, the high and low closures, nosings, or other details will generally be in nearly the same areas on the two kinds of maps.

An approximate comparison of vertical-gradient and second-derivative patterns for the same area is shown by Fig. 12-8. The solid contours are from the second-derivative map of the Puckett area, taken from Fig. 15-18. The dashed lines are from a published vertical-gradient map of the same field. It will be noted that the general contour pattern and the sharpening of local features are quite similar in the two sets of contours.

The horizontal gradient can be measured in marine operations by adding a second magnetometer to the towed cable, with a suitable separation between the two instruments. A second recording is made which is the instantaneous difference between the two, which, of course, is a measure of the horizontal gradient. This gradient can be very useful in recognizing diurnal effects (see page 359).

13

MAGNETIC FIELD OPERATIONS
AND DATA REDUCTION

This chapter takes up first the general field-operation procedures, primarily as related to airborne surveys for oil prospecting. It includes brief reviews of the various location systems, since they are an integral part of the data-processing and mapping procedures. The measurements necessarily include unwanted magnetic effects, i.e., those of the normal earth's field and diurnal disturbances, and methods for their determination and removal are covered in the second part of the chapter.

GENERAL PLAN OF FIELD OPERATIONS

A schematic diagram of a sample from an actual aeromagnetic survey, showing the ground surface, the sedimentary section, the basement, the flight lines of the survey, and the resulting aeromagnetic map, are given in Fig. 13-1. The upper part of the figure shows a magnetic profile and geologic cross section along the front edge of the block diagram below. This figure was prepared from a portion of a flux-gate

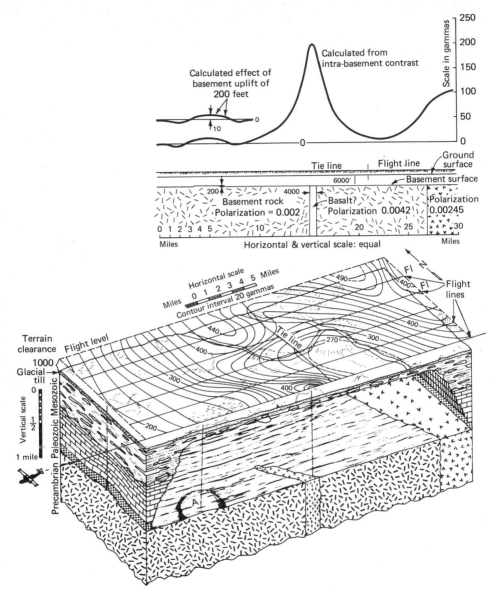

FIGURE 13-1
General plan of operation of airborne magnetic survey, showing flight-line pattern, flight-line profile on line indicated by airplane, magnetic contour map, and geological section as inferred from interpretation of magnetic data. (*From Nettleton, 1950; by permission of Oil in Canada.*)

magnetometer survey in the plains of western Alberta, Canada, and its interpretation. In the following paragraphs we shall review the steps necessary to produce the magnetic contour map.

We shall not consider here the details of the operations in the airplane or of the magnetic instruments except to say that it is presumed that the equipment is properly maintained and that analog or digital records (or both) are produced which reflect all the useful detail of the changes of the total magnetic intensity at the flight level.

To make full use of the high-sensitivity instruments requires digital recording. This permits making an analog record if desired, at any horizontal scale needed, particularly that of the resulting map. It also permits computer operations of various kinds, e.g., those involved in determination of critical points and distances used in estimation of depths to sources or other applications to interpretation procedures.

A major part of the overall operation is the procedures and instrumentation by which the positions of the flight lines are determined so that the observations in the airplane can be properly related to their positions on the ground, as described later (page 347).

The flight pattern (Fig. 13-1) usually consists of a series of primary flight lines at a fixed spacing, which, ideally, should be about one-half the depth from the height of the airplane to the basement but usually is not changed within a given survey even when it is known that large variations in basement depth may be found. These main flight lines are tied by crosslines at greater distances (Fig. 13-1). Common dimensions of the rectangles formed by the flight lines and tie lines are 1 by 6 miles, 2 by 10 km, 2 by 10 miles, etc. In special circumstances the flight-line pattern may be entirely different. In sectionized, usually farming, areas, most of the roads are on section lines 1 mile apart. Many of the earlier surveys were flown over these roads, and therefore the flight patterns had lines at 1-mile spacings.

For reconnaissance surveys over very large areas, with a limited budget of miles to be flown, it is usually preferable to use "band flying" rather than widely spaced single lines. In band flying two (or preferably three) parallel lines are flown at normal spacing, such as 1 mile or 2 km, with a much greater distance, say 25 miles, between bands. The magnetic field can be contoured within the bands and the trend directions established for making azimuth corrections in the interpretation process (see page 411) to give much more accurate basement depths than can be derived from single widely spaced lines. Thus, for a large new basin, a given number of miles of line flown in bands will give a better overall regional picture of the major geologic features than the same number of lines in a uniformly spaced pattern with more widely spaced lines.

An example of well-used band flying was the initial exploration of the huge Canning Basin in Western Australia. The area of the basin is roughly 250,000 square miles, approximately that of the state of Texas. A limited budget of some 150,000 line miles of magnetic surveys was assigned to the project. About half of this budget was used for three-line bands and the other half for more detailed (2- by 10-mile) coverage of the areas indicated by the interpretation of the bands as having thick sedimentary section or broad deformation of the basement surface. As proved by later, much more detailed gravity and seismic surveys and drilling, the first results from the reconnaissance magnetics were very useful.*

During flight on a given line, it is highly desirable that the airplane be held on a straight line at constant elevation, and much modern flying is done with auto-pilots to hold direction and elevation. For basement mapping it is the usual practice to keep an entire survey, where feasible, at a constant elevation, commonly about 1000 ft above the average surface elevation. If the area includes a wide range of topographic elevations, the survey may be divided into blocks, each at a fixed elevation, with some overlap between blocks to make the map continuous. This causes a hiatus in the contours at the elevation-block boundaries because of the difference in values at the two elevations, i.e., the effect of the vertical gradient. The magnitude of this vertical gradient is quite small. Grant and West (1965, p. 307) give an approximate average value of $-0.047H_0$ in gammas per meter, where H_0 is the value of the total intensity in oersteds. Thus, for example, where two surveys, with a flight-level difference of 1000 m come together, and where the total intensity is, say, 0.5 oersted, the hiatus in total magnetic intensity would be approximately $0.047 \times 0.5 \times 1000 = 23\ \gamma$. To make a continuous map, a line can be drawn on the map at the boundary between the two elevation blocks and contours on each side drawn to this line with a discontinuity there corresponding to the vertical-gradient effect.

FLIGHT-LINE ORIENTATION

The principal, i.e., more closely spaced, lines should be oriented roughly perpen-dicular to the expected geological strike or tectonic trends of the area. These may not be accurately known in advance of planning the survey, but usually they can be determined roughly from regional geologic or tectonic maps. This alignment means that magnetic features of geologic origin are crossed to the best advantage, so that the azimuth corrections, which are part of the basement-depth calculations (see page 411), are smaller and determined more accurately. When the angle between anomaly trend and flight line becomes less than about 60°, the interpretation deteriorates quite seriously.

---

* A great deal of airborne magnetic work also has been done for the mining industry, usually in much closer detail than for petroleum exploration. Commonly, mineral surveys are made with a line spacing of about ¼ mile. In many instances the work is done from helicopters to bring the observations nearer the ground surface and to permit lower speeds for better recording of very sharp anomalies from shallow sources. The Canadian government has had under way for several years a systematic magnetic survey of a large part of the country, including closely spaced (¼-mile) surveys in the continental-shield areas. These have been found useful, not only for mining prospecting and the location of individual anomalies possibly associated with mineral accumula-tion, but also in routine geologic mapping. It is found that, in the Precambrian areas, the general trend pattern of the magnetic contours is often closely parallel with that of the outcrops, and the magnetic pattern can be used to interpolate contacts over covered areas.

In low magnetic latitudes where a simple source produces a complex anomaly with a low on the north side (see Fig. 14-7) it may be preferable to fly approximately north-south no matter what the geologic strike because the magnetic character is better determined. A magnetic east-west line between the maximum and minimum would not reveal the essential nature of the anomaly.

LOCATION SYSTEMS

To make a map of the airborne-magnetometer observation, it is necessary, of course, to be able to locate the flight lines and to correlate the readings at any moment with a position on the ground. To retain the detail of high-sensitivity recordings, as mentioned later (page 352), the control of the aircraft and its positioning must be very exact.

Several different location systems are used to locate the flight lines. Some of these were considered in connection with the measurement of gravity at sea (page 121) and also are used for the location of marine seismic operations.

Positioning by Aerial Photography

Many of the earlier airborne magnetic surveys on land were positioned by aerial photography, where controlled photomosaics were available from previous photo-mapping. A simple procedure is to draw the flight lines on such a photomap and then cut it into long strips of such a width that when they are rolled into a scroll, they can be handled by the copilot. He then unrolls the strip as the flight proceeds, calling course corrections to the pilot and making any necessary notes. With a skillful crew, this procedure can produce a quite satisfactory line pattern.

The horizontal position of the airplane is determined by positioning photographs made during flight. They may be either 35-mm frames taken at intervals of 1 or 2 sec or a continuous 35-mm strip-camera film of the ground passing under the airplane. Later, in the mapping office, the individual pictures or the continuous-strip record are compared with the photomap (rephotographed at about the same scale as the position photographs) and each frame or segment carefully matched with the picture of the ground. Then recognizable points on the flight line are spotted on the photomap. Each unit picture carries a fiducial number, and the corresponding number is marked on the magnetometer record so that when these numbers are shown on the magnetic map, the exact relation of the fiducials on the magnetic record to the points on the ground is established. Total magnetic field values, usually after removal of the earth's normal field, can then be transferred from the record to the map.

The preparation of a magnetic map from photographic control as outlined above is a tedious and relatively slow process, particularly the locating of individual points of control on the photomosaic, and requires a substantial crew in a central office to keep up with a single airplane.

Electronic Positioning Systems

Flight lines are commonly positioned by one of several electronic location systems. Some such system is necessary when flying over water or for areas where photo-mapping is not available.* The first such system applied used high frequencies (3000 to 10,000 MHz). Radio-frequency systems (2000 to 3000 kHz) may be operated either in a circular (range-range) or hyperbolic mode. Doppler systems operate on signals sent out by the aircraft; they are reflected back from the surface, and their phase shifts are measured. The operating principles of these systems will be described briefly.

Radar   Ordinary radar, operating in the frequency range of 3000 to 10,000 kHz, has been used as a positioning system in marine operations where suitable identifiable reflecting objects are available. Direct observation of clearly recognizable objects is rare, so that nearly all offshore radar has used special targets, such as towers or floating buoys using corner reflectors.

Shoran   This system, essentially a modification of the radar system, depends on transponders which respond to a signal from the moving vehicle. The transponders are mounted on towers at fixed shore positions or, for more distant marine operations, on anchored boats. The equipment on the moving vehicle sends out a very sharply pulsed signal. When this signal reaches a transponder, it activates a responding signal, which is picked up by the moving-station equipment. Thus the time of transmission from the moving station to the transponder and back, with proper allowance for the time between the receipt of the incoming signal and transmission of the outgoing signal at the transponder, is a measure of the distance between the moving vehicle and the transponder. With at least two transponders which can be interrogated simultaneously from the moving station, the distances from the two fixed locations are determined. The intersection of circles with the measured distances as radii from two or more fixed transponder locations gives the instantaneous position of the moving vehicle. Since the system uses very high frequencies it cannot operate over the horizon of the curvature of the earth. For a flight elevation of, say, 1500 ft, the practical operating limit is some 40 to 50 miles.

The Shoran system was used extensively in the earlier days of airborne magnetic operations, and locations for one of the first large surveys, covering some 85,000 square miles over the Bahama Islands, were controlled by this method (Bemrose et al., 1950).

Radio-frequency systems   Beginning about 1955, the Shoran systems began to be replaced by the radio-frequency systems for operations over water. This was because the latter are not limited to line-of-sight operation, and also there were

---

* In the total geophysical use of electronic positioning systems, their use in airborne magnetics is relatively minor. A much larger number of applications are an essential part of nearly all marine geophysical operations, including those for seismic, gravity, and towed-magnetometer observations.

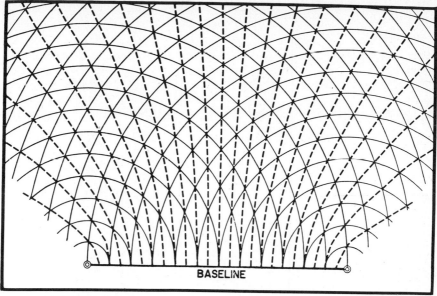

FIGURE 13-2
Standing-wave patterns (circles) and hyperbolic lanes for radio-location system.
(*Off-Shore Navigation Co.*)

difficulties in obtaining licenses in the restricted Shoran frequencies because of their military use.

Several radio-location systems, using frequencies in the 1500- to 4000-kHz range, were developed more or less simultaneously, particularly Lorac and Raydist, in this country, and Toran in France. These systems all depend on patterns of standing waves maintained very accurately with respect to fixed shore stations.

There are two modes of operation using the same basic equipment. In the "hyperbolic mode" two sets of circular standing waves in space are formed by interference of signals between two base stations. Lines of equal phase difference (or "lanes") form a set of hyperbolas (the dashed lines of Fig. 13-2). A second set of such hyperbolas is formed from a second pair of base stations, one of which may be at a central location in common with that of the first pair (Fig. 13-3). Two phase meters on the moving vehicle measure phase changes within the lanes, each making a complete revolution of 360° in crossing one lane. Thus, the two phase meters on the moving vehicles together with a count of the whole number of lanes give a set of location coordinates.

In the "circular" (or range-range) mode, the standing waves are formed by interference between a signal from a fixed base station and a transmitter on the moving vehicle, which form a set of circles of equal phase centered at the base station. Two sets of such circles, centered at two base stations, give location coordinates.

The hyperbolic system has the advantage that since the receiving equipment

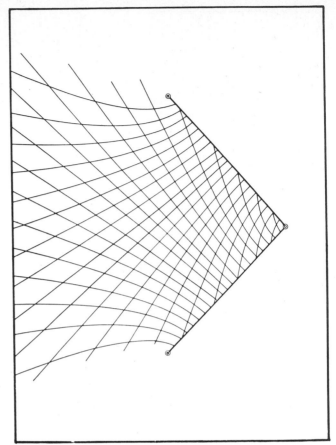

FIGURE 13-3
Hyperbolic lanes from two intersecting nets with common center point to form a coordinate grid of "red" and "green" lanes. (*Off-Shore Navigation Co.*)

on the moving vehicle is entirely passive, any number of operators can work in the same system. It has the disadvantage that three (or four) base stations are required to set up the two systems of hyperbolas. Also, the shape and size of the lane units change with distance from the base stations and become badly distorted near the outer part.

The circular system has the advantage of a simpler geometry made up of two sets of circles with equal lane width throughout the area covered. Also the relative coverage of the system for the same total base-station spread and transmission range is greater, as shown by Fig. 13-4. The principal disadvantage is the limited number of separate users.

Operations of all the radio-frequency systems depend on ground waves transmitted over or near the surface. There can be serious limitations due to "sky waves," which are reflected from the ionized layer and can interfere with the phase measurements and severely distort them. In extreme cases, the phase meters will

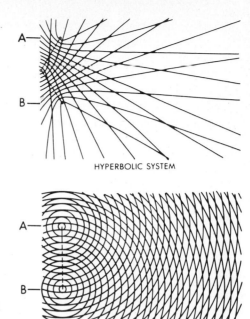

HYPERBOLIC SYSTEM

CIRCULAR SYSTEMS

FIGURE 13-4
Comparison of lane patterns for spread
of base stations for hyperbolic and circular location systems.

move very erratically or even spin around several times, so that the lane count is lost and must be reestablished. This interference is most common near sunrise or sunset but severe enough to prevent any work at night (which is much more important in marine operations).

Maintaining the lane count is greatly aided by a phase recorder. Its record shows the phase within a lane and jumps from one edge of the record to the other as a lane boundary is crossed. Counting the number of those jumps gives the lane number from an initial starting point. The lanes can be counted through moderate interference, but the count may be lost at times of severe sky waves.

Several hyperbolic location networks have been established and maintained along the Gulf Coast for many years and used to position magnetic, gravity, and seismic operations, drilling locations, and offshore-survey block boundaries.

In general, the hyperbolic systems are used where permanent service for several or many users is required, such as those covering the very active Texas-Louisiana offshore. For more remote or one-time exploration programs, where the limited number of users is not a disadvantage, the range-range system is used because of its somewhat simpler instrumentation, better coverage, and uniform lane width (see Fig. 13-4).

Doppler Navigation Systems

The radar-Doppler system is self-contained within the airplane and does not depend on auxiliary reference stations. It is widely used for airborne magnetic operations over land and to a limited extent over water relatively near shore. It depends on the Doppler principle that waves traveling between two locations which are moving in relation to each other change in frequency in proportion to the relative velocities of the source and receiver.* In the application to navigation high radar frequencies are used (the k_a band, for example), and differences in frequency between signals reflected from various points on the ground (forward and astern of the aircraft) give a component of velocity. For aerial surveying, the source is the aircraft, and the receivers are the ground traces of the transmitted beams. In turn, the ground as a reflector becomes a source and the aircraft the receiver. The phase difference between transmitted and reflected signals is a measure of velocity. Dispersion of the frequencies of the reflected beams is kept within tolerable limits by causing the illuminated pads on the ground to be sections of hyperbolas, to take advantage of the fact that the figure of constant Doppler shift is a hyperbola.

Two sets of transmitters and receivers measure frequency differences between the transmitted and reflected signals. These are fed into analog or digital computers to give components of velocity. Various airborne Doppler equipments measure different components of the motion of the airplane, which can then be resolved into components parallel and perpendicular to the aircraft axis.

The velocity components are integrated with time to give distance components. Thus the equipment can plot the track of the airplane or register its position components on magnetic tape. Fiducial numbers simultaneously registered on the Doppler and magnetometer records serve to correlate details of the record with position on the ground. Precision of Doppler positions is generally within 1 to 2 percent of distance traveled from a control point.

It is common practice for the paper speed of the magnetometer recorder to be controlled by the Doppler ground speed. This makes it possible for the magnetic record from the airplane to be at the same linear scale as that used in the mapping of the results. This is a very practical convenience in preparing and interpreting the magnetic map.

A variation of the Doppler principle, using sound waves in water (Sonar Doppler), is used for navigation at sea. One pair of sound-wave beams directed fore and aft and another directed starboard and port are reflected off the bottom. Phase differences between the reflected beams of each pair give speed components. Usable reflections are obtained in maximum water depth of 500 to 1000 ft.

DATA-REDUCTION SYSTEMS

The production of a contoured map from airborne magnetic data is a complex process. It begins with the recording, in the airplane, of magnetic, elevation, and position data. In the mapping process various sources of error have to be taken into account and corrections made. For instance, apparent crossing points of observation lines and tie lines may not be exactly where first mapped and may not

---

* The familiar example, for sound waves, is the increase in pitch of a sound from an approaching source (such as a train whistle) and its sudden decrease in pitch as the source passes and recedes in the opposite direction.

be at the same elevation. Magnetic differences around loops then will not close (add up to zero) as they should. Minor corrections in flight-line positions may be necessary.

Full use of high-sensitivity magnetometers requires very high precision in the control of aircraft. To keep short-time magnetic changes due to variation of aircraft position to 0.1 γ requires flying within an elliptical tube with a vertical dimension of 20 ft and horizontal of 100 ft (Jensen, 1965). This is perhaps a more stringent control requirement than for any other aircraft operation. Also, full realization of the possible precision and detail of high-sensitivity magnetometers requires that operation be confined to times of relatively quiet magnetic "weather."

None of these methods is adequate to make diurnal corrections with sufficient precision and detail to isolate all the possible geologically caused components of recordings by high-sensitivity magnetometers. A much more complete procedure using measured vertical gradients and rather complex computer operations on the various digitally recorded data elements is described by Hartman, Tesky, and Friedberg (1971).

In modern data processing the various steps in eliminating errors and making final maps are done by digital computers. The original data are recorded digitally and the resulting tapes checked by programs which detect errors, such as erratic or missing values, etc. Many other programs and computer processes are used in various steps so that, to a large extent, the entire procedure from original data recording to final contouring is computerized. Somewhat different methods are used by different operators and have now reached a high degree of development. A quite complete description of one such system is given by Bhattacharyya (1971). So far as this writer knows, none of these processes is without some interruptions along the complex chain of operations, and some human intervention always occurs.

CORRECTIONS TO THE MAGNETOMETER OBSERVATIONS

A magnetometer reading is affected by the sum of all contributions to the magnetic field at the time and place of observation. The object of the survey is to map the anomaly part of the total magnetic field (page 320). Thus it is necessary to remove those effects which are extraneous. This is absolutely essential to the attainment of a satisfactory picture of magnetic anomalies of small relief. The magnetic expressions of geologic structures that may be economically important in the search for oil are often very weak. Therefore, careful attention to the various corrections to magnetometer readings is much more important in the application of the magnetometer to oil prospecting than to the exploration for iron ores or igneous intrusions or contacts where the magnetic anomalies are much stronger and usually much more local and definite. It is a relatively simple procedure for a ground-magnetometer operator to set up an instrument and make a reading to a precision of 2 or 3 γ. Similarly, properly installed and adjusted airborne equipment produces a record of the time variations of the total magnetic field traversed by the airplane. It is quite a different matter to be sure that when the final map is made the magnetic values thereon will truly represent to the same precision the local variations in the magnetic intensity that are of interest.

It has been pointed out (Chap. 11) that the total field measured at any one time and place consists of three components: (1) the inner field of the earth due to the basic properties of the earth as a whole, (2) the external field due to sources above the surface, the so-called diurnal variations, and (3) the anomaly field due to sources within the outer part of the earth's crust. Only the last component is used in magnetics for petroleum exploration. Therefore, if the maps are to show only this component, the other two must be removed and the field operations carried out in such a way that these corrections can be made in the data processing and mapping.

Normal Corrections

Normal corrections are made to remove the spatial variation of magnetic intensity due to the earth's inner field. This corresponds in a general way to the latitude correction of gravity values. However, any mathematical expression by which normal magnetic effects can be calculated is very complex, and the corrections are determined empirically from magnetic charts of tables (see page 320 and Figs. 11-1 and 11-2).

For ground surveys the corrections can be made by drawing contours of normal variations at a convenient interval, such as 10 γ, over the area of the survey. The direction and distance between contours are estimated from the magnetic chart. Alternatively, the north-south and east-west components of the normal rate of change of intensity can be estimated from the magnetic chart and the corrections figured from the coordinates of the station. These methods may be used for relatively small areas but are not satisfactory if the survey area is large or is connected to other surveys because the normal variation factors are not linear.

For large or connected surveys the normal corrections should be determined from smooth normal contours, based on the standard magnetic charts or tables. For mapping of airborne surveys the normal correction can be determined from the IGRF charts. If the map is compiled from the analog records, the normal magnetic change from the tables can be drawn as nearly straight lines on the records. The values mapped are then the departures from these lines. If the data processing is by computer, the correction component can be included in the computer operations by relating the values from the tables to the digitized location data. Computer programs available from the appropriate U.S. government agency (NOAA) contain the basic numeric data and can be incorporated into the data-reduction operations.

Diurnal Corrections

It has been pointed out (Chap. 11) that the magnitude of the diurnal variations of the earth's magnetic field is from 10 to 100 γ or more. Therefore, corrections must be made for their removal if the results of a magnetic survey are to be accurate to better than these magnitudes. Since the local anomalies of interest in oil prospecting are commonly of much lower relief than the magnitude of the daily variation, magnetic mapping for such prospecting always includes the removal of the diurnal effects.

The nature of the diurnal changes   Daily records of the principal components of the magnetic field have been made for many years by the U.S. Coast and Geodetic Survey at Cheltenham, Pennsylvania, and at Tucson, Arizona, and, more recently, by NOAA at many other stations in the United States. There are magnetic observatories in many other parts of the world. These records show the nature of the daily or diurnal changes, which usually consist of a somewhat irregular decrease of magnetic intensity during the daylight hours from a more regular level at night. However, there are many short-time irregularities, and the general form of the daily curves varies from day to day. Measurements near continental-shelf areas must be used with care because of local changes caused by variations in crustal conductivity. For comments on the source of the diurnal changes see page 326.

The necessity of making corrections for these diurnal changes was recognized early in the history of magnetic prospecting, and studies of their nature and the geographic variations of their details were carried out by Soske (1933) and Vacquier (1937). In both studies simultaneously recorded curves of magnetic field variation at different locations were compared. Soske compared two fixed locations (at Pasadena and at the U.S. Geological Survey station in Tucson) with one movable station. Vacquier compared simultaneous observations at five field-magnetometer operations and at Tucson, with a separation of up to 1500 miles.

Samples from Vacquier's curves are shown in Fig. 13-5a and b. It is quite evident that accurate corrections for diurnal effects require continuous or quite frequent observations of these changes. Note, for instance, the sharp peak on curve 1 for the upper part of Fig. 13-5a between 14 and 15 hr, where a change of 10 γ occurred in less than 15 min, and the very steep change of over 80 γ in Fig. 13-5b.

A careful study in Ireland by Riddihough (1971) is based on one continuously recording station on the southeast coast and recordings of several days at a time at seven other stations, mostly near the coast around the island with maximum separation of about 350 km. A statistical study was carried out of the differences between each test station and the base. The two principal conclusions were (1) that there is, in general, a phase difference corresponding to change in sun time but no definite statistical improvement was gained by taking it into account and (2) that amplitudes are related to geographical location. These relations can be contoured and have a generally north-south trend. This relation probably is caused by geological and structural features on a large scale, e.g., conductivity anomalies in the crust, oceans, deep sedimentary basins, etc. A statistical study of the continuous records at the single base station using all the data from the several field stations to derive a method of correction indicates that the standard error at any local area would be of the order of 5 γ or less for days that are relatively quiet, magnetically, and 6 γ or more for moderately "stormy" days.

Examples of diurnal variations measured with high-sensitivity (rubidium-vapor) magnetometers are given by Hartman, Tesky, and Friedberg (1971). These comparisons are between three stations in northwestern Canada with distances between them of 85 to 290 miles. The detail is greater than that of the other records discussed above, both in sensitivity and in time scale, by one or two orders of magnitude, as variations with magnitudes of less than 1 γ and periods of 15 sec or less are shown. Consideration of such details is necessary to eliminate extraneous effects so that all the possibly significant detail of the high-sensitivity recordings can be preserved.

It is evident from these studies and others that accurate correction of magnetometer observations requires careful attention to the diurnal effects.

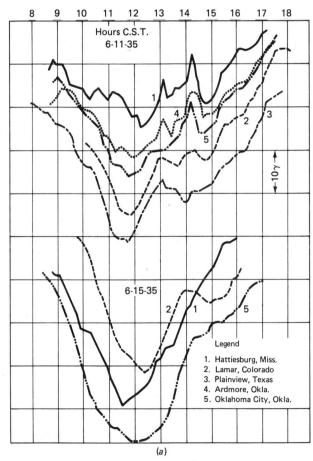

(a)

FIGURE 13-5
Samples of simultaneous curves at different localities of daily variation of vertical component of earth's magnetic field. All curves made from continuous observations of a fixed base-station ground magnetometer. (*Nettleton, 1940.*)

Diurnal correction procedures   We can now consider the field procedures which are carried out to provide data for the removal of the diurnal effects.

For ground magnetometer surveys the data for diurnal corrections can be provided by returning the instrument at intervals during the day to the same point (or to points between which the magnetic intensity differences have been determined previously). The readings at the same station would check if the magnetic intensity were constant. (This is similar to the observations for instrument drift and tidal effects in gravity operations.) The variation in these readings therefore is a measure of the magnetic change during the day. If readings are made at intervals of not over 2 or 3 hr, a fair measure of the daily variation can be obtained. However, an inspection of the sample daily-variation curves in Fig. 13-5 shows that readings at an interval as great as 2 hr could miss details of the daily variation as great as 10 γ. Therefore this method is not adequate for magnetic work in which a precision of a few gammas is attempted.

A much better alternative is to use an auxiliary base instrument, which may be either an ordinary field magnetometer read at frequent intervals (preferably not

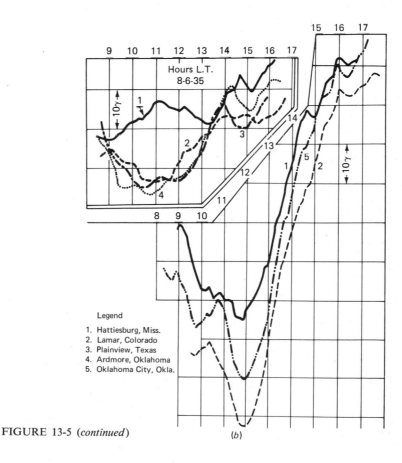

Legend

1. Hattiesburg, Miss.
2. Lamar, Colorado
3. Plainview, Texas
4. Ardmore, Oklahoma
5. Oklahoma City, Okla.

FIGURE 13-5 (*continued*)

(b)

over 10 to 15 min) or a recording instrument. By either method a continuous curve is obtained of the daily variation at the base location. This curve can be used for correcting as many field instruments as are operating in the general vicinity of the base instrument. Just how far afield it is safe to depend on the curve at a given point is not definitely known. Vacquier (1937) shows the statistically probable variation in magnetic intensity as a function of distance between stations to be less than 3 γ for station separations up to 500 miles and only 4 γ for station separations up to 1500 miles, but these statistical conclusions may not be valid for short-period details. From the rather incomplete data available up to the present, it is probable that a given diurnal curve may be used quite safely for corrections to a few gammas for distances of 50 miles, and usually for much greater distances, from the point at which it is determined.

For airborne surveys, as pointed out above, the flight pattern is nearly always laid out with sets of normal flight lines at a standard survey spacing tied by cross-lines at a spacing several times greater. This gives a series of rectangles in the flight-line pattern. In normal flying practice with airplane speeds of 150 mph or more the distance between tie lines, of the order of 6 to 10 miles apart, is covered by the airplane in a time of the order of 3 to 5 min. The diurnal change within this time is usually quite small and is ordinarily removed by the closure adjustments for the rectangular closed loops.

While short-time diurnal changes may be absorbed in the closure adjustments, it is common practice for a continuously recording monitor magnetometer to accompany each airborne survey. Its primary purpose is to show the general nature of the magnetic weather. When the diurnal changes exceed certain rates, the operation is discontinued. This ensures that records will not be used for the final mapping that were made during times of magnetic storms or at other times when changes are too irregular or too rapid to be properly removed in the normal data-reduction procedure. During severe magnetic storms field operations may be shut down for many days, particularly in high magnetic latitudes.

It is possible to make diurnal corrections by digitizing the monitor record and automatically subtracting its variations from the record observed in the airplane in the data-processing procedure. This is only partially successful, probably because of geographic differences in the magnetic changes like those shown by the records of Fig. 13-1. A simpler procedure, which can be quite helpful when a distant monitor record must be used, is to digitize that record so that it can be printed out, with proper time relation, along with the recording being interpreted. Then diurnal effects can readily be recognized as such when they appear nearly simultaneously on both records.

The vertical-gradient measurement (page 341) is insensitive to diurnal changes and can be used for their recognition because any change in the field intensity from distant sources above the airplane affects the two magnetometers simultaneously and equally and thus is eliminated in the difference. Thus an anomaly may appear on a total-intensity record which can be recognized immediately as a diurnal effect if its vertical derivative does not appear simultaneously on the gradient record. This is a particular advantage in high magnetic latitudes, where diurnal effects are strong.

In principle, a vertical-intensity map can be calculated from a vertical-gradient map by a surface-integration procedure, or an interpretation can be made directly from a vertical-gradient map. Thus, useful operations can be carried out by vertical-gradient measurements even at times and locations where strong diurnal disturbances make a total-intensity map almost useless.

The diurnal problem is particularly serious in marine operations when the recording is on isolated lines far from shore. Then there is no nearby reference, either from a fixed recording station on land or intersection with another line from which the diurnal change can be recognized as such and not misinterpreted as being from a geological source. Measuring the horizontal gradient is one method of meeting this problem. A diurnal change affects both instruments simultaneously and has no horizontal-gradient effect, while any disturbance from a fixed source below the ship would have a corresponding horizontal-gradient effect. In practice, this may not be as simple as it seems because of noise on the records. If the magnetometers are too near the ship, there may be disturbances from short-period lateral motions which carry with them a remanent of the magnetic field of the ship itself. This effect, of course, will be more severe in rough weather.

14

THE SOURCES AND NATURE OF MAGNETIC ANOMALIES

Rocks become magnetized because they contain magnetic minerals. There are several such minerals, including magnetite, hematite, pyrrhotite, ilmenite, maghematite, and leucoxene, but magnetite is by far the most magnetic and the most common of these minerals. The other rock-forming minerals are essentially nonmagnetic. Therefore, for most practical purposes, we can say that rocks are magnetizable if they contain magnetite, and their magnetic properties depend on the amount of magnetite disseminated among the nonmagnetic minerals which make up the principal material of the rock. As pointed out later (Table 14-2), this means that, in general, the igneous or basement rocks are so much more magnetic than the sediments that most magnetic effects observed by airborne magnetometers are essentially the same as they would be if the sediments were not present at all.*

* There is some evidence for slight negative magnetic anomalies from salt domes which result from the practically zero or slightly negative, i.e., diamagnetic, polarization of the salt contrasted with slightly magnetic sediments. For a specific example, a long salt column with a diameter equal to twice its depth below the surface and having zero polarization contrasted with an assumed polarization of the surrounding sediments of 50×10^{-6} would have a magnetic effect over its center of -9γ. On the other hand, Lynton (1931) has shown that certain sediments in California have strong magnetic effects because they contain the magnetic mineral vivianite ($Fe_3P_2O_8 \cdot 8H_2O$).

As discussed in more detail on page 365, the magnetism of a rock may either be that induced by the earth's field or remanent ("frozen-in") magnetization from a previous condition, i.e., during cooling or during deposition. While there are many exceptions (some very important), magnetic anomalies, worldwide, have the character and orientation consistent with magnetization by the earth's field in its present orientation.

THE MAGNETIZATION OF DISSEMINATED MAGNETIC PARTICLES

Since magnetization of disseminated particles of magnetite in the earth's field is the principal source of magnetic anomalies of interest in petroleum prospecting, we shall consider the fundamentals of this process in some detail (Nettleton and Elkins, 1944).

Consider that a magnetizable rock is composed of individual particles of magnetite disseminated through a nonmagnetic matrix so that the particles are separated by distances several times their diameter. To the extent that the individual particle can be approximated by an ellipsoid, the magnetization I_i of a single such particle will be determined by the applied earth's field H_e, by the susceptibility of the material k_0, and also to a very great degree by the demagnetization factor* λ, as shown (see, for instance, Grant and West, 1965, p. 318) by the relation

$$I_i = \frac{k_0 H_e}{1 + k_0 \lambda}$$

If a rock has a volume fraction P of magnetite, the magnetization I of the rock as a whole will be

$$I = PI_i$$

Then the effective susceptibility k of the rock as a whole will be

$$k = \frac{I}{H_e} = \frac{PI_i}{H_e} = \frac{Pk_0}{1 + k_0 \lambda} = Pk' \qquad \text{where } k' = \frac{k_0}{1 + k_0 \lambda}$$

* The demagnetization factor can be considered as the result of the poles induced on the surface of a body when it is magnetized. These poles are of opposite polarity to the direction of the magnetizing field and therefore tend to reduce the magnetization within the body.

For a uniform magnetizable space, the demagnetization factor is $\lambda = HT/3$, and the induced field within the body is

$$I_i = \frac{k_0 H}{1 + (4\pi/3)k_0}$$

Thus (for the sphere) the apparent susceptibility is reduced from k_0 to $k_0/[1 + (4\pi/3)k_0]$ or by a factor of $1/(1 + 4\pi/3) = 0.19$. The value of the demagnetization factor varies greatly with shape, from zero for very long bodies magnetized along their axis to 4π for a flat plate magnetized transversely. For a rock body as a whole containing disseminated magnetite the demagnetization effect is small because of the low value of k_0 (see Table 14-1).

It is evident that if we knew the volume percentage of magnetite in a rock together with some way to evaluate k', we could calculate the effective susceptibility.

Grenet (1930) apparently was the first to consider the geometric and demagnetization factors carefully and to point out that for disseminated magnetite in rocks the effective susceptibility k' is only slightly dependent on the susceptibility of the magnetite itself, i.e., on k_0. Grenet gives a table of this relation for spherical particles, which (slightly extended) is given as Table 14-1.

If the particles are not spherical but elongated, the value of k' will be larger than the figures in Table 14-1 if the magnetization is in the direction of the long axis and less if in the direction of the short axis. For instance, Grenet calculates that for an ellipsoid of revolution with a ratio of long to short axes of 10, the value of k' is 2.08 (for $k_0 = 10$) for magnetization along the long axis and 0.150 for magnetization along the short axis. For the same ellipsoid, the mean value of k' for the three axes is 0.797.

On the basis of these figures, Grenet calculates that as a rough average the effective susceptibility of magnetite disseminated in small particles, i.e., the value of k', should be about 0.25.

The susceptibility of magnetite, i.e., the value of k_0, appears to be quite variable and also is highly dependent on the strength of the field in which it is measured (Heiland, 1940, p. 311; Slichter, 1929, p. 245), but values in the range 1 to 2.5 in low fields are indicated. However, as shown by Table 14-1, the susceptibility of a body composed of magnetite disseminated in a nonmagnetic matrix, that is, k', is very insensitive to the susceptibility of the particles, that is, k_0, and any susceptibility value above 1.0 gives a rock susceptibility of a little over 0.20 (for spherical particles).

The foregoing considerations have indicated that we should expect to be able to calculate the approximate susceptibility of a rock containing a volume proportion P of disseminated magnetite as

$$k = Pk'$$

Table 14-1 RELATION BETWEEN MINERAL SUSCEPTIBILITY k_0 AND EFFECTIVE SUSCEPTIBILITY OF ROCK k' FOR DISSEMINATED SPHERICAL PARTICLES

k_0	k'
0.01	0.0096
0.1	0.0705
1	0.193
5	0.228
10	0.233
100	0.238
∞	0.239

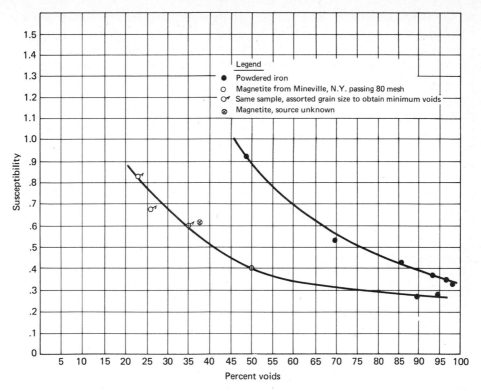

FIGURE 14-1
Specific susceptibility of powdered magnetite and powdered iron in various concentrations in a nonmagnetic matrix. (*After Slichter, 1929.*)

with k' having a value in the neighborhood of 0.20 to 0.25 (200,000 to 250,000 $\times$ 10^{-6}). Thus a rock with 1 percent magnetite magnetized by the earth's field of 0.5 oersted should have an intensity of magnetization, or polarization, I of approximately

$$I = Hk' = 0.5 \times 0.01 \times 0.25 = 0.00125$$

which is an order-of-magnitude figure for polarization of basement rocks commonly used in magnetic-model calculations.

Slichter (1929) made fundamental studies of the properties of magnetite, including a study of the effective susceptibility* of powdered magnetite when disseminated through a nonmagnetic matrix. Figure 14-1 is, in effect, a measurement of the value of k' for various proportions of magnetite. The value decreases regularly to about 0.27 for very small magnetite concentrations. Almost as low a

---

* The ordinate (susceptibility) on Slichter's figure is specific susceptibility, i.e., the measured value divided by the fraction of magnetite.

value is shown when the magnetic material is powdered iron, for which the susceptibility for the material, that is, k_0, must be much larger than for magnetite. These curves are just what would be expected from the theoretical conditions outlined above, as the conditions postulated in calculating Table 14-1 should be approached at the right side of Fig. 14-1. The fact that at low concentrations the value of k' for iron is almost as low as that for magnetite is explained by the small effect of the value of k_0 on k' when k_0 is larger than about unity (see Table 14-1). The slightly higher value of k' than that calculated for spherical particles may be explained either by the particles' being somewhat elongated with a small degree of alignment parallel to the field or by the aggregation of individual particles into groups elongated parallel to the field, or both. We can consider that Slichter's work on powdered magnetite has given an experimental value of k' of about 0.27. Thus the theory pointed out by Grenet (1930) has given a good explanation of Slichter's experiments.

A number of other and later experiments to relate magnetite content to susceptibility in general show a wide scattering when susceptibility is plotted against magnetite content. The methods used for determining magnetite content, i.e., magnetic separation of pulverized rocks, areas in thin sections, and chemical analysis, are not very exact. However, the values derived from averages of such measurements are similar to the values derived above. Slichter (1929) has two values, 0.29 and 0.28. From a large number of values Grenet has an average of 0.156. Mooney and Bleifuss (1953) give an average value (in the units used here) of 0.29 with average values for different rock types ranging from 0.12 to 0.38. Puzicha (1941) gives an average value of 0.25 to 0.30.

All the above shows wide quantitative variability in the relation of magnetite to susceptibility, but the averages are generally consistent with a value around 0.25, which means that 1 percent magnetite gives a polarization in the earth's field of around 0.001 cgs unit. For application to magnetic prospecting, the useful result of the above is that the magnetization of rocks is of the general nature and magnitude to be expected from magnetization of disseminated magnetite by the earth's field.

MAGNETIC POLARIZATION OF ROCKS

In the geophysical literature there are a number of tables of susceptibility of rocks (Heiland, 1940, pp. 310–315; Jakosky, 1950, p. 164; Nettleton, 1940, p. 201; Dobrin, 1960, p. 269). Many of them, especially the older measurements, were made in relatively strong fields, which give lower values than a weak field like that of the earth.

For practical application to petroleum prospecting we must consider general averages rather than specific values for a given rock or locality. In general, sedimentary rocks have very low magnetite content and low susceptibility, metamorphics are intermediate, and igneous rocks are much higher, with the granites and light-colored rocks generally lower than the basic rocks. Table 14-2, from averages of measurements with a considerable range in individual values, illustrates the two essential factors on which magnetic interpretation for petroleum exploration is based:

1 The susceptibility (and therefore the polarization when magnetized in the earth's field) is very much less for sediments than for basement rocks, which is the basis for computing basement depths from magnetic maps.

2 The wide range in individual values indicates a basis for blocks or cells of magnetic material within the basement rocks which are the sources of the anomalies from which basement depths are determined. Also the wide range of values indicates that the expected susceptibility (and polarization) contrasts between magnetic units should be almost as great as their total values.*

Remanent Magnetization

The above discussion has considered that the magnetization of rocks is entirely that induced by the earth's present field. However, there may be a second component representing the magnetization which remains from a previous state. A probably common source of remanent magnetization is alignment of magnetic particles as they cool through the Curie temperature (page 322) so that the rock becomes magnetized in the direction of the earth's field at the time of cooling. On cooling, this direction is "frozen" and becomes a permanent component of the total magnetization, along with the component due to induction by the earth's present field.

* This is quite different from the relations of density to gravity anomalies, as the larger density contrasts are usually only relatively small fractions (one-tenth to two-tenths) of the density of the rock units in contact.

Table 14-2*

Rock type	No. of samples	Susceptibility $\times 10^6$ cgs units		
		Low	High	Average
Sedimentary rocks:				
Dolomite	66	0	75	8
Limestone	66	2	280	23
Sandstone	230	0	1665	32
Shale	137	5	1478	52
Basement rocks:				
Metamorphics	61	0	5824	349
Acid igneous	58	3	6527	647
Basic igneous	78	44	9711	2596

* From Dobrin (1960, p. 270).

The importance of remanent magnetization in the analysis of magnetic anomalies is a matter of some debate. A strong remanent component will modify both the magnitude and direction of the magnetization of a body of magnetized rock and would make a material difference in the form and boundaries of the causal rock body derived from the analysis of its magnetic effect. Magnetization along the same vector axis but in the opposite direction is more common, as mentioned (page 324) in connection with reversals of the earth's magnetic field. The direction of the earth's field need be known only roughly in making basement-depth estimates, which is the usual oil-prospecting application of magnetics.

As a practical matter, in the analysis of magnetic maps for petroleum prospecting, these questions are of relatively small consequence because from world-wide experience with magnetic maps the patterns of anomalies quite generally change with magnetic latitude in the manner expected if the rock units are magnetized in the direction of the earth's present field.

The Principle of Infinite Detail

The magnetization of igneous rocks depends on the distribution of magnetite within them and may be very erratic. Airborne magnetic traverses are always sharply anomalous when flown low over igneous terrain. Profiles and maps become smoother as the distance above the igneous surface increases because of higher flight levels or because the basement is buried by a thick section of sedimentary rocks. This is illustrated by Fig. 14-2, which shows magnetic curves calculated from point sources in a north-south vertical plane with vertical magnetization and with a single dipole at each dot.

The upper curve, with high relief and much detail, is calculated for an elevation of 0.1 unit, corresponding to the depth to the first row of dots. The lower curve, which is much smoother and of lower relief, is calculated for an elevation of 0.5 unit. At the low flight level the influence of individual sources is distinguishable as small variations or waves in the total effect. At the higher level above the surface these effects merge together, and the magnetic field is practically the same as it would be if the sources of the magnetic disturbances were simple blocks (or sheets) of magnetized material which have the boundaries indicated by dashed lines and which correspond to the divisions between areas with different degrees of concentration of sources. This is analogous to the way in which a picture made up of small dots (as in a coarse-grained newsprint halftone) changes when viewed from different distances. Very close up, the individual particles or dots are separately distinguishable. At greater distances certain of them merge into the smaller details of the picture. At still greater distances these merged areas in turn coalesce into larger components of the picture. Similarly, as the magnetic field is observed from successively higher flight levels (or as depths became successively greater), the sharp details which are apparent at low levels merge into broader and broader features at higher levels (or greater depths). Thus it is usually possible by simple inspection of a magnetic map over a broad area covering a wide range of depths to separate the map into areas of shallow, intermediate, and large basement depths

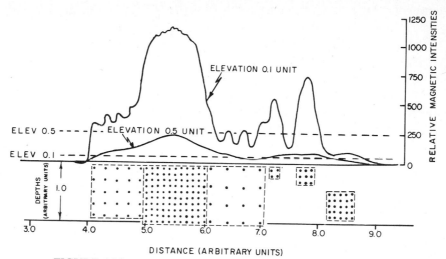

FIGURE 14-2

Calculated magnetic effects of point sources. Each dot represents a unit dipole magnetized vertically ($i = 90°$). The two curves are for simulated north-south flight lines at elevation 0.1 and 0.5 distance unit. Note that the various groups of poles act as blocks of magnetized material, as indicated by dashed outlines. At the very low flight level the effects of individual sources are detectable; at the higher level their effects are much smaller and merge into a very smooth curve. (*Courtesy of the Society of Exploration Geophysicists.*)

by noting the areas of sharp, intermediate, and broad magnetic anomalies. In making model calculations of the effects of magnetized bodies, it is usually assumed that these bodies have regular geometric boundaries, corresponding to the blocks of dots outlined by dashed lines in Fig. 14-2. The fact making basement-depth calculations possible is that these blocks are not homogeneous. They always contain internal detail which produces anomalies appropriate to successively shallower depths as the magnetometer flight lines approach the surface of the basement more closely. Thus, a shallow basement produces sharp anomalies relatively close together, and a deep basement produces broad anomalies farther apart. The whole process of basement-depth determination depends on devising quantitative measurements of this relationship of sharpness to depth.

A field example which illustrates the principle of infinite detail is available from some early experimental airborne magnetic surveys by the U.S. Geological Survey near Mangum, Oklahoma. Figure 14-3 shows four magnetic maps over the same area. The one made on the ground (upper left) is quite similar to the one made in the air at 300 ft above the ground surface.* The survey at 1300 ft shows consider-

---

* These two maps are not strictly comparable because (1) the ground survey was contoured from point-by-point observations while the other was mapped from continuous profiles and (2) the ground survey measured the vertical component (where the inclination is about 67°) while the airborne survey (with a flux-gate magnetometer) measured the total intensity.

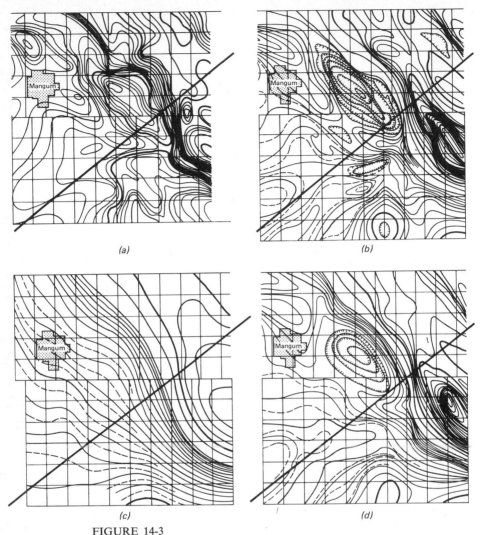

FIGURE 14-3
Magnetic maps of the same area, near Mangum, Oklahoma. *A*: vertical
magnetometer at surface. *B*, *C*, and *D*: airborne magnetometer at elevations 300,
1300, and 6300 ft, respectively, above surface. (*From Vacquier et al., 1951.*)

able smoothing and loss of detail and also has lower total relief. Finally, the picture
at 6300 ft shows much more smoothing and still lower relief.

The same general characteristics are illustrated by the profiles of Fig. 14-4,
which are made from the same diagonal line as shown on the four maps of Fig.
14-3. Note the decrease in amplitude of the two sharp peaks just right of the center
of the line between the 300- and 1300-ft curves and their complete disappearance
at the 6500-ft level.

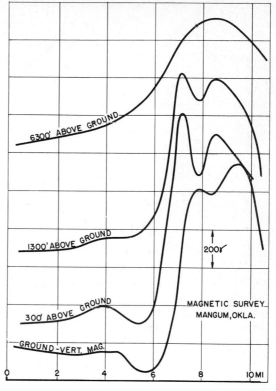

FIGURE 14-4
Profiles from magnetic maps of Fig. 14-3 on the northeast-southwest lines shown. Note loss of amplitude and detail with increasing elevation. Difference in detail between the ground survey and 300-ft-elevation airborne map results because the ground survey measures at isolated points and misses features shown by the continuously recording airborne instrument.

Natural Rock Bodies and Magnetic Anomalies

The preceding paragraphs have described the fundamental details of the magnetization of rocks. The actual basement rocks may be compared with a mildly stirred-up matrix of components with varying magnetite content. The condition may be generally likened to an old-fashioned marble cake with dark and light batter lightly stirred together (Fig. 14-5) and covered with a frosting. The two colors of batter correspond to basement materials with relatively high and low magnetite concentrations and the frosting to the overlying sediments.

Such concepts would seem to make numerical calculations from geometric models not very relevant. This is not true, however, apparently because at depths comparable with the horizontal dimensions of units with different magnetite concentration the magnetite effects from bodies with irregular boundaries are effectively simulated by models with simple geometric forms. Therefore calculated

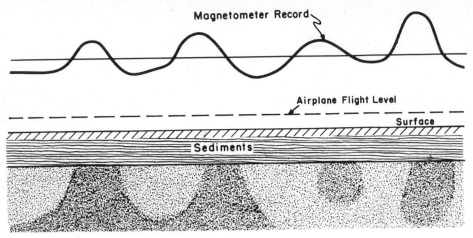

FIGURE 14-5
Marble-cake analogy of gross intrabasement features and magnetic anomalies.

effects from simple models can be very useful in understanding the magnetic effects observed in nature.

THE NATURE OF MAGNETIC ANOMALIES

For a given magnetized body magnetic anomalies are very much more complicated than the gravity anomalies for a body with the same boundaries. In gravity the anomaly produced by a given body of rock or as calculated from a given model depends only on the geometry and density contrast of the body. In the magnetic case, the total intensity of the field, which is the quantity measured by nearly all the magnetic instruments used in petroleum prospecting, is in a direction which varies from near vertical near the magnetic poles to horizontal at the magnetic equator (the zero-degree contour of Fig. 11-2) to near vertical again near the south pole. Furthermore, although the body usually is magnetized in the approximate direction of the earth's field, this is not necessarily so. Therefore, the form of the magnetic anomaly from a given body depends on all the following factors:

1 The geometry of the body
2 The direction of the earth's field at the location of the body
3 The direction of polarization of the rocks forming the body
4 The orientation of the body with respect to the direction of the earth's field
5 The orientation of the line of observation (flight line) with respect to the axis of the body

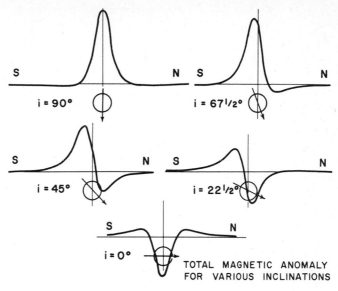

FIGURE 14-6
Variation in form of anomaly in total magnetic intensity of a sphere with change in magnetic latitude. (*From Nettleton, 1962; by permission of the American Association of Petroleum Geologists.*)

These complications mean that the calculation of models to account for magnetic anomalies is much more complex than using simple normalized characteristic curves for evaluating the general nature of gravity anomalies.

This is illustrated by the models for a sphere. For the gravity anomaly the curve is a simple maximum (for a positive density contrast), its magnitude depending only on the radius, density contrast, and depth of the body. The corresponding curves for a magnetized sphere (equivalent to a magnetic dipole) are shown in Fig. 14-6. The magnitude and nature of the anomaly are greatly modified by the direction of magnetization. The curves are calculated on the assumption that the body is magnetized in the direction of the earth's field, and the curves are for a north-south line over the center of the body with north to the right. These are the forms that would be measured by a total-intensity magnetometer (flux-gate, nuclear-precession, or pumped-vapor instrument). The change in character of the anomaly corresponding to changes in the direction of the earth's magnetic field is dramatic. At high magnetic latitudes, where the magnetic inclination is nearly vertical, the theoretical anomaly is a simple maximum with very weak symmetrical flanking minima (amplitude less than 5 percent of that of the maximum) on the two flanks. For the next diagram, with an inclination of $67\frac{1}{2}°$, the figure begins to become unsymmetrical, the maximum amplitude is slightly lower, its peak is shifted slightly south of a point over the center of the sphere, and a negative component

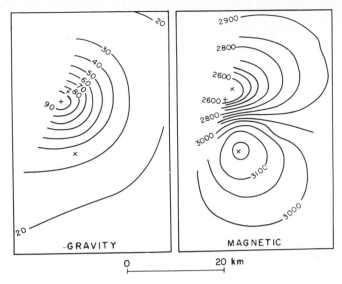

FIGURE 14-7
Comparison of gravity and magnetic maps in the same area. (*From Nettleton, 1962; by permission of the American Association of Petroleum Geologists.*)

begins to develop on the north side. At 45° the maximum amplitude is further reduced, the peak is shifted farther to the left, and the negative anomaly is quite strong. At $22\frac{1}{2}°$ the curve becomes nearly symmetrical with positive and negative parts of roughly equal amplitude and with the point over the center of the sphere quite close to the maximum negative amplitude. At the magnetic equator, where the direction of polarization is horizontal the anomaly has become a strong minimum with symmetrical flanking positive features. Thus the magnetic curve, which is a simple positive anomaly at high magnetic latitudes, is approximately turned inside out to become a negative anomaly at very low magnetic latitudes. The series would be repeated with left-to-right reversal if the diagram were continued to the south magnetic pole.

An example taken from an actual survey is shown in Fig. 14-7. This is from an area where the inclination is about 20° and where both a gravity and an airborne magnetic survey were carried out. The diagram on the left shows a simple, nearly circular positive gravity anomaly (the left half of the anomaly is offshore and was not observed). The magnetic map over the same area shows a strong negative anomaly to the north (upward on the map) and a positive anomaly to the south. The plus sign on each of the two diagrams shows the center of the gravity anomaly and the two x's on each figure show the centers of the magnetic negative and positive anomalies. Figure 14-8 shows profiles of the gravity and magnetic anomalies on a north-south line together with calculated effects from a heavy sphere and a magnetized sphere. The approximate agreement in size and location of the two spheres, with reasonable assumptions for the density and magnetic contrasts,

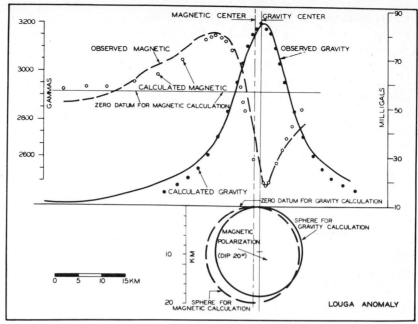

FIGURE 14-8

North-south gravity and magnetic profiles from Fig. 14-7 with calculated values and cross section of possible anomalous heavy magnetized body. (*From Nettleton, 1962; by permission of the American Association of Petroleum Geologists.*)

indicates strongly that these two anomalies come from a magnetized heavy mass with nearly the same boundaries for the magnetic and density contrasts.

MAGNETIC EFFECTS OF GEOMETRIC MODELS

Since the complexities of calculated magnetic effects for geometric bodies due to variations with direction of magnetization make any calculation of models to use as guides in interpretation much more complex than for gravity effects, such calculations now are made almost entirely by digital computers. There are a number of useful compilations of magnetic effects of geometric models.

Vacquier et al. (1951) give a series of 83 contoured anomaly maps which were laboriously calculated by hand before the advent of electronic computers. These maps show total magnetic intensity and its vertical derivative (or curvature) for different rectangular bodies with unit depth to top and infinite depth to bottom, with various orientations with respect to the earth's field, with various ratios of length to width, and with magnetic inclinations of 0, 20, 30, 45, 60, 75, and 90°. The calculations of the total intensity and its curvature are all for magnetization in the direction of the earth's field. Since their initial publication and their reprinting in 1963, these models and the accompanying theory and examples have been widely used in the interpretation of aeromagnetic surveys. A series of north-south profiles across the 6 by 8 depth-unit body is given by Fig. 14-9. The change in character of the magnetic effects is generally similar to that for a spherical body shown by Fig. 14-6 but different in detail. The curves also show profiles of the curvature of the magnetic field. This parameter is useful in magnetic-map interpretation and is discussed later.

Andreasen and Zietz (1969) have given another and much more complex series of model fields, calculated and contoured by an electronic computer and plotter. The entire series of 825 magnetic fields is for a single model with horizontal dimensions of 4 by 6 depth units, with thicknesses varying from 0.1 depth unit to infinity, and with various directions of the earth's field and directions of magnetization (which, in most cases, is different from that of the earth's field because of assumed remanent magnetization of the rocks). With these several parameters, the variety of situations becomes very large, and therefore these calculations are not as simply applicable to routine magnetic interpretation as those of Vacquier et al. (1951).

A compilation by Reford (1964) is for a series of north-south profiles over thin sheets (dikes). There are sets of curves for magnetic inclinations of 0 to 90° by 15° steps. For each set, curves are given for profiles over sheets with dips of 0, 45, 90, −45, and 0° and for strike directions of east-west, northeast-southwest (or northwest-southeast), and north-south. There is a wide range of magnitudes and slopes of the anomalies, depending on the three variables of magnetic inclination, dip of the sheet, and its strike direction relative to magnetic north. In particular, the anomaly near the magnetic equator (inclination zero) completely disappears in some cases.*

Bhattacharyya (1964) gives the theory and some examples for calculations of fields for prism-shaped bodies. The examples include the total field and its second vertical derivative for several bodies with the same boundaries as those of Vacquier et al. (1951) but with the polarization direction different from that of the earth's field, as would be the case for remanent magnetization of the block.

The Inclined Two-dimensional Dike

It has been mentioned just above that the magnetic effects from the dipping thin sheet are the basis for one method of direct interpretation. In other computer systems, also, the fundamental mathematical properties of the thin dike and its expansion to form thick bodies and semi-infinite bodies (faults) are of fundamental importance. Thus, the total magnetic field of the thin dike and its horizontal and vertical derivatives are fundamental to the theory of computer operations for

---

* There is much repetition in Reford's curves as the entire series can be derived by horizontal (left to right) or vertical inversion of the basic curves for polarizations between 45 and 90° (Bean, 1966, p. 963).

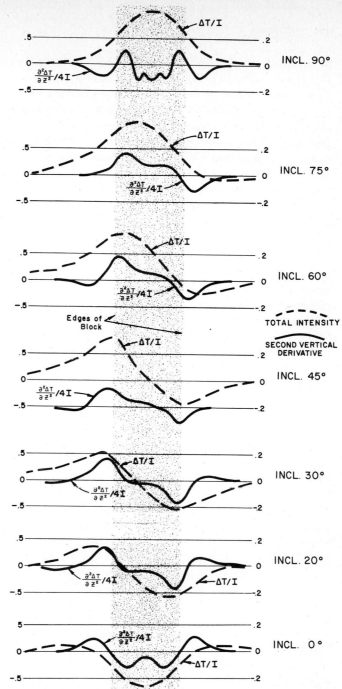

FIGURE 14-9
Profiles of magnetic and vertical derivative (curvature) fields on a north-south line across a prism with top at 1 unit depth, bottom at infinity, and for the various angles of inclination shown. All curves are for a body 8 depth units long (represented by the shaded area) and 6 units wide north to south. (*From Vacquier et al., 1951.*)

basement-depth determination, e.g., those of Hartman, Tesky, and Friedberg (1971) and of Bean (1966), but also for determination of location, depth, and form of individual magnetic bodies in mineral prospecting.

The mathematics of the magnetic field of a dipping body is given by Hutchison (1958). A family of characteristic curves, with certain normalizing factors, is plotted on transparent log-log paper, which can be laid over a log-log plot of the curve to be analyzed. When the characteristics of the field data are reasonably within the fundamental assumption (particularly the criteria of two-dimensionality) the depth, thickness, and slope of the body can be estimated. The paper also gives some commonsense comments on the commonly used depth-estimate methods.

Gay (1963) gives the theory for the thin dike in detail as the basis for a curve-matching system for determining depth, dip, and thickness of the source body. The standard curves to which the field curve is matched or for determination of body parameters are quite numerous.

The Hutchison and Gay papers are concerned primarily with the determination of geometric parameters of individual magnetic bodies. The basement-depth factors of locations of maxima, minima, maximum slope, half-slope, amplitude ratios, etc., are derivable from the equations of total intensity and its derivatives and are readily applied to computer operations. Thus, any attempt to develop a routine computer process for automatic depth determination probably will involve the basic mathematics given in these papers.

Relation between Magnetic and Gravity Effects

In the study and interpretation of magnetic maps it would be useful to have normalized form curves for simple bodies, such as spheres, cylinders, dikes, and faults, as we have for gravity (see Chap. 7). Because of the much greater complexity in the magnetic case, as pointed out in the previous section, such simple treatment is not possible. However, in spite of the wide range in form depending on magnetic inclination, the order of magnitude of magnetic effects can be approximated by calculations from vertically magnetized bodies of simple form. Furthermore, for vertical magnetization, the effects can be calculated quite simply from the relation between the gravitational and magnetic potentials. Also, it is common to have magnetic and gravity maps of the same area or to have simultaneously recorded gravity and magnetic profiles from marine surveys, so that it becomes important to understand the relations between gravity and magnetic fields of the same body.*

In the illustrations of the following section, curves are given for the vertical magnetic intensity of simple bodies. The illustrations also include curves for gravity effect and for the second vertical derivative of gravity ($\partial^2 g / \partial z^2$). These are given to facilitate comparison of magnetic maps with gravity maps and with the commonly available second-derivative maps (see page 140). The process of reduction to the pole gives a quantity corresponding to the first vertical derivative of gravity,

---

* The relationships are applicable to the evaluation of common features on the gravity and magnetic maps. Such common features must originate within the basement. We must remember that many gravity anomalies have their origin in density contents within the sediments and have no counterpart in magnetic effects.

which would give a curve corresponding to the vertical intensity of a vertically magnetized body and thus correspond to the magnetic-intensity curves of the figures. It is evident, from the wide divergence in form between the gravity curves and either the magnetic or second derivative of gravity, that a magnetic map will have much more in common with a map of the second derivative of gravity than with the gravity map itself. This, of course, is true only to the extent that the magnetic and density contrasts have approximately common boundaries.

Just as the gravitational force in a given direction is the derivative of the gravitational potential in that direction, the magnetic force in a given direction is the derivative of the magnetic potential in that direction. Poisson has shown that there is a simple relation between the gravitational and magnetic potentials for an important special case.

Let U be the gravitational potential due to the mass of a body with uniform density σ. If this same body is magnetically polarized uniformly in a direction i with an intensity of magnetization I, and if W is the magnetic potential and G is the gravitational constant, then, according to Poisson,

$$W = -\frac{I}{G\sigma}\frac{\partial U}{\partial i}$$

The magnetic force in any direction is the negative of the derivative in that direction of the magnetic potential. Thus the magnetic force in any direction S is

$$-\frac{\partial W}{\partial S} = \frac{I}{G\sigma}\frac{\partial}{\partial S}\left(\frac{\partial U}{\partial i}\right)$$

For the special case where the polarization is vertical (downward in northern latitudes) and we wish the vertical component V of the magnetic field,

$$V = -\frac{\partial W}{\partial z}$$

where z is measured vertically downward. Then

$$V = +\frac{I}{G\sigma}\frac{\partial^2 U}{\partial z^2}$$

Note that the magnetic field is one higher derivative than the gravity field.

In many areas where magnetic measurements are made, the vertical component is the principal part of the magnetic field. Therefore, a very simple and rough but often useful approximation is to consider the vertical magnetic effects of bodies vertically polarized. In this special case the effects can be derived quite simply from the gravitational effects.

The Sphere

The gravitational potential of a sphere (Fig. 14-10) is

$$U = \frac{Gm}{r}$$

and

$$\frac{\partial^2 U}{\partial z^2} = Gm\frac{2z^2 - x^2}{r^5}$$

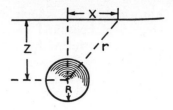

FIGURE 14-10
Vertical magnetic effect of a vertically
polarized sphere.

If the same sphere is uniformly polarized in the vertical direction, the vertical magnetic intensity is

$$V = \frac{I}{G\sigma} Gm \frac{2z^2 - x^2}{r^5}$$

$$= \frac{I}{G\sigma} G(\tfrac{4}{3}\pi R^3 \sigma) \frac{2z^2 - x^2}{r^5}$$

$$= \tfrac{4}{3}\pi R^3 I \frac{2z^2 - x^2}{(x^2 + z^2)^{5/2}}$$

$$= \frac{4\pi R^3 I}{3z^3} \frac{2 - (x/z)^2}{[1 + (x/z)^2]^{5/2}}$$

$$= Kf\left(\frac{x}{z}\right)$$

where
$$f\left(\frac{x}{z}\right) = \frac{1 - 1/2(x/z)^2}{[1 + (x/z)^2]^{5/2}} \quad \text{and} \quad K = \frac{8\pi R^3 I}{3z^3}$$

A plot of $f(x/z)$ is given by curve 1 of Fig. 14-11.

The similar normalized form curve for the gravity effect of a sphere has been added for comparison. Also, for later reference the normalized curve for the second derivative of the gravity effect of a sphere is shown.

The Horizontal Cylinder

The gravitational potential of a horizontal cylinder (Fig. 14-12) is

$$U = 2Gm \log \frac{1}{r}$$

where m is now the mass per unit length, and

$$\frac{\partial^2 U}{\partial z^2} = 2Gm \frac{z^2 - x^2}{(z^2 + x^2)^2}$$

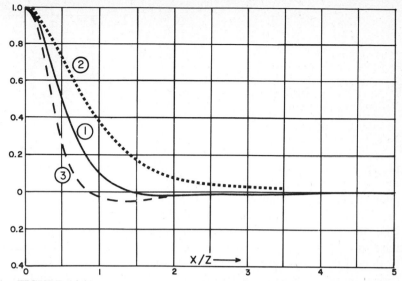

FIGURE 14-11

Curve for vertical magnetic effect on a north-south line over a vertically magnetized sphere (*solid line*), with curves for gravity effect (*dotted line*) and second vertical derivative of gravity effect (*dashed line*) shown for comparison.

If this same cylinder is polarized in the vertical direction, the vertical magnetic intensity is

$$V = \frac{I}{G\sigma} 2G\pi R^2 \sigma \frac{z^2 - x^2}{(z^2 + x^2)^2} = \frac{2\pi R^2 I}{z^2} \frac{1 - (x/z)^2}{[1 + (x/z)^2]^2} = K'f'\left(\frac{x}{z}\right)$$

where

$$f'\left(\frac{x}{z}\right) = \frac{1 - (x/z)^2}{[1 + (x/z)^2]^2}$$

which is plotted as curve 1 of Fig. 14-13, and

$$K = \frac{2\pi R^2 I}{z^2}$$

As for the sphere (Fig. 14-11), curves for the gravity and second derivative of gravity have been added for comparison.

The Fault

A very common anomaly is a low-relief feature attributed to a fault or edge of a plate or area of local relief on the basement surface. For vertical magnetization, and as approximated by a thin sheet, the anomaly is given (Nettleton, 1940, p. 212) as

$$2I\frac{t}{z}\frac{x/z}{(x/z)^2 + 1}$$

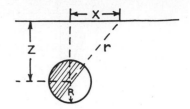

FIGURE 14-12
Vertical magnetic effect of a vertically
polarized horizontal cylinder.

The form of the anomaly curve is given by curve 1 of Fig. 14-14, and the total magnetic relief (ΔV in the figure) is $\Delta V = 2It/z$. As in the other figures, the curves for gravity and for second derivative of gravity are shown also. Here gravity and magnetic curves are quite different in character, as the gravity curve (for a positive density contrast) rises from zero to approach asymptotically a maximum value, while the magnetic curve is zero over the edge with maximal and minimal values on the flanks. The general character of the second-derivative curve of gravity is similar to that of the magnetic curve but a little steeper. In this case, even more than as shown by the curves for spheres and cylinders (Figs. 14-11 and 14-13), the comparison of the gravity second derivative with a magnetic map would be much more effective than comparison of gravity and magnetic maps for indicating features with common boundaries.

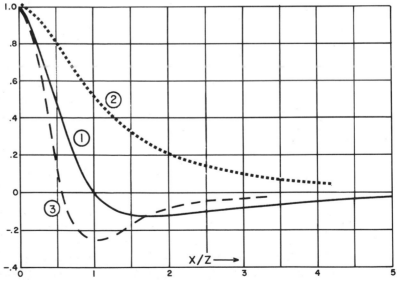

FIGURE 14-13
Curve for vertical magnetic effect on a north-south line over a vertically magnetized horizontal cylinder (*solid line*), with curves for gravity effect (*dotted line*), and second vertical derivative of gravity effect (*dashed line, Elkins, 1951*) shown for comparison.

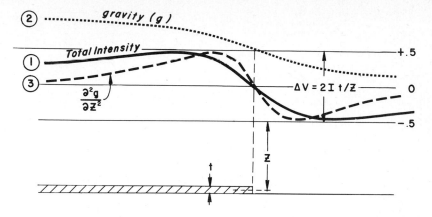

FIGURE 14-14

Curve for vertical magnetic effect on a north-south line over magnetized thin sheet or fault (*solid line*), with curves for gravity effect (*dotted line*), and second vertical derivative of gravity effect (*dashed line, Elkins, 1945*) shown for comparison.

This comparison is valid only for relatively high magnetic latitudes where the magnetization angle is greater than around 45° (see Fig. 14-6) but, of course, is valid for any angle of magnetization after reduction to the pole (see Fig. 15-4).

End Corrections

The foregoing expressions for magnetic effects of the horizontal cylinder and fault are for two-dimensional models. Also many of the algorithms for computer determination of magnetic effects from geometric models assume infinite length perpendicular to the plane of the calculation. Such effects, of course, are larger than they would be if the length in the third dimension were limited.

In the similar gravity case, the determination of an end correction to estimate this effect is quite simple (page 207 and Fig. 7-14). For the magnetic case it is much more complicated because of effects of direction of magnetization and of length-to-width ratios for the body.

Figure 14-15 gives curves for two special cases which are of some help in estimating the amount of deficiency when models are of less than infinite length. Both are on a north-south profile across the center of an east-west rectangular body with infinite depth to the bottom. One is for a body with width of 1 depth unit, the other for a width of 2 depth units. The calculations are for total intensity at 90° magnetic latitude and vertical polarization. The curves would be different in detail but of similar form for other magnetic latitudes.

As for the gravity case, the curves of T/T_∞ are for the ratio of the effect of a model of finite length to that of one of infinite length. The curves for the two

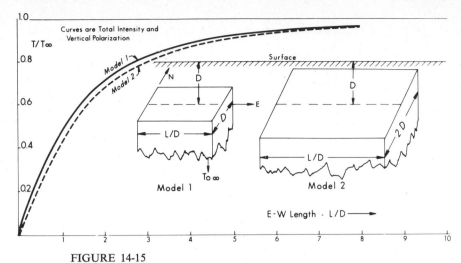

FIGURE 14-15
End correction for estimating deficiency of effects for a body of finite length compared with one of infinite length. The curves show the ratio of the magnetic effect of a body of finite length to that calculated for finite length (two-dimensional approximation).

models are not very different, and we see that when the total length perpendicular to the profile is greater than about 3 times the depth, the magnetic effect is more than 80 percent that for infinite length.

COMPARISON OF GRAVITY AND MAGNETIC QUANTITIES

It has been noted that since the magnetic effects are one higher derivative than the gravity effects, the quantities in the numerator and denominator of the expressions for the magnitudes (the amplitude factor k) are of the same order, that is, R^3/z^3 for the sphere, R^2/z^2 for the horizontal cylinder, and t/z for the fault. Therefore the magnitudes of the magnetic effects do not depend on the scale, as they do for gravity. Thus the magnitude of a magnetic anomaly for a small body at a shallow depth is the same as that of a large body at a large depth as long as the ratios of dimensions are the same. Only the horizontal scale of the anomaly will be changed.

The fact that magnetic effects are one derivative higher means that the characteristic magnetic curves are considerably sharper than those for gravity effects. This is illustrated by the normalized curves for gravity effects (curves 2), which have been added to the charts for the magnetic form curves for the sphere, cylinder, and fault. The second-vertical-derivative curve for gravity has also been added to the figures because second-derivative maps are often made for gravity surveys as an aid in interpretation (as discussed in Chap. 6).

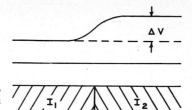

FIGURE 14-16
Maximum vertical magnetic anomaly, resulting from a vertical contact between rocks with polarization I_1 and I_2.

LIMITING MAGNITUDES OF MAGNETIC QUANTITIES

Present magnetic anomalies are caused by the present state of polarization of the rocks however it may have been acquired. Because of our uncertain knowledge of the contributions of remanent polarization and the indications that it is smaller than the induced polarization, especially for the older rocks, about all we can do is estimate the polarization as that calculated from the susceptibility and the present strength and direction of the earth's field. On the basis of the foregoing estimates of magnetite content (page 361) a general figure for probable polarization of an average igneous rock is around 0.001 to 0.002 cgs unit. Its direction is probably near that of the present earth's field (or its reversed direction). However, it must be remembered that in any particular case the values for polarization may differ widely from this figure, which is particularly true in some of the applications of magnetic work to mining problems in which iron-bearing ore bodies are of interest. As an extreme example, the famous Kursk magnetite ore body in Russia has a susceptibility estimated as 2.7 and a magnetic anomaly of 200,000 γ.

For simplicity the few examples which follow are calculated for vertical polarization. However, as is shown by the theoretical curves for the sphere (Fig. 14-6) and the rectangular block (Fig. 14-9), the radical changes in shape of the anomaly curves resulting from differences in direction of the magnetic field are accompanied by magnitude changes of only about 2:1. Therefore, the approximations of magnitudes calculated for vertical polarization are generally applicable in spite of the variations in anomaly form resulting from the worldwide variation in the angle of inclination of the polarization.

Intrabasement Block

In large-scale magnetic prospecting for oil it is seldom that even very large and broad anomalies have relief of more than 1000 γ. The greatest possible magnetic relief would be produced by a sharp contact of two large masses of rock with different polarizations and with infinite depth to bottom (Fig. 14-16). The relief so produced (for vertical polarization) would be

$$\Delta V = 2\pi(I_1 - I_2)$$

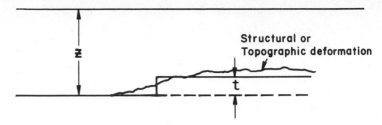

FIGURE 14-17
Simulation of low-relief basement surface feature or structure by edge of faulted block.

Therefore, if we say that the regional magnetic anomalies seldom have a relief greater than 1000 γ, we can calculate that a common upper limit for $I_1 - I_2$ is given by

$$I_1 - I_2 = \frac{0.01}{2\pi} = 0.0016$$

Because the probable range of polarizations is quite large, we should expect the individual values of I_1 and I_2 to be of the same order of magnitude as their difference. Therefore, the foregoing consideration also indicates a common order of magnitude for polarizations and polarization contrasts of igneous rocks in the neighborhood of 0.002.

Suprabasement Block

The magnetic relief produced by topography or by structure which deforms the basement surface without a change in material is very much smaller than that due to intrabasement contacts. If the deformed area is broad compared with its depth, the magnetic effect is mostly over the edge and can be approximated by a step, or fault, as indicated by Fig. 14-17. The total relief of the magnetic anomaly for vertical polarization (page 385 and Fig. 14-18) is

$$2I\frac{t}{z}$$

In this case, the magnetic-polarization contrast is the total magnetization I, since the contrast is that of the uplifted basement with respect to the nonmagnetic sediments. For a basement with polarization of, say, 0.001 cgs unit and for an uplift or displacement of one-tenth the depth, the magnitude of the anomaly is

$$2 \times 0.001 \times 0.1 = 0.0002 = 20 \ \gamma$$

This is quite a small anomaly compared with the anomalies of hundreds to a 1000 γ or more resulting from intrabasement sources. This is illustrated by Fig.

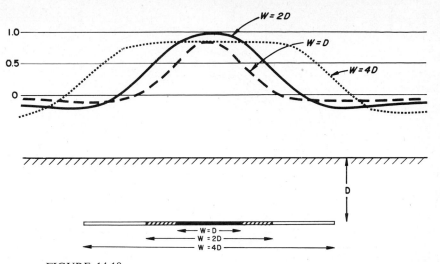

FIGURE 14-18

Curves for north-south profile of the vertical component for vertically polarized strips of different widths.

13-1, where the basement uplift A produces a much smaller effect than the intra-basement fault and dike. For relief up to about one-fourth the depth, the supra-basement anomaly magnitude in gammas is approximately

$$20 \frac{I}{0.001} \frac{t}{z}$$

The surface, or structure, anomalies may be hard to recognize particularly if the basement surface deformation results from geological deformation along a boundary which corresponds with an intrabasement magnetic contact. Then the low-relief anomaly due to the surface deformation is superimposed on a very much larger relief intrabasement effect and can be recognized, if at all, only by very subtle deviations of the anomaly curve.

The forms and amplitudes of the anomaly curves for platelike bodies not infinitely wide vary widely, depending on the width of the body. They can be calculated quite simply by adding two curves like curve 1 of Fig. 14-14, one positive and one negative and displaced by the width of the body. Curves so calculated for · vertical polarization and for widths of 1, 2, and 4 times their depth are given in Fig. 14-18. The magnitude from the curve multiplied by $2It/z \times 10^5$ gives the anomaly magnitude in gammas.

An example calculated to illustrate the relative contributions to a magnetic anomaly from basement-magnetization changes and from basement structure is shown by Fig. 14-19. This has been calculated (for vertical polarization) for a cylindrical body extending to great depths, on which are superposed the calculated

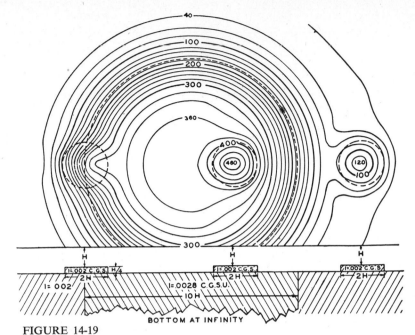

FIGURE 14-19

Contours of calculated magnetic effects, to show variation of appearance of the effect of a given basement feature depending on its relation to boundaries of an intrabasement polarization contrast.

effects of several flat, cylindrical disks, intended to simulate structure in the basement surface. It is evident that the appearance of the picture given by the structures depends on their position relative to the larger regional feature. If this regional is properly estimated and removed, however, the local features will show up in their proper form and relief.

15

THE INTERPRETATION OF MAGNETIC MAPS

The discussion in this chapter is in terms of airborne magnetic profiles and maps. The same principles are applicable to maps made from ground magnetometers and profiles made from them. The improvement in continuity and precision of continuously recorded, high-sensitivity airborne profiles over those which can be made from point-by-point ground observations or from contours made from them is so great that the latter are of only historical interest or used by necessity as the only data available.

In the application of magnetic prospecting in the petroleum industry, the objective, in nearly all cases, is the determination of the depth to the basement. This application is, of course, based on the fact that the magnetite content of the basement rocks is so much greater than that of the sediments that, for practical purposes, the airborne magnetic record can be considered the same as it would be if the sediments were not present.* Therefore, in most applications to oil exploration, the individual magnetic anomalies are not of interest in themselves but only as they contribute to details of the map which aid in the calculation of depths to

---

* In some cases certain filtering or selective processes may be applied to remove instrumental noise or shallow effects from intrasedimentary sources.

magnetic sources. The depths are nearly always calculated directly from the magnetic record as made in the airplane or a computer reproduction of that record made from a digital recording in the airplane instrumentation.*

MAGNETIC MAPS

The interpretation, of course, begins with the observed field, which is usually the total-intensity map.† In the earlier airborne magnetic maps there often were many local irregularities caused by errors in the recording or in the mapping of the positions of the flight lines. These often led to herringbone contours on the map, caused either by the datum of a line or a portion of a line being in error or by a line being inaccurately located. Modern airborne maps are almost completely devoid of such errors, as they are eliminated by the much more accurate locations mapped by electronic positioning systems or by systematic reduction of tie or closure errors in the data-reduction process.

Second-Derivative Maps

It is common practice in magnetic mapping to prepare a second derivative from the observed total-intensity or vertical-intensity map. The procedures used and the variations depending on the choice of coefficients or weighting factors have been discussed in some detail in connection with their application to the interpretation of gravity surveys (pages 140 to 144). The second-derivative map serves much the same purpose as in its application to gravity surveys, in that it acts as a filter, emphasizes the expressions of local features, and removes the effects of large anomalies or regional influences. The principal usefulness of such maps in magnetic interpretation is to indicate the outlines of individual intrabasement blocks or the edges of suprabasement disturbances. These indications serve to suggest features on the individual flight-line profiles which can be used for basement-depth determinations and, in particular, to determine the strike of the edges of the bodies and the angles between the flight lines and those edges which are used to make the azimuth correction to the depth determinations (see page 411).

---

* A great deal of magnetic exploration has been carried out by the mining industry, where the objectives and interpretation processes are quite different from those in the petroleum industry. In mining exploration the objective usually is the location and delineation of a magnetic ore body. Then the interpretation procedures may be concentrated on using special graticules and charts or computer processes from which the depth, thickness, and dip of an ore body can be determined. Such procedures commonly use models of geometric forms from which magnetic effects are computed and compared with those observed (see, for instance, Grant and West, 1965, Chap. 11).

† The older surveys made with mechanical magnetometers on the ground, of course, measured the vertical component of the magnetic field. In magnetic latitudes higher than about 60°, which includes all except the southwestern part of the United States, the qualitative difference between the vertical-component and total-intensity maps is relatively small. At lower magnetic latitudes these differences become very conspicuous, and near the magnetic equator the vertical- and total-intensity maps look entirely different.

As in their application to gravity maps, the appearance of second-derivative maps varies widely, depending on the spacing of the grid system used and the weighting coefficients. The airborne magnetic maps, being based on data 1000 ft or so above the ground, may be much smoother (if the mapping is accurate) and not subject to small superficial disturbances such as often occur in gravity maps where the surface material is not homogeneous. This means that for magnetic maps second derivatives with sharper filters, such as equations (2) and (3) of Table 6-1, can be applied to magnetic maps. This is because they are relatively free of local disturbances, so that the small differences within the calculation pattern are reliable, whereas, for a gravity map, the principal effect may be to emphasize irregularities due to instrument or elevation errors or inhomogeneities in the shallow subsurface.

Continuation Calculations

The general nature of continuation calculation and an example of downward continuation for gravity interpretation are given on page 144. Such calculations have been applied more in airborne magnetics than in gravity interpretation.

Upward continuation is used in magnetic interpretation as a filter to remove anomalies from shallow sources. It has also been used to compare the field continued upward to a higher level with that measured at that level. When the calculation is made from an adequately mapped field, the continued and measured fields at the higher level are almost identical.

Downward continuation can be used to sharpen anomaly definition and to separate sources with overlapping effects in much the same way as described in the gravity example. Although the quantity mapped is mathematically different, it has much the same general effect in localizing anomaly boundaries and trends as a second-derivative or grid-residual calculation.

All such calculations or high-pass filters tend to sharpen errors or near-surface disturbances. However, since airborne surveys are made at some distance (commonly around 1000 ft) above the ground, they can be operated upon by sharper coefficients (or narrower filters) without degenerating into a result dominated by errors or shallow-source interference, which would result if the same operations were applied to measurements at the ground surface.

All operations of continuation, second derivative, filtering, and reduction to the pole (page 139) are based on some form of sampling, usually a regular grid of values from a map. The result is very much dependent on the adequacy of this sampling. The discussion of this and other factors of filtering in connection with gravity interpretation in Chap. 6, particularly pages 139 to 143, is equally applicable to magnetic data.

Reduction to the Pole

Another mapping procedure helpful in the interpretation of magnetic anomalies is "reduction to the pole," first developed in France by V. Baranov. This somewhat involved mathematical procedure is carried out on a grid of values written from a contour map or computer-derived by mathematical interpolation, similar to that of contouring procedures, from observed values not on a grid. Its purpose is to correct for the variation in the appearance of an anomaly depending on its magnetic latitude and the corresponding variation of the dip angle of the magnetization vector in the body. The result of the operation is to produce a "pseudo-gravity anomaly," so called by Baranov.*

The principles of the method and the mathematical development are given in considerable detail by Baranov (1957). In examples of the coefficients used and the development of their numerical values given by Baranov and Naudy (1964) the grid is on a trigonal pattern, so that the successive distance circles at which input values are read and reduced are calculated on circles with equal radial increments. The numerical coefficients used are variable, of course, depending on the angle of inclination, and different sets must be derived which are applicable within a limited range of inclination angles. Figure 15-1 shows the magnetic-contour pattern for a sphere magnetized in the earth's field at an angle of 60° and a corresponding pattern for the same anomaly reduced to the pole. It will be noted that the asymmetry of the original anomaly is completely removed and that the reduced anomaly is a set of circles corresponding to the magnetic field of a sphere magnetized vertically. Figure 15-2 shows a similar set of original and reduced magnetic anomalies, this time for a rectangular body with infinite depth to the bottom.

Examples of the numerical calculation of coefficients, for a trigonal grid, are given by Baranov and Naudy (1964). The coefficients are for an inclination of 16°30′. Application to a theoretical sphere is shown in Fig. 15-3. An application to magnetic and gravity data from the west coast of Senegal is shown in Fig. 15-4.†

The calculations for reduction to the pole can be carried out quite readily in the wave-number domain. The operation also uses, as input, magnetic field values on a uniform grid of points. These are transformed into the wave-number domain in a manner similar to that described earlier in relation to gravity-anomaly separation (page 157). The variations due to different dip angles are taken into account

* The term pseudo-gravity anomaly is unfortunate. As pointed out below and also in Baranov's publications, the anomaly produced is similar in location relative to the body and in general form to the gravity anomaly which would be produced by a body having a density contrast at the same boundary at which the magnetic contrast occurs. However, the characteristics of the anomaly are those of a magnetic rather than a gravity feature and correspond to the vertical derivative $\partial g/\partial z$ of gravity and therefore are labeled as pseudo-gravimetric gradient in Baranov's figures (see Figs. 15-1 and 15-2). For these operations the word "pseudo-gravimetric" should not be used without including "gradient" or its equivalent. It would be still better if no reference to gravity were made at all and reduction to the pole or equivalent vertical intensity were used to describe the result of this operation. The characteristic curves are distinctly sharper than those for gravity. This is shown by Figs. 14-11, 14-13, and 14-14, the sphere, horizontal cylinder, and fault, respectively. As discussed above (page 116), this means that comparison of gravity and magnetic maps (or as reduced to the pole) are better if made with a second-derivative map of gravity than with the gravity map itself.

† This application is to the same area which was used for a comparison of gravity and magnetic maps (Figs. 14-7 and 14-8) but apparently from different maps of the original data, as some of the details are slightly different.

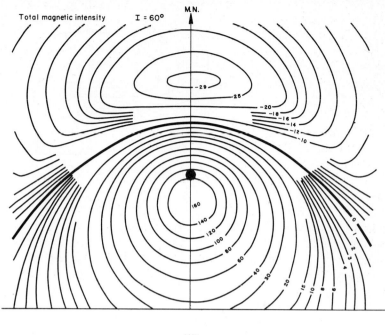

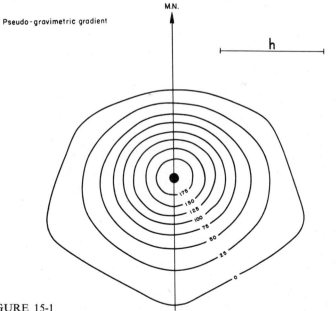

FIGURE 15-1

Total magnetic intensity for a sphere magnetized in the earth's field with inclination 60° (*upper*) and the same field reduced to the pole (*lower*). The length of the bar *h* is the depth to the center of the sphere below the heavy dot. (*Modified from Baranov, 1957.*)

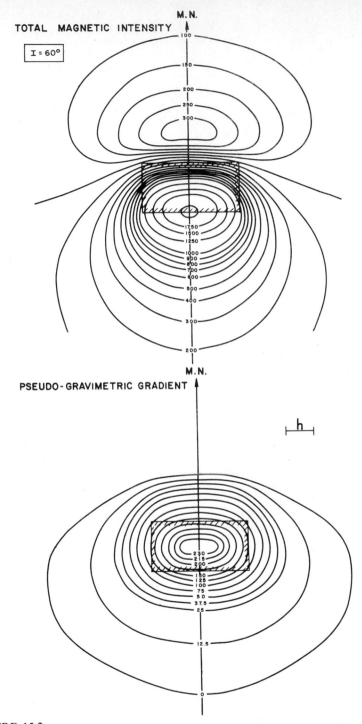

FIGURE 15-2
Total magnetic intensity for a rectangular block with base infinitely deep, magnetized in the earth's field with inclination 60° (*upper*) and the same field reduced to the pole (*lower*). The bar *h* is the depth to the top of the block. (*Modified from Baranov, 1957.*)

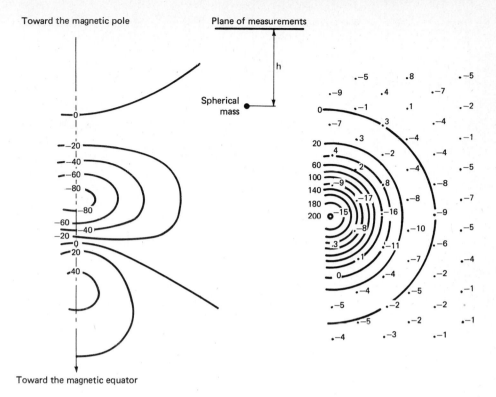

FIGURE 15-3
Total magnetic intensity for a sphere magnetized in the earth's field with inclination 16° 30′ (*left*) and the same field reduced to the pole (*right*). The vertical bar *h*
is the depth to the center of the sphere. (*From Baranov and Naudy, 1964.*)

by using appropriate factors in the Fourier transform, which are readily included
in the computer programming. Some practitioners prefer the wave-number system
as being simpler than the method described by Baranov. The mathematics of
application to two-dimensional profiles in terms of the Hilbert transforms is given
by Shuey (1972).

   With modern computer techniques the application of the rather involved
mathematical procedure for reduction to the pole (either in the spatial or frequency
domain) becomes quite practical. The procedure makes for considerable simplification of the magnetic maps in low magnetic latitudes, as illustrated by Figs. 15-3
and 15-4. Comparison of such maps with gravity maps of the same area should be
much more straightforward. Again it should be kept in mind in such a comparison
that if density and magnetic contrast have the same boundaries, such a comparison
is of value but not otherwise.

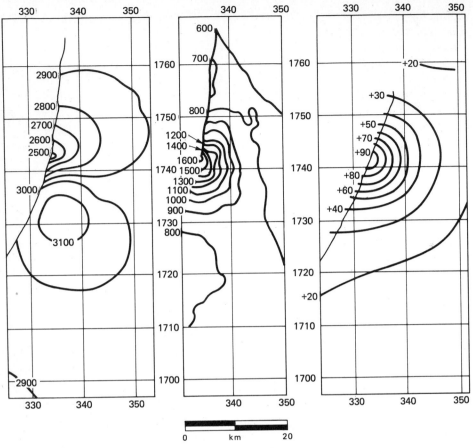

FIGURE 15-4
Total magnetic field (*left*), field reduced to the pole (*center*), and gravity field (*right*) over an area on the east coast of Senegal. The same area is used for the comparison in Figs. 14-7 and 14-8. (*From Baranov and Naudy, 1964; by permission of the Society of Exploration Geophysicists.*)

DETERMINATION OF BASEMENT DEPTHS

The primary purpose of magnetic exploration in the petroleum industry is for the determination of the depths to the basement surface. As has been mentioned, most basement rocks have more magnetite and are more magnetic than nearly all sedimentary rocks. In a general way (and subject to many minor and some major exceptions) the magnetic effect recorded by an airborne magnetometer (and to a lesser degree by magnetic measurements on the ground surface) is substantially the same as it would be if the sediments were not present. Therefore, determining the depth to the source of the recorded magnetic anomalies is generally equivalent to determining the thickness of the sedimentary section. Since oil occurs only in sedimentary rocks, a reliable determination of the depth to the basement rocks gives a measure of the volume of sediments available in a given basin and is a first limitation on its potential as a source of oil.

With modern magnetic surveys of high precision and close control, magnetic measurements can be complementary to seismic data in the interpretation of a body of geophysical data in a given area. This is especially true in complex areas in relatively deep basins, where reflection seismic surveys are subject to alternative interpretations. With a complex pattern, reflections from the basement may be difficult to recognize, and a basement profile from a careful interpretation of magnetic data may be very helpful in identifying a basement reflection.

With adequate magnetic surveys it is often possible to determine basement depths in sufficient detail for local structure on the basement surface to be reliably delineated. In addition to individual determinations of depths to the magnetic source this may involve modeling of an individual basement feature and determination of its major geometric parameters. If the geological deformation which caused the local basement feature, either originally or by rejuvenation, is later than the time of deposition of the overlying sediments, the basement disturbance may have caused deformation of those sediments which could be a major factor in oil accumulation. Some examples of such interpretations are given later (pages 413 to 422).

Some of the methods of calculation of basement depths are quite old. The pioneer work by Peters (1949) published a rather complete system of quantitative magnetic interpretation which involved downward continuation to the basement surface and an estimation of local relief of that surface.* Peters' paper introduced the concept of the measurements of maximum-slope and half-slope distances as depth criteria. Since these methods have been very widely used, their history, uses, and limitations are discussed in some detail.

SLOPE AND HALF-SLOPE METHODS FOR ESTIMATING BASEMENT DEPTHS

The principles of the half-slope and maximum-slope distance parameters are indicated in Fig. 15-5. The application to a portion of an actual magnetic profile is shown in Fig. 15-6.

On a magnetic profile (which may be a portion of the actual record from an airborne survey or drawn from contours on a map) a line of maximum slope is drawn through the point of inflection O (Fig. 15-5). A drafting triangle is set with one-half of this maximum slope and slid along another triangle to determine the

---

* Peters (1949) reviewed interpretation processes which had been in use by the Gulf Research and Development Company beginning in about 1930. It is of interest that Peters' acknowledgment of those contributing to the development mentions particularly John Bardeen; this was very early in the career of an outstanding scientist who is now the only person to have received two Nobel prizes for physics.

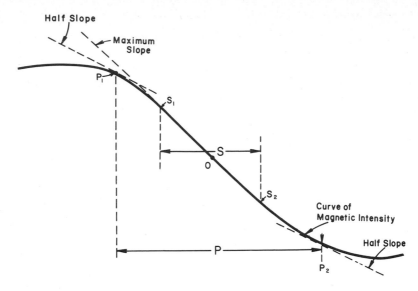

FIGURE 15-5
Idealized magnetic profile showing measurements for slope (S) and half-slope (P) parameters used for basement-depth estimation.

points of tangency P_1 above and P_2 below the point of inflection. The half-slope parameter P is the horizontal distance between these two points.

It is quite usual in actual practice to find that the maximum slope coincides very closely with the magnetic profile for a certain distance and then breaks away from the straight line at consistently measurable positions, S_1 and S_2. The distance S between these points is the slope parameter.

The principal objective of Vacquier et al. (1951) was to determine the factors by which these and some other parameters should be multiplied to give the depth to the magnetic source. The system is very straightforward: a series of total-magnetic-intensity and second-derivative contour maps was calculated for prismatic bodies of rectangular form with their tops at unit depth and bottoms at infinity. The lengths of the interpretation parameters could then be measured on the calculated maps and the ratios of these lengths to the unit depth determined. From these measurements and from extensive practice in their application it has been determined that, as a general rule, the depth to the prism source is approximately equal to S and also approximately equal to $P/2$.

The application of the principles outlined above is complicated by the nature of the source and by interfering bodies. The above discussion has been entirely for magnetic sources reaching to infinite depth. In Fig. 15-7 magnetic curves are given for simple prismatic bodies with unit depth to the top and with varying depths to the bottom. For curves A, B, and C, the essential shape of the magnetic curve is nearly the same but with the amplitude changing from about 500 γ to 150 as the

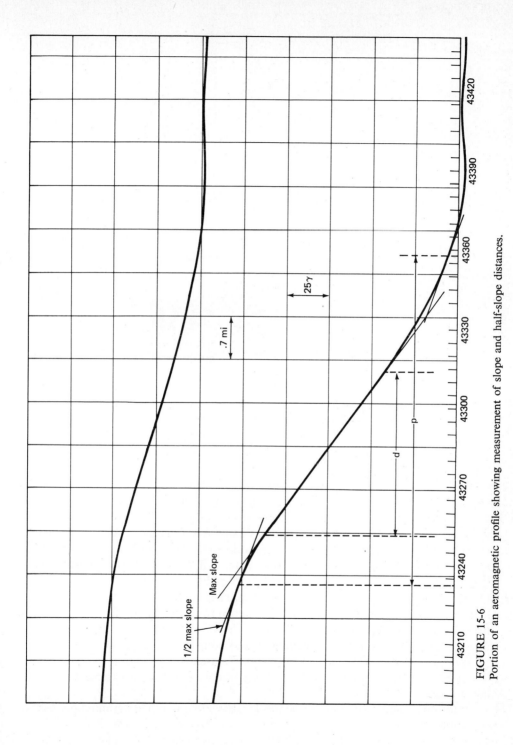

FIGURE 15-6

Portion of an aeromagnetic profile showing measurement of slope and half-slope distances.

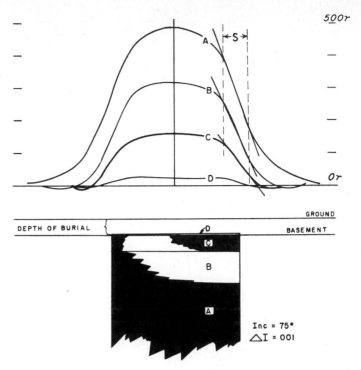

FIGURE 15-7
Calculated magnetic effects for body with variable depth to bottom. Also shown is slope measurement S for determination of depth. (*From Steenland, 1965; by permission of the Society of Exploration Geophysicists.*)

depths to the bottom decrease from infinity to twice the depth to the top. Curve D is for a body with a thickness of only one-tenth of its depth of burial.

For the three curves A, B, and C the slope parameter S is substantially the same, but for the curve D the distance S is smaller and a larger depth factor must be used. This relationship has been discussed in some detail by Steenland (1962). He calls thick basement blocks, corresponding to A, B, or C of Fig. 15-7, intrabasement sources. Thin blocks, corresponding to D of Fig. 15-7, which really amount only to irregularities on the basement surface, such as faults or edges of uplifts, are called suprabasement sources. The difference in character of anomalies from intrabasement and suprabasement sources is shown in Fig. 15-8. By recognizing the low-amplitude suprabasement anomalies and using the proper depth parameters it is possible to determine many more points of basement-depth control and thus materially increase the amount and reliability of data on which a basement-depth map is made.

The slope and half-slope principle for basement-depth determination has been applied very widely and, with experience, can be quite effective. The parameters are not really constant, as is shown in Vacquier et al. (1951, plate I) by the

N–S AEROMAGNETIC PROFILES
CROSSING 2 X 6 E–W BLOCKS

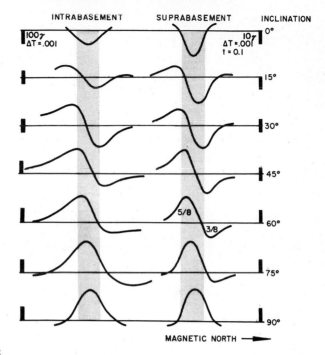

FIGURE 15-8

Calculated magnetic profiles over an intrabasement prism (bottom at infinity) and suprabasement source (thickness one-tenth of depth). Note that the scale on the left is 10 times greater than that on the right. (*From Steenland, 1962; by permission of the Society of Exploration Geophysicists.*)

curves for several parameters as functions of the length and width of the prismatic bodies and of the angle of polarization. Some of the parameters vary by a factor of over 2:1.* However, these apparent variations, particularly those of the rather subjective slope measurement, have not been clearly established, and the indicated variations have been largely ignored in practice.† However, some practitioners have their own empirical rules for depths to structure anomalies, depending on

---

* Peters (1949, p. 310) gives the depth as approximately $P/1.6$ but with variations from $P/1.2$ for very narrow bodies to $P/2.0$ for bodies infinitely wide. Am (1972, fig. 4) gives a range of 0.5 to 2.0 for width-to-depth ratios from 1.0 to 10.0. The curve is nearly linear from a value of 0.8 for a ratio of 1.3 to 1.6 for a ratio of 3.2.

† Vacquier et al. (1951, plate II) gives curves for another set of depth parameters based on the calculated second-derivative maps. Since in practice measurements are usually made directly on the aeromagnetic record of total-intensity variations, there appears to be little interest or application of these characteristics to basement-depth estimation.

Standard deviation 6.5%

Average error 7.4%
+ 9.0%
− 4.2%

Number of cases 169
+ 78
− 74
0 17

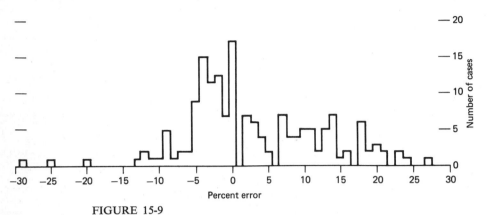

FIGURE 15-9

Histogram of errors in basement-depth estimates, Peace River, Alberta, at 169 wells drilled after magnetic survey and interpretation. (*From Steenland, 1963; by permission of the Society of Exploration Geophysicists.*)

judgment of the nature of the anomaly and particularly the probability of the apparent slope distances being shortened by interference.

To a considerable extent, the determination of basement depths by slope and half-slope measurements and by use of somewhat variable parameters for different situations becomes an art rather than an exact science. The slope distance (S in Fig. 15-5) has no mathematical or theoretical reality. If it did, the horizontal derivative of the magnetic-intensity curve would be constant over that distance and would have a more or less definite change of character where the slope changes; but such is not the case. In spite of this, however, an experienced interpreter will make consistent measurements and will determine reliable basement depths from such measurements, as illustrated by the following examples.

An opportunity for comparison of basement depths determined by the methods outlined above with drilling depths was provided by a fairly detailed and adequate magnetic survey over the Peace River area of northwestern Alberta (Steenland, 1963). An interpretation was originally made in 1950 over an area of some 27,000 square miles and a basement map constructed. The map was based almost entirely on depths determined from slope and half-slope distances and parameters for intra-basement sources. At the time of the survey a few drilling contacts on the basement were available for general control and checking of the interpretation. After the

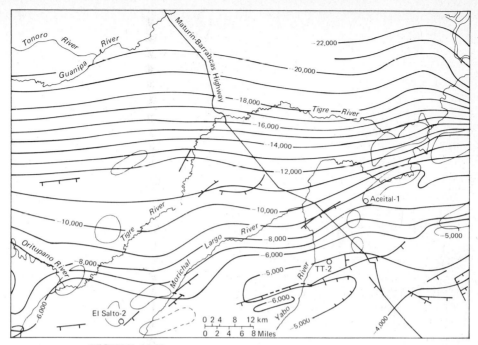

FIGURE 15-10

Basement map, eastern Venezuela Basin, from magnetic interpretation. (*From Jacobsen, 1961; by permission of the Society of Exploration Geophysicists.*)

survey and interpretation were carried out, the area became an important oil and gas province and a great many wells were drilled either actually into the basement or so far into the sedimentary section that a reasonable extrapolation to the depth of the basement could be made. There were 169 new wells, and each such well provided a comparison between the actual depth of the basement and that predicted from the magnetic analysis. Figure 15-9 is a histogram that shows the differences between the basement depth from the magnetic analyses and that actually found by drilling. The differences are expressed in percent of the total basement depth from the magnetometer flight level.

The histogram is unsymmetrical, with more values to the right of the center (magnetic depths too shallow) than would be expected from a normal distribution curve. This is now thought to be due to using some depth indices attributed to intrabasement sources when those for suprabasement sources should have been used.

Another comparison to demonstrate the effectiveness of magnetic basement mapping in the Orinoco Basin in eastern Venezuela is available from published material (Jacobsen, 1961). In this case the survey was made, in part, to evaluate the method. The results of a basement-depth interpretation are shown by the map of Fig. 15-10. As control on the interpretation, basement depths were provided at

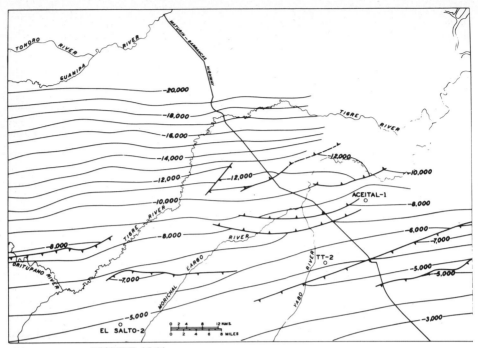

FIGURE 15-11
Basement map, eastern Venezuela Basin, from subsurface and seismograph
control. (*From Jacobsen, 1961; by permission of the Society of Exploration
Geophysicists.*)

the three drilling locations shown. The survey itself was not well planned, as the
primary control was on east-west lines with north-south tie lines in an area where
the tectonic strike and magnetic trends are nearly east-west. As a result, the basement-depth determinations depended, to a considerable extent, on the tie lines,
which were much more nearly perpendicular to the tectonic trends but, of course,
more widely spaced.

The principal features shown by the basement map are (1) a northerly dip
of the basement surface from depths of around 4000 ft in the southeast part of the
map to 22,000 ft at its northern edge; (2) a fault with about 1000 ft displacement
in the southeastern part of the map and with its downthrown side to the south; i.e.,
it is an antithetic fault downthrown in the direction of the regionally rising basement surface; (3) a number of smaller fault indications shown by lines with
hachures on their downthrown sides; and (4) a number of subcircular or elongated
outlines indicating areas of possible uplift on the basement surface.

The actual basement surface is shown in Fig. 15-11. These contours were
derived from extensive drilling in the southern half of the basin and from refraction
and reflection seismograph work, which extended the basement surface northward
beyond the area where it was defined by drilling. This information was not available to the interpreters making the map of Fig. 15-10. This map, of course, is not

identical with the one from the magnetic interpretation, but when depths are compared, the differences are rather small. Some specific features which may be compared are:

1 Where the 17,000-ft contour from the basement interpretation crosses the highway, the actual basement contour is at about 18,000 ft.

2 The fault just south of the TT-2 well has almost the identical position where it crosses the highway, nearly the same strike, and approximately the same 1000 ft of displacement to agree quite closely, except in minor detail, with the fault interpreted from the magnetic survey.

3 The more local features (faults and local uplift) generally are not confirmed definitely by the basement map although about half of the faults are near, parallel with, and have displacements in the same direction as faults shown by the map of the actual basement surface.

OTHER METHODS OF DEPTH ESTIMATION

While the slope and half-slope distance measurements may be the most widely used at the present time, in the geophysical literature there are many other systems for basement-depth or model-depth estimation. The general principles of selected examples of these systems are described in this section.

The systems described are based on the properties of magnetic curves calculated for geometric bodies. Some of these have been developed for determination, in some detail, of the dimensions and depth of ore bodies, and others have been developed primarily for estimation of depth to the top of the body. This, of course, is the basement depth if the top of the magnetic body is at the basement surface (which is usually the case).

Am (1972)

This very extensive paper gives a mathematical review of magnetic effects of the thin and thick dike and derives characteristics which can be applied to depth estimation. Thick dikes to and including a half-sheet or fault are obtained by integrating the thin-sheet anomaly. The paper includes a series of normalized curves for various angles of polarization and of body width-to-depth ratios. These are similar to those of Gay (1963), who also gives a set of standard curves for interpretation.

Am's paper also reviews the usefulness and limitations of the "depth estimators" of several other systems, and, quite significantly, this section begins (Am, 1972, p. 80) with the statement, "No ideal depth estimator has been found." Several charts show wide variation of depth parameters with body width-to-depth ratios and polarization angles. A simple statement of the problem is quoted: "Some of these methods are sound in theory but unsuitable for general use. Others, although not really sound in theory, often give satisfactory results." A good example is the straight-slope method. As Am states (p. 69), "After twenty years, this work [Vacquier et al., 1951] perhaps still presents the best approach to the problems of making depth estimates from aeromagnetic maps."

Comments on several of probably the most useful depth-estimation systems are given in the following paragraphs.

Smellie (1956)

This method uses theoretical anomalies for single pole, line of poles, single dipole, and line of dipoles. Ratios of distances, e.g., from peak to half-width of each side, are used. Curves for variation with angle of polarization allow application to a complete range of magnetic latitudes. Application to test anomalies were on magnetic ore bodies at shallow depth and gave good results. The forms used would seem poor approximations for broad anomalies, usually the type available for routine depth estimates.

Bean (1966)

Bean has given a graphical method of depth estimation based on inflection and half-slope points. The basic measurements made are indicated in Fig. 15-12. The upper part shows the essential parameters for a two-sided anomaly. The lower part shows them for a broad, or one-sided, anomaly such as a zone of steep gradients on a map without any other obvious boundary.

The defining points are the inflection points O and O' and the half-slope points P_1, P_2 on the right and P'_1, P'_2 on the left. The measurements made are those of the distances l, m and n, as indicated. There is one value for n, but m and l can be measured on both sides. It is pointed out that if there is a regional magnetic gradient, such as the normal northward (or southward) increase in total intensity, it must be taken into account in measuring the maximum slope.

From the measurements, the ratios $A = l/m$ and $C = n/m$ are determined, C being calculated for both sides, if possible (local disturbances or distortions from complex sources may ruin one side or the other). For the single-sided anomaly only the ratio A is available.

The ratios A and C are subject to modification by the width-to-depth ratio of the body and by the angle of inclination of the magnetization vector. Charts for corresponding modification of the depth indices have been prepared from a series of calculated magnetic curves (Bean's figs. 1 to 3). Charts are also included for determination of widths of two-sided blocks and for location of the center of the block. The principal variable in these charts is the angle of the magnetization vector.

According to Bean (personal communication), the necessary measurements can be made routinely and quite rapidly directly on airborne magnetic records and have been programmed for automatic computer interpretation of digital data.

Grant and Martin (1966)

Grant and Martin describe a system which uses certain depth "estimators" derived from measurements on a map and on a profile across the anomaly. These are used

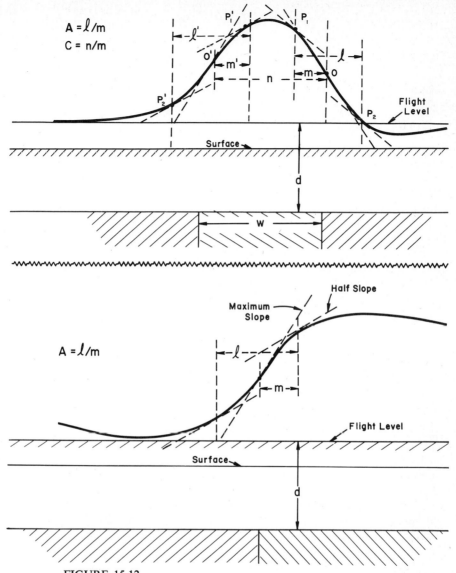

FIGURE 15-12
Assumed magnetic profiles showing quantities measured in Bean's method of depth estimation.

with charts of the variation of certain ratios of these estimators, depending on the form of the models used to approximate the magnetized body (see also Grant and West, 1965, p. 346). An example of one estimator is given by Fig. 15-13. This is applicable to an intrabasement prismatic block such as those of Vacquier et al. (1951). The example is for near-vertical magnetization.

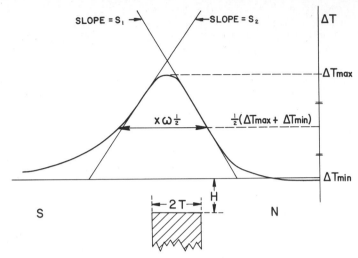

FIGURE 15-13
Schematic magnetic profile with quantities measured in Grant and Martin's system of depth estimation. (*From Grant and Martin, 1966.*)

The quantities measured are the maximum slopes, S_1 and S_2 on the two sides of the anomaly, the half-width $W_{1/2}$, and the total amplitude $\Delta T_{max} - \Delta T_{min}$.

One estimator, here called E_1, is the ratio of apparent length to width of the anomaly, with these dimensions determined by the distances between half-amplitude contours in the long- and short-axis directions of the anomaly.

Another estimator, here called E_2, is the average slope multiplied by the half-width and divided by the amplitude, i.e.,

$$E_2 = \frac{(S_1 + S_2)W_{1/2}}{2(\Delta T_{max} - \Delta T_{min})}$$

Auxiliary charts are made for ratios of certain estimators. On one such chart curves are plotted of E_1 versus E_2 for a range of values of H/T and L/T which make an approximately orthogonal family of curves, where H, L, and T are the depth, half-length, and half-width of the model body.

The estimators for a particular anomaly determine a position on the chart of H/T and L/T values. From an auxiliary chart these values determine the ratio of half-width to depth value, from which the depth can be determined.

The example given is relatively simple, as it is for near-vertical polarization. A rather complex series of charts (about 30) is required for a reasonably complete accommodation to a wide range of polarization angles and of strike directions of the model with respect to magnetic north. The system would seem to be better adapted to the analysis of individual anomalies, as in mining exploration for determination of depths of isolated ore bodies, than to the routine application to many anomalies or magnetic profiles, as in petroleum exploration.

Moo (1965)

This paper gives an extensive analysis of the characteristics of theoretical magnetic-anomaly curves for a two-dimensional block, which may have sloping sides. The analysis considers horizontal distances between critical points such as maxima, minima, points of inflection, and half-slopes and also the tangents of slopes at inflection points and presents three depth-estimation methods.

One depends on the half-slope (Peters) method, and Moo's fig. 6 gives depth indices for variations in width of the body and of the ratio of slopes of the flanks of the anomaly curve. The extreme range of the half-slope indices is from 1.3 to 3.0, but most of the chart is in the range 1.3 to 2.0, which is in approximate agreement with the range of 1.2 to 2.0 given by Peters (1949, p. 310).

The second method uses the distance between the maximum and minimum of the anomaly curve (a quantity which often can be measured on a map, but a good minimum may be hard to find). This parameter varies over a large range (2 to 10) depending on the width of the body and the ratios of slopes and critical distances on the anomaly curve.

The third method is based on the measurement of the horizontal distance from the anomaly maximum to the inflection point on the north flank. Here the extreme range of depth indices is from about 0.6 to 2.0.

All the Moo methods require the use of auxiliary charts to determine certain other factors of the anomaly curves to give the depth index for a particular case. The general effectiveness is difficult to judge without experience, but it would seem that the system would be relatively slow in routine applications to large masses of airborne data for basement-depth estimation.

Koulomzine, Lamontague, and Nadeau (1970)

This rather elaborate mathematical analysis is based on the two-dimensional dipping-dike model. The model curve is decomposed into symmetrical and anti-symmetrical components, a process which depends on locating certain points and distances on the anomaly curve. The critical parameters of the model (its width and depth to top) are determined from the measurements on the anomaly together with a single auxiliary chart which contains six "master curves."

The mathematics of the paper is quite involved, but the authors state that its actual operation takes only a few minutes for a given curve. Test applications to two relatively simple ore-body anomalies gave derived depths which agree quite well with those known.

Other Magnetic-Model Methods

The basement-depth calculation schemes described above are based on magnetic anomalies from rather generalized models of the form of the magnetized body. There are many other model calculation systems in the geophysical literature, e.g., Smith (1959), Powell (1965), and Grant and West (1965, pp. 344–353). For the most part these are given as methods of determining the properties of a single occurrence of magnetized material and are used more in mineral than in petroleum prospecting.

For routine use in oil prospecting a system must be quite simple and readily applied to the magnetic profile. Since a single project may contain many thousands of miles of flight-line traverse with hundreds of anomalies suitable for measurement, it is essential that any system used must be such that the individual measurements can be made quickly. It may be more useful to have many simple measurements of certain parameters than a much smaller number of more elaborate curve matchings for individual anomalies.

In routine use of simple measurements, such as slope and half-slope distances, there is a danger of systematic errors which may come about from using incorrect depth indices because of lack of recognition or of application of the considerable range of values such as shown by some of the Am charts. Another possible source of systematic error is from filtering which may be included in the processing for converting digitally recorded data, e.g., by a Calcomp plotter, into a graphical printout. This is a particular hazard with a marine survey made with a proton-precession magnetometer with the individual measurements or steps being recorded digitally and being filtered to remove the small, short-period variations.

The systems used vary with different interpreters according to their own practices and preferences. Any possible checks should always be used. The best control comes from any wells drilled into the basement or deep enough into a well-known sedimentary section for the basement depth to be reliably extrapolated, as shown by the examples from surveys over the Peace River and Orinoco Basin areas (pages 400 and 402). A good test method, but one rarely used because of the expense, is to fly test profiles at various levels over igneous terrain so that the changes in character of anomalies with height and any changes in indices can be determined as the record changes from complex at low levels to simple at high levels; this is illustrated by Figs. 14-3 and 14-4.*

At best, determining depths from a basement map is painstaking, rather tedious, and to most people, a rather unexciting task. That together with its rather subjective nature are the reasons for the developments outlined in the next section.

AUTOMATIC DEPTH CALCULATIONS

The excessive labor, care, judgment, and subjectivity of the various model methods when applied to basement mapping from extensive aeromagnetic surveys have led to attempts to make the process faster and less subjective by automatic procedures using digital computers. This development has not yet produced a completely satisfactory and objective system and probably never can because of the inherent

---

* Theoretically, the same result could be achieved by upward continuation of a detailed low-level map over igneous terrain and plotting profiles from the continuation results. Two-dimensional upward continuation of a profile would not be reliable.

ambiguity of the direct interpretation problem. On the other hand, it is quite possible for such a system to produce numerical or graphical depth values which, by judicious use, can save the interpreter a large part of the tedious routine of measurements on the magnetic records. Only the general principles of the methods without the mathematical detail will be given here.

The Werner "Deconvolution" Method

This method, along with the general data-reduction and mapping system to which it is applied, are described by Hartman, Tesky, and Friedberg (1971). The system uses a vertical-gradient profile which may be measured or calculated and a calculated horizontal-gradient profile in addition to the recorded total-magnetic-intensity curve.

The calculations are made on the basis of two-dimensional models with flight lines assumed perpendicular to the strike of the model. The elemental model is composed of a single dipping sheet for a body with thickness less than its depth or of a group, or "stack," of such sheets to make a wide block or a contact (one-sided block) of differently magnetized material. To quote Hartman et al. (p. 919):

If thin sheets (dikes or sills) are present their anomalies are recognized and analyzed from the total field data, while interfaces (dipping contacts or changes in slope of a particular surface) are analyzed from the measured vertical gradient data and the computed horizontal derivative of the total field. Typically, an anomaly will produce both a thin-sheet solution and an interface solution.

The elementary Werner equation has six or seven unknowns. This means that six or seven uniformly spaced data points are used to produce that many equations to be solved simultaneously. The solution produces a horizontal distance and depth to a definite point on the surface of the model body. Thus each solution produces a point on a magnetic body which may be on the basement surface but also may be above or below that surface.

A certain amount of depth discrimination is introduced on the basis of the spacing of the individual data points from a minimum of the elemental recording interval ($\frac{1}{40}$ nautical mile) to any multiple of that distance. A short interval will favor solutions for shallow sources and a wider interval that for deeper sources. In practice, solutions are made for several successively wider intervals.

The results of the solutions are plotted automatically on wide (30-in.) paper on a drum plotter. For each solution a mark is made at the horizontal position and depth of the corresponding source. Successive solutions are made as the group of data points involved spans different segments of the anomaly curve; each segment gives an individual mark on the chart. The positions of the individual solutions are printed with different symbols, such as ($\times$) from total-intensity data, ($-$) from horizontal derivative, and ($|$) from the vertical derivative. Ideally, these marks would be at the same point as long as an adequate segment of the anomaly curve is included. Actually the points tend to differ slightly as successive solutions are obtained and form a pattern of some kind. From experience, the form of the pattern, commonly a "hook," or commalike group, or a short vertical group, may be found to be characteristic of a type of source.

Examples of application of the process to synthetic data calculated from simple models give quite good results. The original computer printout from application to routine survey records always contains too many solutions or groups of points to be reasonable. Frequently there are many near the ground which reflect near-surface (often cultural) sources. Frequently some solutions have depths which must be (if real) from sources within the sedimentary section; in some cases these seem to be clearly related to faults or other known or reasonably inferred geological sources within the sediments. The ones of primary interest are those which apparently have sources at the basement surface.

From the standpoint of the primary petroleum industry interest, i.e., mapping the basement surface, the problem of using the source indications as printed out automatically is that of identifying the sources at the basement surface. In some cases they may be clear, and connecting the right ones to indicate the basement surface is relatively simple. In other cases, where both deeper and shallower sources are shown, it may take check solutions by other methods to suggest which level of indication is most probably indicative of the basement.

The Naudy Method

This method, developed in France, is another computer operation which gives automatic indications of depths to magnetic sources (Naudy, 1971). Its general operation is somewhat similar to the Werner-derived system in that it also uses a number of passes over the magnetic curve with sampling at successively wider intervals. The mathematical basis is quite different. The Naudy system uses the anomaly form for an inclined dike for which the thickness can vary from a thin sheet to a thick prism. The curve is separated into symmetrical and antisymmetrical parts and operates on the unsymmetrical part with a reduction-to-the-pole calculation.

The operation with a succession of increasing unit intervals tends to locate the centers and depths to the top of the different anomalous bodies. Calculations may be made and depths plotted by the computer and may be based either on a bottomless dike or a thin plate. The final result is a series of symbols, placed on the chart by the computer, which correspond to the depths to the tops of the corresponding model. The concluding paragraph of the paper is a useful reminder of the place of such operations in magnetic interpretation and is quoted in full:

Depth determination is a job suitable for a high speed computer. In comparison with a manual process, computer processing saves time and gives more objective results. However, interpreting an aeromagnetic survey involves something more than a simple depth determination. The experience of a geophysicist is always necessary whatever may be the part played by the computer.

The Spector and Grant Method (1970)

This system depends on a two-dimensional spectral analysis of a given map or region. The physical basis is that the magnetic map represents the effect of a group of magnetic sources and that the individual sources are rectangular parallelepipeds. By varying the relative sizes of length, width, thickness, depth, polarization angle, etc., any of the various shapes, such as thin plate, bottomless prisms, vertical dike, etc., can be approximated. The group of such blocks is treated by statistical theory and reduced to a power spectrum. The result of the analysis is plotted on a logarithmic scale against the frequency. On such a plot, if a group of sources has a similar depth, they will fall into a line of constant slope. Thus, if there are groups of sources with the individual groups at widely different depths, such as shallow volcanic over a deep basement, the plot will be separable into parts with different slopes and the magnitude of the slope is a measure of depth.

For an example of the application of such a statistical study to the determination of deeper structures under shallow volcanics in southern Algeria, see Curtis and Jain (1975).

The system would seem to have its application primarily to evaluation of general conditions over broad areas and to give relatively objective separation of the sharp and broad anomalies in such a way that multiple depth zones could be recognized. It is not applicable to determination of depths to individual anomalies, as used for mapping a basement surface, but could give an objective confirmation of the general depth of such a surface.

CORRECTIONS TO THE DEPTH ESTIMATES

Most of the methods of basement-depth estimation described above are commonly carried out directly on magnetic records made in the airplane or as reproduced from digital recordings. These give distances on the recording tape, which must be corrected to give the depths used in making the basement map.

Correction for Scale

In most of the operations with flux-gate magnetometers the recordings made in the airplane are on at a constant rate and therefore on a time basis. For such records the scale with respect to ground distances is variable, depending on the ground speed of the airplane, which can vary widely as one flies with or against the wind.

The scale of the record is then determined by simply measuring sample distances on the record corresponding to known distances between certain points on the ground to give a factor for conversion to feet, miles, or other distance units. In modern recording, scale variation can be eliminated by controlling the recorder by Doppler distance measurements (when recorded on the airplane) or by computer coupling with the location data when recorded digitally.

Correction for Azimuth

In general the flight line is not perpendicular to the magnetic contours, and the distances determined from the record are greater than they would be if it were.

BASEMENT DEPTH MAPPING

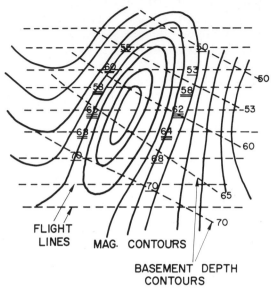

FIGURE 15-14
Schematic diagram of basement-depth mapping showing east-west flight lines,
total-magnetic-intensity contours, depths calculated from flight-line records, and
contours on the basement surface.

This means that distances measured on the record must be shortened by multiplying by the cosine of the angle between the direction of the contours and the flight line. The contour direction of the particular feature being used for the depth calculation is usually better determined from a second-derivative map than from the total-intensity map.

Correction for Flight Elevation

The basement maps are usually made with depths below sea level, while the parameters measured on the record apply at the flight elevation. Therefore, it is necessary, after determining the depth of a given point from the record, to subtract the sea-level elevation of the flight line to give the depth on a sea-level datum.

THE BASEMENT MAP

The general mapping procedure is illustrated by Fig. 15-14. The background contours indicate a large positive magnetic anomaly. The slopes on the flanks of this anomaly correspond to the straight-line portion of a record like that in Fig.

15-6. Depths (below sea level) are determined for each flight line and on each side of the anomaly. In the mapping system from which this example was taken (as used by Gravity Meter Exploration Co.) the underlinings of one, two, or three lines are a quality indication corresponding approximately to *poor* for one, *fair* for two, and *good* for three lines. This evaluation is based, in part, on the consistency of the depth indication between the slope and half-slope measurements and in part on the quality of the record and its freedom from local disturbances. After these depth numbers are placed on the map, the basement surface is contoured, as indicated by the northwest-southeast-trending depth contours shown. Note that in this example the very large magnetic anomaly does not conform to any local relief on the basement surface as the basement depth contours pass regularly across it. Thus this is an intrabasement feature which has its origin entirely in magnetic contrasts below the basement surface and which occurred at a geological time before the development of the basement surface on which the overlying sediments were deposited. Such a feature, in itself, would not be a factor in oil exploration.

EXAMPLES OF MAGNETIC INTERPRETATION

The millions of square miles of basement-depth mapping for petroleum exploration are mostly proprietary with the commercial firms by or for whom the work was done and are not available for publication. There are, however, a few available published examples which can be used as examples to illustrate the effectiveness of magnetics as a very useful tool in the total exploration "package" for determining the general and detailed geology of an area and evaluating its petroleum prospects.

The examples are of two different types. Those covering relatively broad areas are made primarily for mapping depth to basement and thereby determining the general form of the sedimentary section over all or part of a basin area. Those illustrating delineation of local structures may be either from careful detailed analysis of the data from a routine areal survey or from more closely spaced data within such a survey. Examples of both types are given in the following paragraphs or by reference to a previous section.

Areal Surveys

Peace River, Alberta, and Orinoco Basin, Venezuela These areas were discussed on pages 400 to 403 as examples of the reliability of the slope and half-slope methods of determination of basement depth. The Orinoco Basin maps (Figs. 15-10 and 15-11) demonstrate in some detail that the basement was reliably mapped by analysis of airborne magnetic data.

Senegal, West Africa In this reconnaissance survey the general form of a large basin area was revealed by analysis of a reconnaissance survey with widely spaced flight lines. Figure 15-15 is a generalized geologic map of the area. From this map

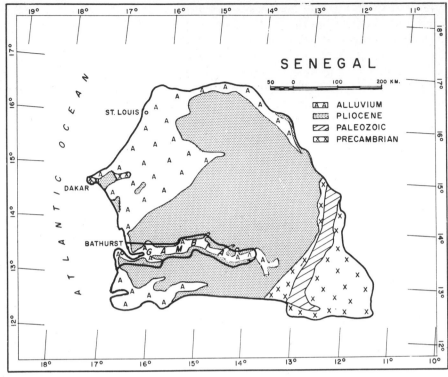

FIGURE 15-15
Geologic map of Senegal. (*From Nettleton, 1962; by permission of the American Association of Petroleum Geologists.*)

alone we can determine very general features only. It can be inferred that much of the map is over a basin area, because there are outcrops of older rocks to the east and there are some outcrops of igneous rocks (not indicated on the figure) in the vicinity of Dakar. This surface geologic information alone would give very little basis for evaluation of the area for its petroleum possibilities.

A rather loose reconnaissance magnetic survey was carried out over the area. This was not closely enough controlled to give a detailed map of the basement surface, but approximate basement depths could be determined throughout the area. Figure 15-16 shows generalized contours, at an interval of 1000 m, on the basement surface. While this reconnaissance has provided little detail, there is now a much more specific and quantitative picture of the gross features of the basin from which to begin an evaluation of its petroleum prospects. An exploration executive having a choice of certain options for selecting limited portions of the area for further exploration would have a much more definite basis for choice than from the surface geologic map alone.*

* Incidentally, the large circular basement feature along the coast in the northwest part of Fig. 15-16 is the magnetic part of the data for the examples of Figs. 14-7, 14-8, and 15-4.

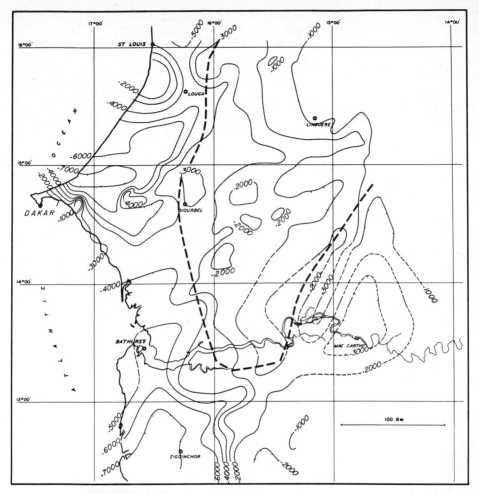

FIGURE 15-16
Basement-depth map, Senegal, calculated from reconnaissance aeromagnetic survey; contour interval, 1000 m. Dashed line is outline of Central Basin Platform. (*From Nettleton, 1962; by permission of the American Association of Petroleum Geologists.*)

The general conditions here may be compared with those of the Permian Basin of West Texas. Note that the area covered by the geologic map is approximately 500 km, or nearly 300 miles, square, which is generally comparable with the broad extent of the Permian Basin from the Central Mineral Region of Texas to eastern New Mexico. From the geologic map alone (Fig. 15-15) just as in the case of the surface geologic map of the Permian Basin, there is no suggestion of the very large structure corresponding to the Central Basin Platform of West Texas. In a further extension of the parallel between the two areas the magnetic basement map of Senegal also shows a sort of central basin platform, as outlined by the

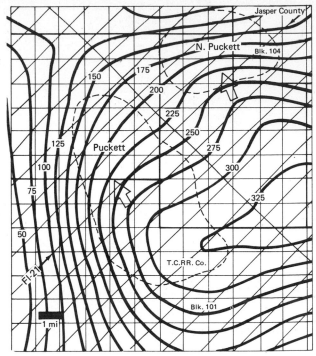

FIGURE 15-17
Puckett oil field; observed aeromagnetic map. (*From Steenland, 1965; by permission of the Society of Exploration Geophysicists.*)

dashed line in Fig. 15-16, which is about 120 km wide and 250 km long or roughly 75 by 150 miles. These dimensions are comparable with those of the Central Basin Platform. The basin at the west with maximum depth of 7000 m, with its axis running northeasterly from north of Dakar is rather comparable in horizontal dimensions and depth with the Delaware Basin. Another shallower basin on the southeasterly side of the platform is crudely comparable with the Midland Basin; and the large basin at the southwest part of the map, again with a depth of over 7000 m (well over 20,000 ft) is rather comparable in size and depth, but not in location, with the Valverde Basin. Thus we see that even a very general basement map derived from a relatively crude "once over" magnetic survey can lend a great deal of general information at small cost to an otherwise almost completely unknown geological province covered by alluvial and recent deposits so that surface geologic studies are almost useless.

Mapping Local Basement Structures

The previous examples are of large areal surveys which have given the general basin form and sedimentary thickness under the area covered. There are available also a number of examples of more detailed surveys and interpretations over areas

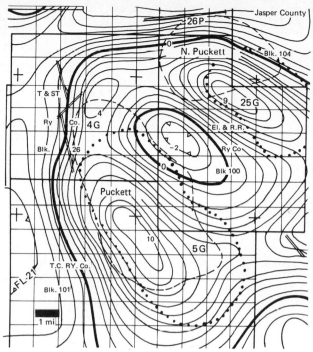

FIGURE 15-18
Puckett oil field; second-vertical-derivative map, contour interval 1×10^{-15} cgs unit. (*From Steenland, 1965; by permission of the Society of Exploration Geophysicists.*)

which include oil fields, so that the actual structure is known from drilling and can be compared with that derived from a magnetic interpretation made in advance of all or most of the drilling. Two such examples, of very different character, discussed and illustrated below, demonstrate that careful analysis of magnetic data can delineate with fair to good precision the location and approximate relief of basement structures under sedimentary deformations which resulted in petroleum accumulation. Basement-depth calculations were from slope and half-slope distance measurements using indices for intrabasement and suprabasement sources. Second-vertical-derivative maps were made to delineate local detail better.

The first illustration is from Steenland (1965). This and several other examples from this paper are included in Nettleton (1971). The second example is from the Zelten Field in Libya, which is a major oil field.

These two examples are very different in geological and magnetic character. Puckett is a structure of very large relief and with a very strong magnetic anomaly. Zelten is in an area where basement polarization is weak and the magnetic expression of the basement deformation which underlies the oil field has a relief of only about 5 γ.

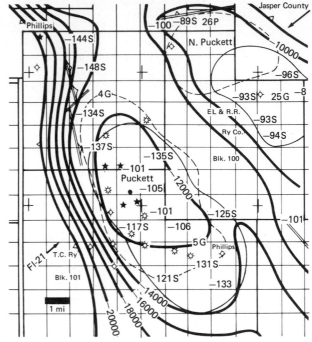

FIGURE 15-19
Puckett Field; magnetic basement map of 1958, contour interval 2000 ft with possible local structures and oil fields indicated by dashed outlines, October 1963. (*From Steenland, 1965; by permission of the Society of Exploration Geophysicists.*)

The Puckett Field, Pecos County, Texas   Here very large magnetic disturbances caused by relief of the basement surface are superimposed on still larger disturbances having their sources in intrabasement magnetic contrasts. Figure 15-17 is the airborne magnetic map of the area from a flight-line pattern of approximately 1 mile spacing with northeast to southwest lines. This shows a very extensive and steep gradient on the westerly side and an irregular high nose in the central part of the map.

The second-derivative map (Fig. 15-18) develops the local anomalies and shows a large closed maximum in the southwesterly part of the map and a smaller closed positive anomaly to the northeast.

The basement contour map developed from the analysis of this survey is shown in Fig. 15-19. The very large change in depth on the western side of the map is from the differences between the depth figures of around 12,000 ft over the central structure and depths of over 20,000 ft just west of the area shown. This magnitude of structural relief on the basement surface is calculated from the local, suprabasement anomaly over the structure, as illustrated by the sample record shown by Fig. 15-20. The resolution of the magnetic profile into the components

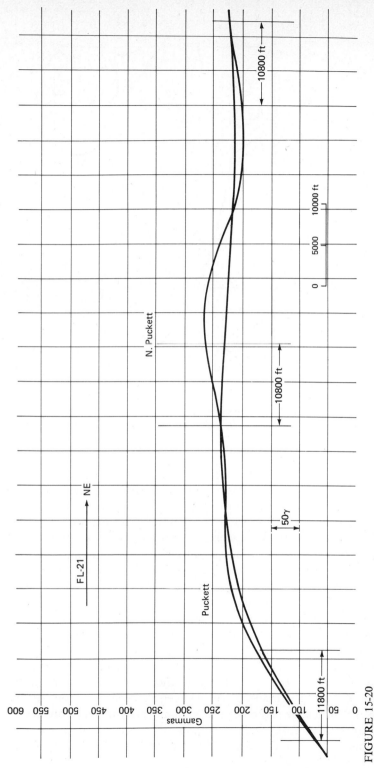

FIGURE 15-20

Total-magnetic-intensity record, flight line 21 of Fig. 15-7, with resolution into intrabasement and suprabasement components. (*From Steenland, 1965; by permission of the Society of Exploration Geophysicists.*)

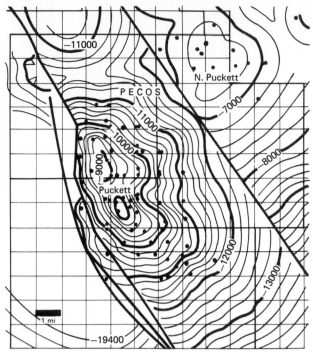

FIGURE 15-21
Puckett Field, subsurface contours on the Ellenburger, showing structure from
drilling. (*From Steenland, 1965; by permission of the Society of Exploration
Geophysicists.*)

due to intrabasement contrasts and those due to the structure itself (suprabasement
contributions) is indicated on this profile.

Finally Fig. 15-21 shows the actual structure of the area as developed by the
drilling with contours on the Ordovician-Cambrian Ellenburger limestone, which
is deep in the sedimentary section and probably reflects fairly well the actual
deformation of the basement surface. The location of the Puckett structure itself
and the large fault, with displacement of some 10,000 ft, on the west side are quite
close to the predictions from the magnetic analysis. The North Puckett structure is
also shown generally, although the agreement in location is not as good as that of
the main Puckett Field. The general configuration with the terracing between the
two producing areas and the fault on the west side of the North Puckett Field are
rather similar to the indications from the magnetic basement contour map.

This is a good illustration of the difference in magnitude of contributions
from intrabasement and suprabasement magnetic anomalies. From the magnetic
profile (Fig. 15-20) we see a general rise of some 200 γ or more due to an intrabasement magnetic contrast. Superimposed on that is a rise of only some 30 γ as
the expression of the suprabasement effect from the great Puckett structure with

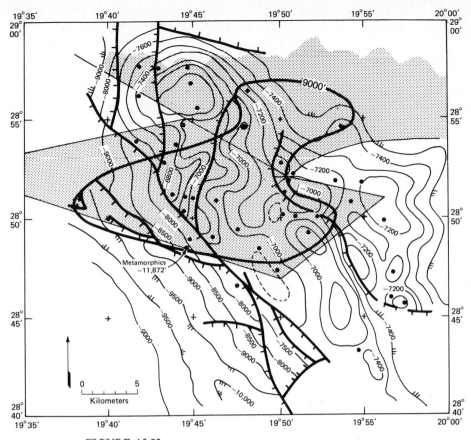

FIGURE 15-22
Zelten Field, Libya; comparison of structure contours (*Exxon Corporation*)
from drilling with interpretation of aeromagnetic survey.

the relief of the order of 10,000 ft. This is a further example of the differences in
magnitude of anomalies arising from intrabasement and suprabasement contrasts.

The Zelten Field, Sirte Basin, Libya   The magnetic interpretation here is of
particular interest as an illustration of the analysis of very subtle anomaly expressions in an area of unusually low magnetization of the basement. The records were
from a high-quality flux-gate airborne survey with east-west lines at 3 km spacing
(Steenland, Hinrichs, and Navazio, 1963). The magnetic interpretation produced
a structure contour map (Steenland, Hinrichs, and Navazio, 1963, fig. 4) which
indicated a large, probably faulted structure with some 3000 to 5000 ft of relief.
The area has since developed into an oil field with many wells. A structure map
on the field is given in Fig. 15-22.

   The principal features from the Steenland et al. map, including the closing

9000-ft basement-depth contour, have been added to the structure map. The shaded "possible local basement structure area" includes most of the oil field, especially when it is noted that the northern limit was not defined because of the limited extent of the more closely spaced control lines of the original survey.

The original interpretation suggested a fault pattern which is not evident from the oil-field structure contours. It is possible, within the control given, to re-construct the contour pattern with some fault suggestions in general conformity with those of the magnetic interpretation. The one contact from drilling indicates a basement depth about 30 percent greater than that of the magnetic interpretation. This is an illustration of a rather common style of systematic error which comes about from use of an incorrect depth parameter. Such an error does not materially distort relative measurements in the same general area, and therefore the pattern, if not the depth level, of the basement interpretation is essentially correct.

The importance of this example is not that the details of the interpretation are not borne out by the drilling but that the general area of one of the great oil fields of the world* was quite closely outlined. Several other oil-field locations in Libya were similarly predicted. They could have been found by magnetic surveys alone at a cost of less than 1 percent of that used for the extensive gravity and seismic surveys which were carried out before the selections for development concessions were made.

---

* Production to the end of 1972 was over 1.5 billion barrels.

REFERENCES FOR PART TWO

ALLDREDGE, LEROY R., GERALD D. VANVOORHIS, and THOMAS M. DAVIS: A Magnetic Profile around the World, *J. Geophys. Res.*, vol. 68, no. 12, pp. 3679–3692, 1963.

AM, K.: The Arbitrarily Magnetized Dyke: Interpretation by Characteristics, *Geophys. Explor.*, vol. 10, pp. 63–90, 1972.

ANDREASEN, GORDON E., and ISADORE ZIETZ: Limiting Parameters in the Magnetic Interpretation of a Geologic Structure, *Geophysics*, vol. 27, no. 6, p. I, pp. 807–814, 1962.

—— and ISADORE ZIETZ: Magnetic Fields for a 4 × 6 Prismatic Model, *Geol. Surv. Prof. Pap.* 666, Washington, D.C., 1969.

BARANOV, V.: A New Method for Interpretation of Aeromagnetic Maps; Pseudo-gravimetric Anomalies, *Geophysics*, vol. 22, no. 2, pp. 359–383, 1957.

—— and H. NAUDY: Numerical Calculation of the Formula of Reduction to the Magnetic Pole, *Geophysics*, vol. 29, no. 1, pp. 67–79, 1964.

BARTELS, J.: Erdmagnetismus, Erdstrom und Polanlicht, in B. Gutenberg (ed.), "Lehrbuch der Geophysik," pp. 378–453, Verlag vom Gebruder Borntraeger, Berlin, 1926.

BEAN, R. J.: A Rapid Graphical Solution for the Aeromagnetic Anomaly of the Two-dimensional Tabular Body, *Geophysics*, vol. 31, no. 5, pp. 963–970, 1966.

BEMROSE, JOHN, J. C. HEGGBLOM, T. C. HOLT, T. C. RICHARDS, and R. J. WATSON: Bahamas Airborne Magnetometer Survey, *Geophysics*, vol. 15, no. 1, pp. 102–109, 1950.

BHATTACHARYYA, B. K.: Magnetic Anomalies Due to Prism-shaped Bodies with Arbitrary Polarization, *Geophysics*, vol. 29, no. 4, pp. 517–531, 1964.

——: Two Dimensional Harmonic Analysis as a Tool for Magnetic Interpretation, *Geophysics*, vol. 30, no. 5, pp. 829–857, 1965.

——: An Automatic Method of Compilation and Mapping of High-Resolution Aeromagnetic Data, *Geophysics*, vol. 37, no. 5, pp. 695–761, 1971.

BLOOM, ARNOLD L.: Principles of Operation of the Rubidium Vapor Magnetometer, *Appl. Opt.*, vol. 1, no. 1, pp. 61–68, 1962.

"Continents Adrift," readings from *Scientific American* with introductions by J. Tuzo Wilson, W. H. Freeman & Co., San Francisco, 1971.

CURTIS, C. E., and S. JAIN: Determination of Volcanic Thickness and Underlying Structures from Aeromagnetic Maps in the Silet Area of Algeria, *Geophysics*, vol. 40, no. 1, pp. 79–90, 1975.

DEMENITSKAYA, R. M., and N. A. BELYAEVSKY: The Relation between the Earth's Crust, Surface Relief and Gravity Field in the U.S.S.R., in Pembroke J. Hart (ed.), "The Earth's Crust and Upper Mantle," *Am. Geophys. Union Geophys. Monogr.* 13, 1969.

DOBRIN, MILTON B.: "Introduction to Geophysical Prospecting," McGraw-Hill Book Company, New York, 1952; 2d ed., 1960.

DUFFUS, H. J.: Techniques for Measuring High Frequency Components of the Geomagnetic Field, in S. K. Runcorn (ed.), "Methods and Techniques in Geophysics," vol. 2, Interscience Publishers, Division of John Wiley & Sons, Inc., New York, 1966.

ELKINS, T. A.: The Second Derivative Method of Gravity Interpretation, *Geophysics*, vol. 16, no. 1, pp. 29–50, 1951.

FABIANO, E. B., and N. W. PEDDIE: Grid Values of the Total Magnetic Intensity, I.G.R.F. (International Geophysical Reference Field), 1965, *ESSA Tech. Rep. C & GS* 38, 1969.

GAY, S. PARKER, JR.: Standard Curves for Interpretation of Magnetic Anomalies over Long Tabular Bodies, *Geophysics*, vol. 28, no. 2, pp. 161–200, 1963.

GRANT, F. S., and LUCIANO MARTIN: Interpretation of Aeromagnetic Anomalies by the Use of Characteristic Curves, *Geophysics*, vol. 31, no. 1, pp. 135–148, 1966.

———, and G. F. WEST: Interpretation Theory in Applied Geophysics, McGraw-Hill Book Company, New York, 1965.

GREEN, R.: The EAEG's Choice of Units, *Geophys. Prospect.*, vol. 16, pp. 1–3, 1968.

GRENET, G.: Sur les propriétés magnetiques des roches, *Ann. Phys.*, 10th ser., vol. 13, pp. 263–348, 1930.

HAALCK, H.: Die magnetischen Methoden der angewandten Geophysik, in W. Wein and F. Harms (eds.), "Handbuch der Experimentalphysik," Band XXV, Teil 3, pp. 303–398, Akademische Verlagsgesellschaft m.b.h., Leipzig, 1930.

HARTMAN, RONALD R., DENNIS J. TESKY, and JEFFREY L. FRIEDBERG: A System of Rapid Digital Aeromagnetic Interpretation, *Geophysics*, vol. 36, no. 5, pp. 891–918, 1971.

HEILAND, C. A.: "Geophysical Exploration," Prentice-Hall, Inc., New York, 1940.

HENDERSON, ROLAND G., and ISADORE ZIETZ: Analysis of Magnetic-Intensity Anomalies Produced by Point and Line Sources, *Geophysics*, vol. 13, no. 3, pp. 428–436, 1948.

HOOD, PETER: Magnetic Surveying Instrumentation: A Review of Recent Advances, pp. 1–31, in Mining and Groundwater Geophysics, *Geol. Surv. Can. Econ. Geol. Rep.* 26, 1967.

HOWE, H. HERBERT, and DAVID G. KNAPP: United States Magnetic Tables and Magnetic Charts for 1935, *U.S. Coast Geod. Surv. Ser.* 602, 1938.

HUGHES, D. S., and W. L. PONDROM: Computation of Vertical Magnetic Anomalies from Total Field Magnetic Measurements, *Trans. Am. Geophys. Union*, vol. 28, pp. 193–197, 1947.

HURWITZ, LOUIS, EUGENE B. FABIANO, and NORMAN W. PEDDIE: A Model for the Geomagnetic Field for 1970, *J. Geophys. Res.*, vol. 79, no. 11, pp. 1716–1717, 1974.

HUTCHISON, RALPH D.: Magnetic Analysis by Logarithmic Curves, *Geophysics*, vol. 23, no. 4, pp. 749–769, 1958.

JACOBS, J. A.: "The Earth's Core and Geomagnetism," Pergamon Press and The Macmillan Company, New York, 1963.

JACOBSEN, PETER, JR.: An Evaluation of Basement Depth Determinations from Airborne Magnetometer Data, *Geophysics*, vol. 26, no. 3, pp. 309–319, 1961.

JAKOSKY, J. J.: "Exploration Geophysics," 2d ed., Trija Publishing Co., Los Angeles, California, 1950; 1st ed., Times-Mirror Press, Los Angeles, 1940.

JENSEN, HOMER: Instrument Details and Applications of a New Airborne Magnetometer, *Geophysics*, vol. 30, no. 9, pp. 875–882, 1965.

KOEFOED, O.: Units in Geophysical Prospecting, *Geophys. Prospect.*, vol. 15, pp. 1–6, 1967.

KOULOMZINE, T., Y. LAMONTAGUE, and A. NADEAU: New Method for the Direct Interpretation of Magnetic Anomalies Caused by Inclined Dikes of Infinite Length, *Geophysics*, vol. 35, no. 5, pp. 812–830, 1970.

LEATON, BRIAN F.: I.G.R.F. charts, pp. 189–203 in Alfred J. Zmuda (ed.), World Magnetic Survey, 1957–1969, *I.A.G.A. Bull.* 28, International Union of Geology and Geophysics, Paris, 1971.

LYNTON, EDWARD D.: Some Results of Magnetometer Surveys in California, *Bull. Am. Assoc. Pet. Geol.*, vol. 15, no. 11, pp. 1351–1370, 1931.

MARKOWITZ, WILLIAM: SI, the International System of Units, *Geophys. Surv.*, vol. 1, pp. 217–241, 1973.

MINTROP, L.: On the Stratification of the Earth's Crust According to Seismic Studies of a Large Explosion and of Earthquakes, *Geophysics*, vol. 14, no. 4, pp. 321–336, 1949.

MOO, J. K. C.: Analytical Aeromagnetic Interpretation of the Inclined Prism, *Geophys. Prospect.*, vol. 13, no. 2, pp. 203–224, 1965.

MOONEY, HAROLD M., and RODNEY BLEIFUSS: Magnetic Susceptibility Measurements in Minnesota, pt. II: Analysis of Field Results, *Geophysics*, vol. 18, no. 2, pp. 383–393, 1953.

NABIGHIAN, MISAE N.: The Analytic Signal of Two-dimensional Magnetic Bodies with Polygonal Cross-Sections: Its Properties and Use for Automated Anomaly Interpretation, *Geophysics*, vol. 37, no. 3, pp. 507–517, 1972.

NAGATA, TAKESI: "Rock Magnetism," rev. ed., Plenum Press, New York, 1961.

NAUDY, HENRY: Automatic Determination of Depth on Aeromagnetic Profiles, *Geophysics*, vol. 36, no. 4, pp. 717–722, 1971.

NETTLETON, L. L.: *Geophysical Prospecting for Oil*, McGraw-Hill Book Company, New York, 1940.

———: Interpretation of Aeromagnetic Surveys in Western Canada, *Oil in Canada*, July 31, 1950.

———: Gravity and Magnetics for Geologists and Geophysicists, *Bull. Am. Assoc. Pet. Geol.*, vol. 48, no. 10, pp. 1815–1838, 1962.

———: Elementary Gravity and Magnetics for Geologists and Seismologists, *Soc. Explor. Geophys. Monogr. Ser.* 1, Tulsa, 1971.

———, and T. A. ELKINS: Association of Magnetic and Density Contrasts with Igneous Rock Classifications, *Geophysics*, vol. 9, no. 1, pp. 60–78, 1944.

PACKARD, M., and R. VARIAN: Free Nuclear Induction in Earth's Magnetic Field, *Phys. Rev.*, vol. 93, p. 941, 1954.

PARASNIS, D. S.: A Supplementary Comment on R. Green's Note "The EAEG's Choice of Units," *Geophys. Prospect.*, vol. 16, pp. 392–393, 1968.

PETERS, LEO J.: A Direct Approach to Magnetic Interpretation and Its Practical Application, *Geophysics*, vol. 14, no. 3, pp. 290–320, 1949.

POWELL, D. W.: A Rapid Method of Determining Dip or Magnetization Inclination from Magnetic Anomalies due to Dyke-like Bodies, *Geophys. Prospect.*, vol. 13, pp. 197–202, 1965.

PUZICHA, KURT: Der Magnetismus der Gesteine als Funktion ihres Magnetitgehaltes, *Beitr. Angew. Geophys.*, vol. 9, no. 2, pp. 158–186, 1941.

REFORD, M. S.: Magnetic Anomalies over Thin Sheets, *Geophysics*, vol. 29, no. 4, pp. 532–536, 1964.

——— and J. S. SUMNER: Aeromagnetics, *Geophysics*, vol. 29, no. 4, pp. 482–516, 1964.

REILLY, W. J.: Use of the International System of Units (SI) in Geophysical Publications, *N.Z. J. Geol. Geophys.*, vol. 15, no. 1, pp. 148–156, 1972.

RIDDIHOUGH, R. P.: Diurnal Corrections to Magnetic Surveys: An Assessment of Errors, *Geophys. Prospect.*, vol. 19, no. 4, pp. 551–567, 1971.

SHERIFF, R. E.: "Encyclopedic Dictionary of Exploration Geophysics," Society of Exploration Geophysicists, Tulsa, 1973.

SHUEY, RALPH T.: Application of Hilbert Transforms to Magnetic Profiles, *Geophysics*, vol. 17, no. 6, pp. 1043–1045, 1972.

SLICHTER, L. B.: Certain Aspects of Magnetic Surveying in "Geophysical Prospecting," *Trans. Am. Inst. Min. Met. Eng.*, vol. 81, pp. 238–258, 1929.

SMELLIE, D. W.: Elementary Approximations in Aeromagnetic Interpretation, *Geophysics*, vol. 21, no. 4, pp. 1021–1040, 1956.

SMITH, R. A.: Some Depth Formulas for Local Magnetic and Gravity Anomalies, *Geophys. Prospect.*, vol. 7, pp. 55–63, 1959.

SOSKE, JOSHUA: Differences in Diurnal Variation of Vertical Magnetic Intensity in Southern California, *Terr. Mag. Atmos. Elect.*, vol. 38, no. 1, pp. 109–115, 1933.

SPECTOR, A., and F. S. GRANT: Statistical Models for Interpreting Aeromagnetic Data, *Geophysics*, vol. 35, no. 2, pp. 293–302, 1970.

STACEY, FRANK D.: "Physics of the Earth," John Wiley & Sons, Inc., New York, 1969.

STEENLAND, N. C.: Gravity and Aeromagnetic Exploration in the Paradox Basin, *Geophysics*, vol. 17, no. 1, pp. 73–89, 1962.

———: An Evaluation of the Peace River Aeromagnetic Interpretation, *Geophysics*, vol. 28, no. 5, pt. I, pp. 745–755, 1963.

———: Oil Fields and Aeromagnetic Anomalies, *Geophysics*, vol. 30, no. 5, pp. 706–739, 1965.

———: Aeromagnetic Study of Coyanosa, *Geophysics*, vol. 32, no. 2, pp. 282–290, 1967.

———, F. W. HINRICHS, and FRED NAVAZIO: Aeromagnetic Interpretation Zelten Field Area, Libya, *Geophysics*, vol. 28, no. 5, pt. 1, pp. 745–755, 1963.

STRANGWAY, DAVID W.: "History of the Earth's Magnetic Field," McGraw-Hill Book Company, New York, 1970.

U.S. NAVAL OCEANOGRAPHIC OFFICE: "The Total Intensity of the Earth's Magnetic Force, Epoch 1965," 3d ed., Washington, D.C., Jan. 1966.

VACQUIER, VICTOR: Short-Time Magnetic Fluctuations of Local Character, *Terr. Mag. Atmos. Elect.*, vol. 42, no. 1, pp. 17–28, 1937.

———: U.S. Patents 2,406,870, Sept. 3, 1946, and 2,407,202, Sept. 3, 1946.

——— and GARY MUFFLEY: U.S. Patent 2,555,209, May 29, 1951.

———, NELSON CLARENCE STEENLAND, ROLAND G. HENDERSON, and ISADORE ZIETZ: Interpretation of Aeromagnetic Maps, *Geol. Soc. Am. Mem.* 47, 1951, reprinted 1963.

VESTINE, E. H., LUCIHE LAPORTE, CAROLINE COOPER, ISABELLE LANGE, and W. C. HENDRIX: Description of the Earth's Main Magnetic Field and Its Secular Change, *Carnegie Inst. Wash. Publ.* 578, 1947.

WEBSTER, A. G.: "Electricity and Magnetism," pp. 370–371, Macmillan, London, 1897.

WHITHAM, K.: Measurement of the Geomagnetic Elements, in S. K. Runcom (ed.), "Methods and Techniques in Geophysics," vol. I, Interscience Publishers, New York, 1960.

WYCOFF, R. D.: The Gulf Airborne Magnetometer, *Geophysics*, vol. 13, no. 2, pp. 182–208, 1948.

ZIETZ, ISADORE, and ROLAND G. HENDERSON: The Sudbury Aeromagnetic Map as a Test of Interpretation Methods, *Geophysics*, vol. 20, no. 2, pp. 307–317, 1955.

——— and ———: A Preliminary Report on Model Studies of Magnetic Anomalies of Three-dimensional Bodies, *Geophysics*, vol. 21, no. 30, pp. 794–815, 1956.

Synergistic Interpretation of Multisensor Geophysical Data

COMBINED GEOPHYSICAL-GEOLOGICAL
RELATIONS IN OIL EXPLORATION

INTRODUCTION

This short Part Three is not concerned with gravity or magnetics as such in oil exploration but in their relations with each other, with seismic prospecting, and with geology. There is a natural tendency, by those concerned primarily with any one of the geophysical methods, to become myopic in their study of the physical or mathematical aspects. This is to be expected in academic study and research, but the large fraction of geophysicists employed in the oil industry will do well to keep in mind that the *objective of their jobs is to find oil*. No one method finds hydrocarbons directly,* and most contributions to new oil fields by any one geophysical method are only a part of a very large body of information behind any new petroleum prospect or discovery.

As eloquently stated by Pratt (1942, p. 52), "oil must be sought first of all in our minds." This idea was expanded much later by Halbouty in a paper entitled Oil Is Found in the Minds of Men (Halbouty, 1972).

---

* The development of seismic data processing to reveal "bright spots" appears to be an exception to this statement which has been true from the beginning of geophysical exploration to the very recent past. Even so, bright spots can be misinterpreted and are not the simple and sure indicators they were hoped to be.

RELATIONS BETWEEN THE THREE GEOPHYSICAL METHODS

From the beginning of geophysical exploration in the petroleum industry in the 1920s, three basic physical principles were used, i.e., the measurement of small variations in the magnetic field, the measurement of small variations in the gravitational field, and the propagation of elastic waves through the earth. While there have been great changes in instrumentation, data-processing, and interpretation techniques over the intervening half-century, these three and only these three physical principles are the basis for practically all the geophysical work up to the present time. Many other methods, particularly electrical, have been conceived and tried in the field in a limited way, but none has persisted to the extent that field operations are carried out on a scale at all comparable with that of the three primary methods.

The reflection seismic method, of course, usually is much more direct in its relation to geology than the potential methods. Reflection zones or horizons frequently are directly correlative with geologic strata and give relatively accurate measures of their depth and form. In many cases, however, correlations with geology may be uncertain or misleading. In such cases, gravity and magnetic data may contribute to geological understanding by establishing limits or alternatives for possible correlations and providing information on physical properties of the rocks.

To some degree, the dominance of the reflection seismic method may be attributed to the relative simplicity of the basic principles applied. The concept of initiating a disturbance at the surface, detecting a return signal from a reflecting layer, and determining the depth of that layer by knowing the speed of the wave propagation is very simple. On the other hand, the magnetic and gravity methods are both potential methods and depend on action at a distance. This means that as measurements are made farther from the disturbing anomalous magnetic or density contrast, the effect becomes more and more attenuated and smoothed out, so that details are less evident. The effect is a little like that when the roughness of objects or other irregularities of the ground surface are covered by a heavy snowfall. Features on the ground which are immediately recognizable under a thin snow cover become more and more rounded out and merged together as the snow gets deeper, until the expressions of complex sources become only rounded bumps on the surface or disappear. In a somewhat similar way the interpretation of magnetic and gravity surveys is inherently ambiguous, and any quantitative control on the nature and depth of the source of the disturbance must come from additional information. Such information can come from drilling contacts, seismic results, or reasonable geologic limitations.

THE OTHER 5 PERCENT

Since the successful development of the reflection method in the early 1930s seismic exploration has so greatly dominated the geophysical search for oil that the expenditures for gravity and magnetic operations have been a small fraction of the total. This is clearly shown by the annual reviews of geophysical activity published since 1944 by the Society of Exploration Geophysicists. For instance, Allen (1971) shows that worldwide expenditures for magnetic, gravity, and seismic field operations in petroleum exploration for 1969 were 10.0, 11.0, and 608 million dollars, respectively. Thus the potential-type exploration attracted only 3.3 percent of the total money spent. The ratios have been somewhat different over the years, but in general we can relegate the combined gravity and magnetic expenditures to "the other 5 percent."

Why is this fraction so small? In part the answer is in the lower field costs and greater rate of coverage of the potential methods. Part also is in a certain lack of appreciation of the value, in terms of geological usefulness and actual oil-finding possibilities, of the gravity and magnetic methods.

The cost data above do not mean that the potential methods make a proportionately small contribution to the overall exploration result. Because of the relatively rapid rate of progress in the field, particularly by airborne magnetics, the total area covered by gravity and magnetic surveys may be greater than that covered by the much larger seismic expenditures. As a very rough rule of thumb, for land operations, the relative costs per unit area covered by magnetic, gravity, and seismic field work with data processing stand in the ratio of 1:10:100. In spite of this very favorable cost ratio the potential methods are not applied as widely or interpreted as carefully and quantitatively as they would be if their oil-finding usefulness were fully appreciated and they were applied in proper perspective in overall exploration programs.

In the earlier days of very general exploration,* it was common to consider that the ideal order of application of the three geophysical methods would be, first, reconnaissance with airborne magnetics, followed by gravity surveys in a degree of detail depending on the geological objectives and the facility of getting over the surface, and, finally, by reconnaissance and detailed reflection-seismograph surveys with the detailing being directed to those areas where gravity or magnetics or both point to a feature of particular interest. This idealized exploration approach is rarely used, partly because of land, concession, or competitive considerations and partly because of a lack of appreciation of the complementary nature of the different geophysical methods. In many instances the argument was used that an area probably will be covered eventually by the reflection-seismograph method and that with such a survey the potential methods would not contribute enough to justify their cost.

In many geological situations the potential methods can contribute independent information which will add critical or decisive data to the geologic knowledge. As a very simple example there have been seismic surveys in areas where domal uplifts are found but where their geologic nature is not known and where they might be caused by sedimentary structural uplift, salt domes, or igneous intrusions. Gravity and magnetic surveys can answer these questions unequiv-

---

* Instructions for field coverage could be very general. The writer knows of a gravity field party which received a one-line order: Your next assignment is to cover Harper County. The party chief had to write back: Do you mean Oklahoma or Kansas? The two counties are adjacent.

ocally. A negative gravity anomaly with no magnetic anomaly or a very slight one would indicate a salt dome. A low-relief positive gravity anomaly with no magnetic anomaly or with a very low-relief magnetic anomaly would indicate a sedimentary uplift with or without (depending on the magnetic expression) an underlying basement uplift. A strong positive gravity anomaly with a strong magnetic anomaly would indicate an igneous intrusive. Examples of such relations are given in the latter part of this chapter.

Seismic interpretations in faulted areas often show a fault by a zone of no reflections and, especially in rather poor record areas, there may be no definite basis for determining the direction of displacement of the fault. Gravity information can often find a corresponding fault anomaly, which will, of course, indicate which is the upthrown side.

Magnetic surveys have their most obvious and economically important application in the exploration of new areas, particularly offshore areas. Rapid and relatively inexpensive airborne magnetics can determine the depth to the basement, usually quite reliably and, by careful interpretation of good data, to a precision of the order of 5 percent or less of the depth below the flight level. Also, as demonstrated by the Puckett and Zelten examples (pages 418 and 421) and by other published cases (Steenland, 1965), it is often possible to determine local relief of the basement which has corresponding oil-field structure in the overlying sediments. In a new area where the thickness of sediments is unknown, a magnetic basement map can determine those basin areas where the sediments are thick enough to make reasonable petroleum prospects and to guide later exploration there. Even when seismograph surveys have been carried out, the basement reflection and its depth may be very uncertain where the sediments are thick and reflections became very poor or unrecognizable. Basement depths from magnetic data together with their confirmation by gravity calculations may give a reliable measure of the thickness of the sediments. Examples of such applications are given later.

A question may be raised whether the magnetic basement and the effective geologic basement are the same. They usually are, but there are exceptions, e.g., where large thicknesses of nonmagnetic metamorphosed rocks are present and where the geologic basement is the top of the metamorphics and the magnetic basement is at their base. An example is a great thickness of nonmagnetic quartzite found in the vicinity of the Uinta Mountains in Utah. However, from worldwide experience in the application of the magnetic method to basement mapping, the instances where the basement depth can be grossly wrong are rare, and usually careful consideration of all the facts will reveal clues which can suggest that a gross discrepancy is present. Therefore, on the whole, magnetic basement mapping has proved to be a very reliable and useful part of overall petroleum exploration.

In addition to the usefulness in locating local structures, gravity surveys are often helpful in evaluating larger geologic units. An example has been given (page 152) of the simple relations of the basin depth to the gravity anomaly in the Los Angeles basin. Similar relations generally apply to the smaller basins with horizontal dimensions not over a few tens of miles, such as the intermontane basins of the basin and range area of Utah, Nevada, and eastern California. The nature of larger basins often can be inferred. For instance, the great valley of California (like other basins in the Rocky Mountains and other parts of the world) is asymmetrical with a foredeep on one side which is distinctly indicated by a regional gravity minimum over the area of thickest sediments. For large basins, however, information on the location and form of gross features within the basin and quantitative information on the thickness of the sediments can be much more reliably determined from magnetic data than from gravity data. This is illustrated by the example from Senegal (page 413). All such definite information of a regional nature, obtainable from relatively low-cost magnetic and gravity surveys, contributes to the regional geologic picture which needs to be taken into account in broad-scale exploration planning.

SYNERGISTIC APPLICATIONS OF MULTISENSOR GEOPHYSICAL DATA

The word "synergistic" has come into rather common use in recent years to indicate a result from a combination of ideas, talent, or other components of a conference, symposium, or combination of information sources which is much greater than the sum of its parts. The term seems particularly apt for combinations of geophysical and geological data as applied to the solution of a problem of a general nature, i.e., the internal characteristics of a broad basin, or a more specific one, i.e., a possible local oil accumulation.

Combinations of independently observed gravity and magnetic data with seismic results have been applied to geologic problems, as mentioned in preceding paragraphs. The same applications are made to marine operations over the continental shelf by combining maps or profiles from marine seismic, marine gravity, and marine or airborne magnetics over the same area. However, when all three types of measurement are made simultaneously on the same ship, the synergistic principle is truly applicable. It is more economical to combine these into a single operation, as, of course, the very substantial costs of the boat and positioning system are shared. If a marine survey is carried out for a seismic operation alone, the total cost (1974) for 24-hr operation including digital recording is of the order of $200,000 per month. A gravity meter with digital recording can be added for about $30,000 per month and a magnetometer for about $10,000 per month. Data-processing costs are additional and are approximately $100 to $175 per mile.

Some contracts are made on a mileage rather than a monthly rate. The primary acquisition costs per mile are (1974) around $150 to $210 for seismic, $22 for gravity, and $5 for magnetic data. The data-processing costs, approximately as given above, are additional.

The obvious cost reductions for a combined operation are never completely realized. A temporary breakdown of a gravity meter or magnetometer may mean permanent loss of that segment of a line because it is too expensive to halt the seismic operation for the time required for repairs. Sea conditions or boat-speed limitations (on account of cable noise) may be more restrictive for good seismic results than they would be for gravity or magnetic data, so that a gravity-magnetic operation alone can be carried out at considerably higher boat speeds.

One practical difficulty arises when the gravity meter is added to what is primarily a marine seismic operation. For the seismograph alone, common practice is to steer the ship to pass over shot points for which position coordinates in an electronic positioning system (Shoran, Raydist, Lorac, etc.) are precalculated. Between these points the course and speed are not critical. This can result in irregular course and speed changes which can produce large Eötvös effects on the gravity meter. For the best gravity data the course and speed should be held as nearly constant as possible, which takes much more careful ship handling in the pilot house. Also, the navigation operator may be required to plot ship positions rather than giving steering directions to the helmsman so that the ship passes over the precalculated shot points. The writer knows of cases where a gravity-meter operator has been told to stay out of the pilot house because his requests for fewer course and speed changes were a nuisance to the helmsman trained in seismograph operation alone. The more constant course desired for the gravity measurements is, of course, no detriment to the seismograph operation. Part of this difficulty is avoided if the ship is controlled by an autopilot and only occasional small course changes are made slowly. This is the modern practice on any ship where good gravity measurements are expected (too often the gravity measurement is a stepchild of a basically seismic operation which is carried out without proper regard for the requirements of good gravity-meter practice).

EXAMPLES OF MULTISENSOR OPERATIONS

There are, of course, very extensive records of reflection seismic, gravity-meter, and towed-magnetometer data obtained simultaneously over many parts of the continental shelf. Nearly all such data are proprietary with the oil company or geophysical company for whom or by whom the work was done. Occasional samples appear in advertising brochures or exhibits.

The following section presents three examples from different sources and in widely separated parts of the world. The writer is greatly indebted to the oil and geophysical companies for the original data used. The original material has been modified to some extent to include additional details of interpretation and calculation.

West Coast of Alaska

This example is from a brochure by the GAI/GMX Division of Edgerton, Gemmerhausen, and Greer. The data were obtained from routine operations but were exceptional in the very large relief of a basement feature and the high-amplitude gravity changes, which correspond closely with the large range of the basement

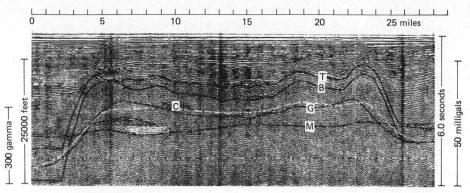

FIGURE 16-1
Seismic record section with gravity and magnetic profiles from combined shipborne geophysical operation, off west coast of Alaska: *T*, seismic interpretation
of bottom of sediments; time scale at right; *B*, bottom of sediments converted
from time to depth; depth scale at left; *G*, Bouguer gravity, scale to right;
calculated gravity (*white line*) is gravity calculated from basement at *B* with
density contrast 0.1; *M*, total magnetic intensity corrected for normal earth's
field; scale at left. (*Original data courtesy GAI/GMX Division of Edgerton,
Gemmerhausen, and Greer.*)

depth. Unfortunately, the basement material is nearly nonmagnetic, and very little
was contributed to the interpretation by the magnetic data.

Figure 16-1 is a portion of the processed record section from the combined
seismograph-gravity-magnetic operations with a four-unit Vibroseis energy source.
The total length of the section is about 28 miles (see mileage scale at top of figure).
The general pattern of the seismic section is an area of shallow basement over the
broad central part of the figure (miles 5 to 23) with a large increase in the thickness
of the sedimentary section near both ends of the figure.*

The line marked *G*, superimposed on the lower part of the record section,
shows the Bouguer gravity from a LaCoste and Romberg gyrostabilized-platform
ship gravity meter after the usual corrections for water depth, latitude, and Eötvös
effects. The water depth is not shown as such but is shallow and nearly constant,
as indicated by the nearly constant first disturbances at the top of the seismic
record. The total magnetic intensity from a towed proton-precession magnetometer
is shown by the line marked *M*. Scales for the gravity and magnetic profiles are at
the right and left margins, respectively, of the figure. The other lines are for the
analysis and interpretation discussed in the following paragraphs.

The time scale for the seismic section (6 sec to the bottom of the chart) is
shown by the $\frac{1}{2}$-sec marks at the right margin. The line *T* was marked by an experienced seismic interpreter as the boundary between the sedimentary section

---

* The zone of sediments is indicated by the more or less continuous nearly horizontal
reflection segments. The zone of basement is indicated by the characterless and discontinuous
pattern below.

above and the basement section below. The time to this change of character varies from a little over 1 sec at the shallowest part (mile 23) to about 4.5 sec at the left margin of the figure.

Line B is the depth corresponding to the reflection times on line T. Average velocities were determined at regular intervals from the processing of the original data. These velocities were used to calculate the depths to the basement, as given by line B, with the depth scale at the left margin of the record. These depths vary from about 3000 ft at mile 23 to 24,000 ft at the left end. The vertical exaggeration of the calculated depths is about 1.8:1.

Gravity calculations A two-dimensional computer program similar to that of Talwani, Worzel, and Landisman (1959) was used for the gravity calculations. The input was based on the depth to the basement as given by line B and density contrast at this depth of 0.1 g/cm^3. The result is shown by the white line marked "calculated gravity."

The overall magnitude and the general form of the calculated gravity are fairly close to those observed. On the right side (mile 24–25) the calculated gravity is about $\frac{1}{2}$ mile farther to the right than the observed, suggesting that the interpretation of the basement surface should be a little farther to the left. It would be shifted to the left if the reflection positions were migrated. The same is true to a smaller extent (because of the smaller slope) at the steep break near the left edge of the figure (miles 1 to 4).

At the top of the rise in the left part of the figure (mile 6) the observed gravity curve is locally higher than that calculated. A 3.2-mgal local residual between the calculated and observed gravity is shown by the insert between miles 4 and 8 with its base at the 3.5-sec time line. This suggests, of course, that there is a disturbance in the subsurface which was not taken into account in the calculations. The area of the suggested disturbance coincides approximately with the shallow irregularity in the reflections centering at about mile 5.5 (but is a little too far to the right).

These two-dimensional calculations assume that the line of the profile is perpendicular to the strike of the subsurface features, particularly the two large faults. The rather close agreement of calculated with observed gravity suggests that the departure from this assumption is quite small.

The magnetic profile of this example is disappointing, as it is quite regular across the entire area and does not indicate the local disturbances expected from the relatively shallow basement or the large relief expected from the great structural relief near the ends of the figures.

This suggests that the basement as interpreted from the seismic reflections is not magnetized to the degree commonly expected for basement rocks. The very weak density contrast used for the gravity calculations also is less than would be expected if this were normal basement material. The combination of a very-weak magnetic-contrast with weak-density-contrast basement suggests quartzite or other nearly nonmagnetic rock with a density of about 2.6. If the sediments just above the basement surface are well compacted, the indicated low value of 0.1 for the density contrast may be reasonable.

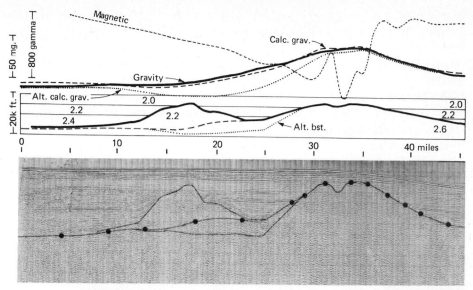

FIGURE 16-2

Seismic record section with gravity and magnetic profiles off the west coast of Africa. (*Lower section*) Twenty-four-fold record section with time scale at left; upper line is interpretation of bottom of sediments; heavy dots are basement-depth values converted to the time scale to give reflection position of interpreted basement; lowest line is alternative basement interpretation from seismic record. (*Middle section*) Model densities used for two-dimensional calculation of gravity effects. Section is true scale. (*Upper section*) Gravity and magnetic profiles; the magnetic basement-depth estimates were made directly from this total-magnetic-intensity curve; "computed gravity" is from the model section with densities shown and with the basement corresponding to the magnetic basement interpretation; the "alternate computed gravity" is from the "alternate basement" of the model section. (*Original data from Gulf Oil Corporation.*)

Continental Shelf off the West Coast of Africa

This example (Fig. 16-2) is from routine operations of the Gulf Oil Corporation's exploration ship R. V. *Gulfrex*. The record section and profiles are similar to, but not identical with, those of Robinson (1971, fig. 2). The instrumentation on the ship is described by Robinson and also by Brown (1970). The computer systems used for real-time reduction of the observations are described by Darby et al. (1973).

The seismic record is a twenty-four-fold stack recording from a 4-unit Aquapulse source with the guns on both sides of the ship. The seismic-record section shows a very large double-peaked uplift, centering at a distance of about 32 miles (see distance scale at bottom of model section) and a smaller disturbance at distance near 17 miles with its top at reflection time of about 1.0 sec. An interpretation based on the seismic section alone would have suggested the same source for both structures.

General inspection of the gravity and magnetic profiles shows that these two seismic disturbances are of very different geological character. The large uplift on the right has a generally corresponding gravity maximum with relief of approximately 40 mgals. Also, the magnetic profile shows sharp anomalies over the shallow part of the uplift. These facts show immediately that this uplift is a basement feature with the gravity anomaly due to density contrasts between the basement rocks and the laterally adjacent sediments while the sharp magnetic anomalies are due to inhomogeneities in the magnetization of the uplifted basement rocks near the surface.

The local seismic feature, at 17 miles, is obviously of a very different geological character. The magnetic curve is much smoother, indicating greater basement depth. There is no obvious local gravity anomaly, although a broad local minimum is detectable by careful inspection of the gravity profile and is analyzed in more detail below (page 439). These facts suggest that this seismic anomaly is caused by nonmagnetic low-density material, possibly salt or low-density high-pressure shale.

Quantitative analysis   The seismic section was analyzed by gravity calculations using the section derived from the magnetic depth estimates.* The basement profile shown on the seismic section was developed from individual magnetic depth estimates. These estimates were made by slope distances and half-slope distances measured on the magnetic profile. As in all such measurements, there is a certain amount of subjectivity in the values obtained, and the numbers are influenced to some extent by the character of the basement profile on the record section.

The depth values are widely variable from less than 3000 ft in the zone of sharp magnetic changes (miles 30 to 35) to over 15,000 ft in the very smooth part of the record (miles 0 to 15). For each depth value the equivalent two-way reflection time was determined from a time-depth chart based on velocity analysis of the seismic record. The times so determined are shown by the heavy dots on the record section. These dots determine a top of magnetic basement as derived from the magnetic interpretation and shown by the line superimposed on the record section. On the whole this magnetic basement follows a fairly regular reflection over the shallower parts of the record section and a very weak but possible zone of reflections in the deeper part and under the uplift at distance 12 to 23 miles.

The next step in the analysis was to carry out a gravity calculation based on a density contrast at the basement-sediment contact as derived from the magnetic depths and seismic data. The model for this calculation is shown in the central part of the figure. This model is true scale, i.e., without the distortion of the seismic-record section caused by the variation of velocity with depth. The general

* The gravity calculations were made at the Houston Technical Services Center of the Gulf Research and Development Co., using a two-dimensional computer program similar to that of Talwani, Worzel, and Landisman (1959). The basement-depth estimates were made at Tidelands Geophysical Co. The writer is very much indebted to these organizations and particularly to Joyce O'Brien at Gulf and Dean Gibbons for making this interpretive analysis possible.

increase of density with depth is approximated by the layers with densities of 2.0, 2.2, 2.4, and 2.6, as shown. The basement is assumed to have a uniform density of 2.6, so that as the large uplift cuts through successively higher sedimentary layers, the contrast becomes +0.2, +0.4, and +0.6. The other structure is assumed to be composed of salt or low-density shale with density 2.2. Its contrast with other layers is −0.2, 0, and +0.2.

The calculated gravity effect is shown by the dashed line in the upper part of the figure, which agrees quite well with the observed gravity profile. This agreement, all within less than 4 mgals and mostly much less than that, is all within less than 10 percent of the total anomaly amplitude.

This general agreement of calculated with observed gravity is not a proof that all details of the model correspond with the actual geologic section, but any reasonable alternative would contain the same general features. The agreement of calculated and observed gravity makes an impressive confirmation of the interpretation.

An alternative seismic interpretation of the basement surface was suggested by a seismologist reviewing the record section. This is shown by the bottom line on the section which follows a fairly continuous zone of reflected events. An alternative calculation of gravity effects was carried out using these depths and the same densities as for the original model. The result is shown by the "alternate computed gravity" line in the upper part of the figure. The wide divergence from the observed gravity anomaly (up to 20 mgals) gives a very strong preference to the first interpretation.

Separation of gravity sources Such a separation with a semiquantitative analysis can show how certain components of the model make subtle contributions to the gravity effects. This demonstrates how careful inspection of the record section and of the gravity profile can give general interpretive results without more elaborate calculations. Also such a simple analysis can contribute to constructing a model for a more elaborate machine calculation.

The left part of the model and the observed and calculated gravity curves are repeated in Fig. 16-3. From the model it is evident that there should be a small positive contribution from the 0.2 density contrast of the small part of the 2.2 density material where it penetrates the 2.0 zone. A quick dot-chart calculation, using a chart, like that of Fig. 7A-10, indicates this should have a magnitude of about 1.5 mgals. Also, there should be a relatively large negative contribution from −0.2 density content from the broad zone of 2.2 material within the normal 2.4 density layer. A quick dot-chart calculation indicates a magnitude of about 10 mgals.

On the basis of these indications, the components were separated as shown in the upper part of the figure. The regional curve is estimated as the gravity which would exist if the low-density body of the model were not present. The difference of the observed gravity from this regional gives the first-residual curve which shows a broad negative effect of 10-mgals amplitude from the broad body of the model with density contrast −0.2.

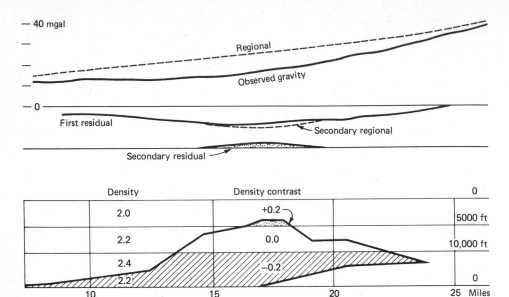

FIGURE 16-3

Gravity effects from details of model section. This figure is enlarged from the gravity curve and model of the left and central parts of Fig. 16-2. The upper regional is drawn to give a first residual with a magnitude corresponding with that expected from the large block of material with density contrast of −0.2 as shown in the model. The secondary regional is drawn to give a secondary residual with a magnitude corresponding with that expected from the shallow material with density contrast of +0.2.

This first-residual curve is not entirely simple but has a relatively positive zone in its central part. We now draw the secondary regional curve which is the estimate of what the curve would be as a simple expression of the deep low-density material alone. Taking the difference again, we have the secondary-residual curve with amplitude about 2 mgals; it gives the positive anomaly which is attributed to the shallow part of the 2.0 density material where it is within the 1.8 density zone.

The process is very similar to that described in connection with the gravity anomalies of salt domes (page 260) which have broad negative anomalies from deep salt within which there may be sharp positive anomaly. This may be caused either by cap rock or by very shallow salt with a positive density content with respect to low-density, very shallow sediments.

Offshore Nova Scotia

This example, furnished by Geophysical Services, Inc. (GSI), is from 65 km (about 40 miles) of ship traverse of a course N16°W and centered about 135 miles east of the east tip of Nova Scotia. The line is within the area of a reconnaissance bottom-meter gravity survey by the Canadian Dominion Observatory (Goodacre, Stephens,

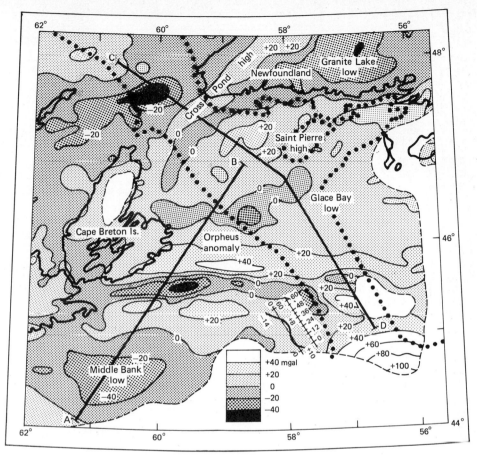

FIGURE 16-4
Gravity map, offshore Nova Scotia, with the location and approximate gravity
profile of the Cabot Straits traverse added. General form of Bouguer gravity
profile from Fig. 15-6 is shown. Distance numbers correspond to those of
Figs. 15-5 and 15-6. (*Modified from Goodacre, Stephens, and Cooper, 1973.*)

and Cooper, 1973). Figure 16-4 shows a portion of that map, slightly modified and
with the traverse line and corresponding gravity profile added.

Observations were carried out by reflection seismograph (air-gun source),
stable-platform ship gravity meter, and towed proton-precession magnetometer.
The results are shown in Figs. 16-5 and 16-6. The lower part of Fig. 16-5 shows
the seismic-record section after computer processing to correct for normal move-
out and conversion to true depth by application of a depth-velocity function for the
area. Calculations were made by a two-dimensional approximation (Talwani,
Worzel, and Landisman, 1959) except for a special treatment of the domelike

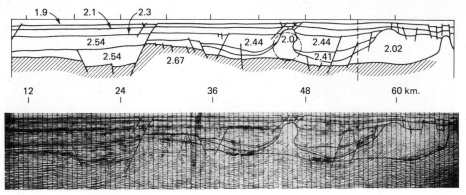

FIGURE 16-5

(*Lower part*) Seismic-record section, Cabot Straits traverse. Corrected for normal move-out and velocity to make vertical and horizontal scales equal. Interpreted geologic contacts, mostly faults, where density contrasts may occur, are shown. (*Upper part*) Model used for calculations. Geologic contacts are repeated from the record section. Identifying numbers, geologic ages, and densities correspond to those of Table 16-1. Circle at salt dome is derived from a simple spherical approximation.

feature at kilometer 46, for which the equivalent of an end correction (see Fig. 7-14) was carried out by insertion of a "salt overlay," as described below.

The Bouguer gravity and total magnetic intensity are shown by the observed-gravity and observed-magnetics lines in Fig. 16-6, with the scales at the left margin. The lines are controlled by values at 0.5-km intervals, each of which is an average of 10 consecutive data points. The smooth dashed lines are from the calculation procedure described below.

Interpretative calculation curves GSI derived these curves by an application of their data-integration procedure. The essential steps in that mathematically complex process are outlined in simplified form in the following paragraphs.

The principal lithological boundaries, as interpreted from the seismic results, are superimposed on the record section (lower part of Fig. 16-5). These boundaries are repeated in the calculation model in the upper part of the figure. As shown by these contacts, the general nature of the section is interpreted as a moderately folded and faulted portion of the sedimentary section which is normal for this area. The section is generally flat from the left margin of the figure (south end) to about kilometer 46, where it is intruded by a dome of Jurassic salt. The detail of the right-hand portion of the section (beyond kilometer 55) was omitted from the quantitative analysis because of irregularities in the field data which could not be readily included in the calculation process.

The calculation procedure is essentially one of finding those gravity fields which are least-squares solutions to a set of simultaneous equations in which the densities of certain of the lithologic units are the unknowns while those of the other

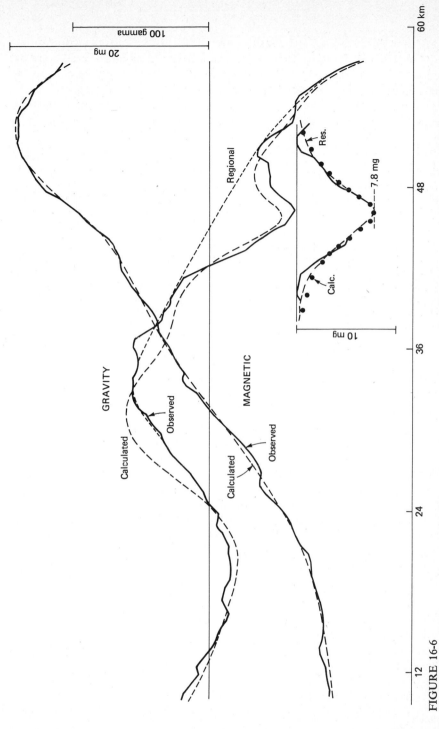

FIGURE 16-6

Observed gravity and magnetic profiles on seismic traverse of Fig. 16-6. Calculated magnetics is derived from a basement below the seismic section except for the infrabasement, shown between kilometers 50 and 55. Calculated gravity is derived from model and densities shown by upper part of Fig. 16-6.

units are considered fixed. The densities are shown in the model section (upper part of Fig. 16-5) and also in Table 16-1, where densities assigned in advance are marked with an asterisk. The calculated effects from the preassigned density and magnetic contrasts were subtracted from the observed gravity and magnetic curves. The remaining anomalies were to be accounted for by the density and magnetization values for the remaining lithologic units. This procedure left a broad 250-γ

Table 16-1 LAYERS USED FOR CALCULATIONS

No.	Age	Density	Density contrast re. 1.90	Magnetization	Mean field mgals	γ
1	Upper Tertiary	1.9*	0	0	0	0
2	Lower Tertiary	2.1*	+0.2	0.000013	4.5	0
3	Upper Cretaceous	2.3*	+0.4	0.00002	14.8	0
4	Lower Cretaceous	2.54*	+0.64	0.00002	32.3	−0.1
5	Lower Cretaceous	2.44*	+0.54	0.00003	6.4	0.2
6	Carboniferous	2.54	+0.64	0.000035	8.0	−0.1
7	Triassic	2.54	+0.64 0.67†	0.00002	7.7	0.0
8	Jurassic	2.59	+0.69 0.63†	0.00005	5.0	−0.1
9	Jurassic	2.41	+0.51 0.60†	0.00046	2.8	1.5
10	Jurassic salt	2.02	+0.12 0.53†	0.00001	−3.3	0.0
11	Basement (Precarboniferous)	2.67	+0.77	0.00006	149.9	−0.2
12	Regional (dip slope 1:12)	2.33	+0.43 0.15†	0.00177	255.9	8.5
13	Infrabasement	3.23	+1.33 0.10†	0.00480	33.3	80.5
14	Dc component				1.07	
15	Salt overlay	2.50*	+0.60	0.00001	13.6	0.1

* Fill-in material to truncate the salt intrusion 2 km each side of the plane of calculation.
† Initial values before inversion to least-squares best fit.

regional magnetic high culminating near the north (right) end of the traverse. This effect was attributed to a deep body with a susceptibility of 0.005 cgs unit and zero density contrast, inserted with an initial depth of 15 km. An iterative calculation was then carried out to determine the depth to the top of this body.

The calculation determined the densities and magnetizations for the seven remaining lithologic units. This procedure consisted of a least-squares linear adjustment of the effects from these units, maintaining the boundaries as given by the interpretation of the seismic section and with the densities and magnetizations as the unknowns.* The calculation of the fields for most of the lithologic units uses a two-dimensional approximation, but this would be seriously in error for a body of limited extent like the presumed salt intrusion. A special modification refined the approximation for this body with a boundary 2 km on each side of the plane of the section (total "length" 4 km). Laterally, beyond this limit, space was filled with the salt overlay with density 2.50.

The final results of the least-squares solutions gave densities and magnetization values for seven units of the sedimentary section. These values, together with those taken as fixed, and with the density contrasts with respect to the Upper Tertiary (density 1.90) are shown in Table 16-1 and also on the model (upper part of Fig. 16-5).

The entire calculation procedure was carried out on a large IBM 370/168 computer, and the resulting structure and field curves presented by Calcomp plotter. The least-squares inversion is an inexpensive matrix operation, while the iterative adjustment scheme is more costly, since the gravity (or magnetization) effect must be recalculated after each cycle of depth estimation.

Comments on the Results

The procedure outlined is, for the most part, a straightforward analytical one. It depends, however, on the interpretive steps of the assignment of lithologic boundaries on the record section, the initial density and susceptibility estimates, and the choice of regional fields.

The fit of the calculated to the observed magnetic curve is generally quite close. This results from the use of an iterative automated interpretation technique, which generally permits fitting to any required tolerance.

---

* The writer used a method similar in principle for the solution of a gravity interpretation problem in the Canadian foothills west of Calgary. A line of gravity stations crossed the steeply dipping outcrops of a thick section of limestones, shales, and sandstones similar to those of the Turner Valley oil field. From the outcrops and also from two or three test wells, two different geologists had drawn their ideas of the probable section under the line of the gravity traverse. A least-squares solution was run, for each of the two suggested sections, with the densities of each of the several geological components as the unknowns. On the basis that one set of densities was much more reasonable than the other it was decided that one of the two geological solutions was much more probable than the other. (This was around 1940 and long before computers were available to carry out the tedious calculations.)

The fit of the calculated to observed gravity curve is not as close. The gross errors may be attributable to incorrect assignment of density boundaries from the seismic record section. The most conspicuous discrepancy is in the zone from kilometers 24 to 32, where the calculated gravity is too high, and in the zone from kilometers 32 to 40, where the calculated gravity is too low. This would be helped if the large fault at kilometers 26 to 28 were moved 2 to 3 km to the right (north), but there is no obvious justification for such a change in the seismic-record section.

The approximate contribution of this large fault to the gravity calculation can be seen quite readily by a simple check calculation. The fault at kilometer 28 has a vertical relief of about 8000 ft and brings 2.67-density material in contrast with 2.54 material to give a contrast of 0.13. The expected gravity effect (see page 193) is $\frac{8000}{80} \times 0.13 = 13$ mgals. On the calculated gravity profile, the gravity increase from kilometers 21 to 31 is about 12 mgals.

Approximate Analysis of Salt-Dome Anomaly

The fit of gravity calculation at the interpreted dome of Jurassic salt is generally quite good. The interpretation can be tested by a simple graphical analysis. As shown in Fig. 16-6, a regional has been added to the observed gravity curve. This is an estimate of what the curve would be if the local salt feature were not present. The residual curve in the lower part of the figure is the difference between the observed curve and this regional. It is irregular because it contains all the irregular local detail of the original observations. The estimated smooth residual leaves out these details.

A simple spherical approximation was made to fit this smooth residual curve. As shown by the figure, the estimated "double half-width" is about 17,500 ft. The corresponding depth to the center of the sphere (page 191) is $17,500 \times 0.652 = 11,400$ ft. Gravity values calculated using Fig. 7A-1 are shown by the small circles. On the whole they are a rather close fit to the smoothed residual curve.

The approximate radius of the body can be calculated from the depth to the center (11,400 ft) and the total gravity relief (7.8 mgals) by Fig. 7A-2. This gives a value of 7800 ft for a density of 0.25. From the densities of the model, the contrast is more nearly 0.5, for which the radius would be $7800/\sqrt[3]{2} = 6200$ ft. The sphere with depth and radius corresponding to these figures is shown by the circle on the model (Fig. 16-6). It is larger than the upper part of the model, because (1) part of the effect comes from the broader lower part of the salt and (2) the length perpendicular to the plane of the section (4 km) is somewhat greater than the sphere diameter in the plane of calculations.

The Dominion Observatory gravity map (Goodacre, Stephens, and Cooper, 1973, and Fig. 6-4) shows very large anomalies, particularly the Orpheus minimum with relief of nearly 100 mgals, which is attributed to low-density Carboniferous and possible younger sediments which fill a deep trench in the Precarboniferous basement. An assumed density contrast of -0.4 for the sediments gives a calculated effect of the correct magnitude for a trench 5 km deep and some 40 km wide.

This contrast would correspond to that between the basement (2.67) and the Tertiary and Cretaceous (1.9 to 2.3) as used for this calculation.

The location and gross features of the gravity profile have been added to the Goodacre map. The maximum at about kilometer 34 of Fig. 16-6 apparently is an eastward continuation of the large maximum which forms the south flank of the Orpheus anomaly.

Goodacre, Stephens, and Cooper (1973, Fig. 6) show a refraction seismograph section which is parallel to, and some 50 km east of, the GSI traverse. This section shows velocities ranging from 1.7 to about 3.8 km/sec for the upper section (above about 2 km) and in the range 5.0 to 6.0 at depths around 4 km. On the basis of density-velocity relations (Fig. 8-9) the corresponding density values are roughly 2.2 to 2.4 for the upper section and 2.6 to 2.7 for the deeper part. These give density differences which are generally in accord with those derived from the above analysis of the gravity data.

Additional Modeling Calculations

After completion of the foregoing normal analytic procedure on the offshore Nova Scotia traverse, further modeling calculations were carried out for both gravity and magnetic data as an experiment. The results are instructive, particularly as being illustrative of a solution which gives a very good fit with observations but is geologically impossible, and of the caution with which the output of the mindless computer must be used.

A gravity or magnetic model may be derived and fitted to observed data by either of two methods of basic control. In the first, the variation in depth to a boundary with uniform magnetic or density contrast is determined. This requires an assumption or determination of depth at one point and a value for magnetization or density contrast at the boundary. This was the method used in the magnetic calculation described above. The "infrabasement" surface is the shallower portion of such a derived magnetic surface.

In the second method, the depths and boundaries of blocks of material are assigned, from other considerations, and the variations in density or magnetization which will give calculated effects to fit the observed data are derived. This was the method of the previous gravity modeling and is further illustrated for both gravity and magnetics by Fig. 16-7.

This example is based on a magnetic basement surface derived from depth estimates made from the observed magnetic profile. Anomalies possibly indicative of faults are present in the magnetic profile, and the details of faulting were further refined by examination of the seismic record section. The top of the faulted magnetic basement was then drawn from the combination of depth estimates and seismic fault suggestions and is indicated in the lower part of Fig. 16-7. The vertical dividing lines between inferred blocks with different magnetization were then drawn and the inversion calculations carried out. The resulting magnetization values for each block are shown by the first figures under the block numbers at the

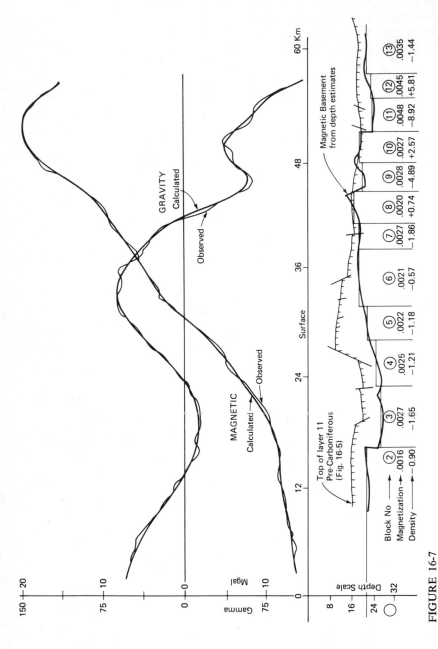

FIGURE 16-7

Alternative basement section with calculated gravity and magnetic effects. Tops of basement blocks conform approximately to magnetic basement derived from depth estimates. Magnetization values were derived from inversion calculations to give calculated magnetic curve above. Density values were similarly determined to give the calculated gravity curve above. These densities are entirely unreasonable physically and geologically.

bottom of the figure. The close fit of the calculated to the observed magnetic profile is indicated in the upper part of the figure. The magnetization contrasts among the blocks are generally reasonable.

A similar calculation was carried out with the gravity data. The complex densities of the previous calculation were ignored, and the gravity effects were assumed to have their sources in density variation among the same blocks used in the magnetic case. The resulting density values are shown by the second figure below each block number. It is obvious that these values are completely unrealistic, but, again, they give a calculated curve in close agreement with that observed. Note the wild gyrations in density values at blocks 8 through 13, where the computer tries to fit detail which is too sharp to have its origin at the depth to the blocks.

In neither the gravity nor the magnetic case is the fit of calculated to observed data a criterion of the validity of the result. Judgment must be based on geologic reasonableness. On that basis the magnetic calculation may be reasonable, but the gravity example fails completely because the densities are impossible, which means that the sources of the gravity effects are largely in the sedimentary section, as used in the first analysis.

In the magnetic case, the principal question is the reliability of the magnetic depth estimates and the basement surface derived from them. The "Precarboniferous" surface from Fig. 16-5 is shown for comparison. In approximately the right half of the figure the agreement is fair, but in the left half the magnetic result is several thousand feet deeper. This may not be a major discrepancy since the "Precarboniferous" surface is not necessarily the magnetic basement.

REVIEW AND CONCLUSIONS

In the early days of gravity and magnetic surveys for oil exploration the work was generally carried out over broad areas by one method and without regard to other geophysical operations. Maps covering millions of square miles of such surveys have rested undisturbed for years in oil company files. Undoubtedly, much could be learned by resurrecting these old maps, studying them by modern methods, and comparing them with any other geophysical or geological data which may be available.

The present pinch on oil reserves is developing some appreciation of the advantages of using all the relevant data, i.e., the synergistic approach. This is receiving more attention in field operations, particularly in marine operations, in that many boats that are equipped and programmed for seismic work also carry gravity and magnetic equipment. Some land seismic crews include a gravity meter to make a measurement at each shot point. This adds very little expense but does require improvement in the usual surveying operation to give better elevation data. Magnetic observations can be added very easily, but in most areas the quality will be inferior to that of airborne surveys because of near-surface disturbances.

There are many geological situations where the potential methods can contribute independent information which will add decisively to the geologic interpretation. As a very simple example there have been seismic surveys which may indicate domal uplifts, but their geologic nature may not be known; they might be caused by a sedimentary structural uplift, a salt dome, or an igneous intrusion. Gravity and magnetic surveys should answer these questions unequivocally. A negative gravity anomaly with no magnetic anomaly would indicate a salt dome. A low-relief positive gravity anomaly with no magnetic anomaly or with a very low-relief magnetic anomaly would indicate a sedimentary uplift with or without (depending on the magnetic expression) an underlying basement uplift. A strong positive gravity anomaly with a strong positive magnetic anomaly would indicate an igneous intrusive. An example is given on page 437.

Seismic interpretations often show a fault by a zone of no reflections, and (especially in areas of rather poor records) there may be no definite basis for determining the direction of displacement of the fault. Gravity information can often find corresponding fault indications and will, of course, indicate the upthrown side.

Magnetic surveys have their most obvious and economically important application in the exploration of new areas, particularly offshore areas. Rapid and relatively inexpensive airborne magnetics can determine the depth to the basement, usually to a precision of the order of 5 percent or less of the depth below the flight level. Also, it is often possible to determine local relief of the basement which has corresponding oil-field structure in the overlying sediments. In a new area where the thickness of the sediments is unknown, a magnetic basement map can determine those basin areas where the sediments are thick enough to make reasonable petroleum prospects and to guide later exploration there.

In addition to the usefulness in locating local structures, gravity surveys are often helpful in evaluating larger geologic units. An example has been given (page 152) of the simple relations of the basin depth to the gravity anomaly in the Los Angeles basin. Similar relations generally apply to the smaller basins with horizontal dimensions of a few tens of miles, such as the intermontane basins of the basin and range area of Utah, Nevada, and eastern California. The nature of larger basins often can be inferred. For instance, in the great valley of California (as in other basins in the Rocky Mountains and in other parts of the world) an asymmetrical basin with a foredeep on one side is distinctly indicated by a regional gravity minimum over the area of thickest sediments. For large basins, however, information on the location and form of gross features within the basin and quantitative information on the thickness of the sediments can be much more reliably determined from magnetic than from gravity data. This is illustrated by the example from Senegal (page 413) and from the Orinoco Basin (page 400). All such definite information of a regional nature, obtainable from relatively low-cost magnetic and gravity surveys, contributes to the regional geologic picture which needs to be taken into account in broad-scale exploration planning.

Careful analysis of local, even very subtle, magnetic anomalies can reveal local relief on the basement surface which has uplifted the overlying sediments to form oil-field structures. Examples are the Puckett and Zelten fields (page 418).

Probably most geologists and geophysicists who have looked at broad-scale gravity and magnetic maps have been tempted to interpret them very simply in terms of broad tectonic features which would have effects on the sedimentary geology and therefore would be of interest in oil exploration. Such temptations must be curbed by the quantitative control which the geophysicist can supply. We may find many cases of close coincidence with regional geologic or tectonic maps. We also find many cases where strong gravity features, some of which are clearly regional and some relatively local, have no counterpart in the geology that we know. It would be a distinct step if we could discover, through broad geological and geophysical thinking, any criteria which would serve as a basis for distinguishing between geophysical features arising from disturbances which may have affected the sediments and those arising from disturbances which have not.

Such criteria, if we find them, will depend upon the quantitative control which can be given by the geophysicists and by the factual control and guessing at possible sources which must be furnished by geologists, together with an active imagination by both and a sympathetic consideration of each other's viewpoint. Also in the interpretation of individual local anomalies, the geophysicist can supply certain limits of location, size, and depth, but the geologist must then give his help.

The overall exploration problem may be compared with a detective story. In a well-constructed story the whodunit has many possibilities at the beginning. As more and more clues are developed, the number of possibilities becomes more and more limited, until finally only one remains. In the exploration analog, the "criminal" to be found and identified is the true geological situation. With a single exploration method (surface geology or any one of the geophysical methods) the number of geological possibilities may be relatively large. As more "clues" are brought to bear by application of more geological and geophysical techniques the number of possibilities is reduced until finally (one hopes) the actual geological situation is revealed by finding the set of geological circumstances which is consistent with all the information available.

When teamwork between geologists and geophysicists has reconciled all the known items of data with each other and with reasonable speculations, we can approach a much clearer picture of the geologic past as reflected in present geophysical and geological facts. In this way we can hope to advance beyond geological and geophysical guesses and let fact keep our feet on the ground when we are tempted to indulge in flights of unrestrained fancy.

> Ah! what avails the classic bent
>   And what the cultured word,
> Against the *undoctored* incident
>   That actually occurred?*

---

* Rudyard Kipling, *The Benefactors*, emphasis supplied. Quoted to this writer many years ago by Paul Weaver, then Chief Geophysicist for Gulf Oil Corp, and one of the pioneer Gulf Coast explorationists.

REFERENCES FOR PART THREE

ALLEN, S. J.: Geophysical Activity 1969, *Geophysics*, pp. 189–210, 1971.

BROWN, DONALD A.: Instrumentation on the *Gulfrex*, *Undersea Technol.*, October 1970, p. 16.

DARBY, E. K., E. J. MERCADO, R. M. ZOLL, and J. R. EMANUEL: Computer Systems for Real-Time Marine Exploration, *Geophysics*, vol. 38, no. 2, pp. 301–309, 1973.

GOODACRE, A. K., L. E. STEPHENS, and R. V. COOPER: A Gravity Survey of the Scotian Shelf, *Earth Sci. Symp. Offshore East. Can., Geol. Surv. Can. Pap.* 71-23, pp. 241–252, 1973.

HALBOUTY, MICHEL T.: Oil Is Found in the Minds of Men, *Trans. Gulf Coast Ass. Geol. Soc.*, vol. 22, pp. 33–37, 1972.

KOLOGINCZAK, JOHN B.: Marine Geophysical Surveys, *Oil Gas J.*, pt. 1, Nov. 16, 1970, p. 204; pt. 2, Nov. 22, 1970, p. 106.

PRATT, WALLACE E.: "Oil in the Earth," p. 110, University of Kansas Press, Lawrence, Kans., 1942.

ROBINSON, W. B.: Geophysics Is Here to Stay, *Bull. Am. Assoc. Pet. Geol.*, vol. 55, no. 12, pp. 2107–2115, 1971.

STEENLAND, NELSON C.: Oil Fields and Aeromagnetic Anomalies, *Geophysics*, vol. 30, no. 5, pp. 706–739, 1965.

TALWANI, MANIK, J. LAMAR WORZEL, and MARK LANDISMAN: Rapid Gravity Calculations for Two-dimensional Bodies with Applications to the Mendocino Submarine Fracture Zone, *J. Geophys. Res.*, vol. 64, no. 1, pp. 49–59, 1959.

INDEXES

NAME INDEX

Affleck, James, 167
Agarawal, R. G., 158
Aldrich, Thomas C., 43, 119
Alldredge, Leroy R., 323
Allen, S. J., 431
Allen, W. E., 273, 277
Am, K., 399n., 403
Andreasen, Gordon E., 374
Athy, L. F., 257

Baranov, V., 390, 392-394
Bardeen, John, 395n.
Barry, H., 44, 253
Bartels, J., 326
Bascom, Willard, 284n.
Bean, R. J., 374n., 376, 404
Berg, J. W., Jr., 172
Berroth, A., 59
Bhattacharyya, B. K., 353, 374
Bible, John L., 93, 247
Birch, Francis, 255
Blakely, Robert F., 145
Bleifuss, Rodney, 364

Bott, M. H. P., 93, 181, 211
Bouguer, P., 11, 20
Bowin, Carl O., 43, 119
Boys, Charles Vernon, 12
Brown, Donald A., 120, 437
Bryan, A. B., 30
Bullard, E. C., 59, 172
Byerly, P. E., 158

Callouet, H. J., 273, 277
Caputo, Michele, 15, 18
Cavendish, H., 11
Chartier, J.-M., 53
Clark, R. H., 260
Clark, Sydney, Jr., 292
Clarkson, Neal, 113
Cook, A. G., 56, 57, 59
Coons, R. L., 48, 172, 173
Cooper, R. V., 441, 446
Cordell, L., 181
Curtis, C. E., 411

Dampney, C. N. G., 157, 181, 185

SUBJECT INDEX